Y0-AAI-959

# Molecular Control of Proliferation and Differentiation

# MOLECULAR CONTROL OF PROLIFERATION AND DIFFERENTIATION

## *The Thirty-Fifth Symposium of The Society for Developmental Biology*

*Asilomar Conference Grounds*

*Asilomar, California*

*June 8–11, 1976*

Society for Developmental Biology.

# Molecular Control of Proliferation and Differentiation


Edited by

**John Papaconstantinou**

*Biology Division*
*Oak Ridge National Laboratory*
*Oak Ridge, Tennessee*

**William J. Rutter**

*Department of Biochemistry and Biophysics*
*University of California*
*San Francisco, California*

**1978**

**ACADEMIC PRESS, INC.** *New York San Francisco London*

*A Subsidiary of Harcourt Brace Jovanovich, Publishers*

*Academic Press Rapid Manuscript Reproduction*



ACADEMIC PRESS, INC.
111 Fifth Avenue, New York, New York 10003

*United Kingdom Edition published by*
ACADEMIC PRESS, INC. (LONDON) LTD.
24/28 Oval Road, London NW1

LIBRARY OF CONGRESS CATALOG CARD NUMBER

ISBN 0-12-612981-9

PRINTED IN THE UNITED STATES OF AMERICA

# Contents

## VII. Factors Effecting Differentiation in Lower Eukaryotes

# List of Contributors

**Boldface Denotes Chairmen**

*John W. Adamson*, Hematology Research Laboratory, Veterans Administration Hospital, Department of Medicine, University of Washington School of Medicine, Seattle, Washington

*T. D. Allen*, Paterson Laboratories, Christie Hospital and Holt Radium Institute, Manchester, M20 9BX, England

*Edward A. Berger*, Department of Biology, University of California, San Diego, California

*James E. Brown*, Hematology Research Laboratory, Veterans Administration Hospital, Department of Medicine, University of Washington School of Medicine, Seattle, Washington

*Graham Carpenter*, Department of Biochemistry, Vanderbilt University, Nashville, Tennessee

*John M. Chirgwin*, Department of Biochemistry and Biophysics, University of California, San Francisco, California

*Stanley Cohen*, Department of Biochemistry, Vanderbilt University, Nashville, Tennessee

*T. M. Dexter*, Paterson Laboratories, Christie Hospital and Holt Radium Institute, Manchester, M20 9BX, England

**Michael Feldman**, Department of Cell Biology, The Weizmann Institute of Science, Rehovot, Israel

*Gideon Goldstein*, Sloan-Kettering Institute for Cancer Research, New York, New York

*Denis Gospodarowicz*, The Salk Institute for Biological Studies, San Diego, California

*John D. Harding*, Department of Biochemistry and Biophysics, University of California, San Francisco, California

*L. G. Lajtha*, Paterson Laboratories, Christie Hospital and Holt Radium Institute, Manchester, M20 9BX, England

*S. Lan*, Ontario Cancer Institute, Toronto, Ontario, Canada

*E. A. McCulloch*, Ontario Cancer Institute, Toronto, Canada

*Raymond J. MacDonald*, Department of Biochemistry and Biophysics, University of California, San Francisco, California

*Vivian L. MacKay*, Waksman Institute of Microbiology, Rutgers University, New Brunswick, New Jersey

*Anthony L. Mescher*, The Salk Institute for Biological Studies, San Diego, California

*Trudy O. Messmer*, The Salk Institute for Biological Studies, San Diego, California

*Malcolm A. S. Moore*, Sloan-Kettering Institute for Cancer Research, New York, New York

*John S. Moran*, The Salk Institute for Biological Studies, San Diego, California

*U. Otten*, Department of Pharmacology, Biocenter of the University, Basel, Switzerland

*Dieter Paul*, The Salk Institute for Biological Studies, San Diego, California

*Raymond L. Pictet*, Department of Biochemistry and Biophysics, University of California, San Francisco, California

*G. B. Price*, Ontario Cancer Institute, Toronto, Ontario, Canada

*Alan E. Przybyla*, Department of Biochemistry and Biophysics, University of California, San Francisco, California

*Hans J. Ristow*, The Salk Institute for Biological Studies, San Diego, California

*H. T. Rupniak*, The Salk Institute for Biological Studies, San Diego, California

**William J. Rutter**, Department of Biochemistry and Biophysics, University of California, San Francisco, California

*H. Chica Schaller*, European Molecular Biology Laboratory, Heidelberg, West Germany

*M. Schwab*, Department of Pharmacology, Biocenter of the University, Basel, Switzerland

**Eric M. Shooter**, Department of Neurobiology, Stanford University School of Medicine, Stanford, California

*Solomon H. Snyder*, Department of Pharmacology and Experimental Therapeutics and Psychiatry and Behavioral Sciences, Johns Hopkins University School of Medicine, Baltimore, Maryland

*H. Thoenen*, Department of Pharmacology, Biocenter of the University, Basel, Switzerland

**J. E. Till**, Ontario Cancer Institute, Toronto, Ontario, Canada

# Foreword

The 35th Symposium, "Molecular Control of Proliferation and Differentiation," was held at Asilomar, Monterey Peninsula, California, June 8–11, 1976. The Society gratefully acknowledges the efficiency and hospitality of the host committee and the financial support from the National Science Foundation. Approximately 300 people attended and enjoyed the science and the fellowship. Special thanks are due Claudia and John Foret who carried out the detailed arrangements for the meeting.

*The Society for Developmental Biology*

# I. Introduction

# Cell Communication in Embryological Development: The Role of Distal and Proximal Signals

William J. Rutter

*Department of Biochemistry and Biophysics*
*University of California*
*San Francisco, California 94143*

## I. INTRODUCTION

### *A. Cells in a multicellular system interact in various ways:*

A fundamental aspect of the division of metabolic labor existing in pluricellular organisms is the production by certain cells of compounds that are utilized or manipulated by other cells. For example the lactic acid produced by anaerobic glycolysis during muscular exercise is a substrate for gluconeogenesis in the liver and the kidney. End products of metabolism are usually detoxified or modified for excretion by cells other than those which produce them. In these instances the specificity of the interaction is derived from the complementarity of metabolic enzymes present in various differentiated cells.

The functions of particular cells may also be regulated by specific chemical signals produced by other cells. In nervous systems cell communication is mediated by neurotransmitters, some of which are listed in Table I. The specificity of the regulation is provided by the neuronal network formed by the effector cells that release neurotransmitters; these in turn trigger target cells containing an appropriate surface receptor. Thus, stimulation of a variable number of cells is achieved and the magnitude of the response is determined by the number of cells involved. Hormone mediated regulation is similar in some respects to nerve mediated responses but different in others. Hormones are produced in cells usually at a distant site from the target organs and are transported via the tissue fluids to the site of action. The target cells must contain a specific receptor, and the magnitude of the response (as reflected in the typical dose response curves) is determined by the degree of saturation of available receptors.

In addition to communication through the extracellular fluid, direct cell communication via gap junctions between neighboring cells has been demon-

**TABLE 1**

*Neurotransmitters Function as Mediators in Neuronal Networks*

| |
|---|
| Acetylcholine |
| Epinephrine, Norepinephrine |
| Dopamine |
| $\gamma$-Amino Butyric Acid |
| Serotonin |
| Glycine |
| (Glutamate) |

strated (1). Only relatively small molecules can exchange readily through gap junctions: There is effective exchange of metabolic intermediates, pleiotypic regulators such as cyclic AMP, cyclic GMP and even small hormones but little if any exchange of large molecular weight molecules. The biological function of junctional complexes may be to integrate the physiological activities of cell populations. Many of the cells in an early embryo are electrically coupled (2), thus implying direct cell communication by junctional complexes. Later in development, electrical coupling is more segregated, and may be restricted to functioning units of similar cells. The direct transmission of information molecules such as DNA or RNA to cells in tissue culture has also been reported (3-6). The function if any of such nucleic acid transfer is unknown.

## II. EVOLUTION OF CELL COMMUNICATION

A consideration of the possible evolutionary origin of cellular communication is instructive since it may provide a basis for correlating structural and functional relationships among regulatory molecules. Primitive unicellular organisms may have interacted largely at the nutritional level. The end products of a given metabolic pathway in one species, may be the substrates for a key pathway in another species. Metabolic dependence between cells may then be stabilized by the production of a needed metabolite, e.g., a needed precursor or coenzyme. In addition to metabolites cells may have secreted hydrolytic enzymes such as proteinases or toxins in order to utilize the products of other biological systems. Hormones may have been derived from these proteins. Indeed, a structural resemblance has been proposed for the serine proteinases and certain hormones (7). It is also possible that other cell surface proteins, or partially hydrolyzed peptides might be employed as signals.

The differentiation of cell surface structures may have occurred relatively early in single cell systems. The individuality of the organism is frequently expressed in its relationship with the external environment and the perception of that environment can be altered by modification of the surface. The aggregation of cells might alter the environment; for example, cell aggregates with complementary metabolic processes should be more nutritionally efficient since great dilution of metabolic intermediates would be precluded. In such systems control of cell proliferation or of function may at first have been related to the balance required for efficient metabolism of the available substrates; but ultimately not only metabolic functions but motility, perception of environmental signals, etc. were segregated in various cells in the population. In such multicellular systems communication between cells would confer a selective advantage in order to control the relative proportions of cells as well as the functions of those cells. In a system having n cell types the number of signals required for independent non-redundant communication among each cell is $n!/n\text{-}2! = n(n\text{-}1)$. Thus the number of regulatory signals in a complex organism might be very large. A multicellular system containing 100 cell types would require 9900 regulatory signals. Additional regulatory events associated with integration of functions, development and redundancy of crucial regulatory functions could increase the number substantially, such that a large proportion of the genome might be devoted to regulatory as opposed to direct physiological functions. There is no evidence that this is the case, in fact there are arguments against it (8-10). Furthermore, there is no evidence suggesting that each cell type communicates directly with every other cell type. A more simplified strategy involving subsets of cells is, more likely, involved. Some of the hormones regulate metabolic processes in many cell types, others regulate function in restricted types.

The strategies involved in regulation of mature cell functions may be quite different from those involved in differentiative and morphogenetic processes. In the former, regulation of the activities of groups of cells may be effectively accomplished from centralized control tissues by neuronal and hormonal mechanisms. In the differentiation and morphogenesis, the organization of the cells into functioning units is a dominant consideration. Some kinds of proximal regulation at the level of individual cells seems involved. Distal regulation may also be involved especially if the cells are dispersed throughout the body, e.g., blood cells.

## III. DISTAL AND PROXIMAL REGULATION IN EMBRYOLOGICAL DEVELOPMENT

The implementation of the developmental program in embryos is crucially dependent upon cell interactions. The requirement for normal development of complex culture media or the presence of other tissues or specific extracellular factors leads to the thesis that both proliferation and differentiation is controlled by molecular signals provided by other cells. In this manner the relative abundance of a given cellular function can be regulated in relationship to other complementary functions. These cell interactions could be mediated by hormones, by nerve cells or by proximal interactions such as those illustrated in Figure 1. Some of the developmental effects can be initiated by single or mixtures of hormones (see Table II). For example, it is well known that thyroid hormones induce amphibian metamorphosis; sexual steroids apparently imprint sexual behavioral and metabolic characteristics on the neonatal animal (11,12) as well as control the development of organs of secondary sex characteristics later in the development. In these processes sequential treatment with hormones (e.g., progesterone and estrogen) may be required. In *in vitro* systems, insulin, cortisone and prolactin are required for development of functional mammary gland acinar cells (13). In addition there are many hormone-like "factors" that specifically control developmental processes (see Table III). The effects of nerve growth factor, epithelial growth factor and erythropoietin are well known to developmental biologists, but a large number of additional factors have now been reasonably well identified and partially characterized. Many of these components specifically effect cell

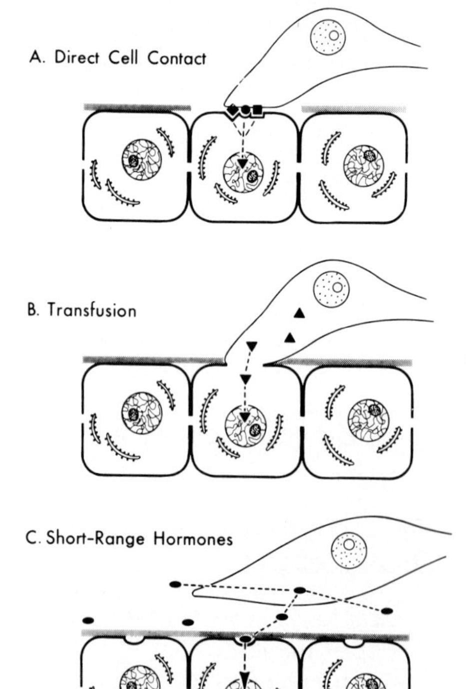


*Fig. 1. Proximal interactions in cyto-differentiation.*

**TABLE 2**

*Some Hormones have Developmental Functions*

| |
|---|
| Thyroid Hormones |
| Glucocorticoids |
| Progesterone |
| Estrogens |
| Androgens |
| Prolactin |
| Insulin |
| Growth Hormone |

proliferation; and their effect on development may be related to this effect. The activity of individual cells may be regulated by several factors. The concentrations required for activity are frequently exceedingly low (in the nanomolar to picomolar range). Thus biologically effective concentrations in plasma can hardly be detected by molecular assays. Therefore, the identification, isolation and characterization of these molecules can be difficult (14). Like hormones, they are produced in selected cells (although they may not be produced in large quantities) and soluble in tissue fluids and act on specific target cells.

Innervation itself appears to play little if any role in the differentiation itself, although it obviously plays a role in the development of the mature functioning tissue (15).

## IV. STRATEGIES OF REGULATION OF DIFFERENTIATION

How are the various differentiative transitions orchestrated? Are the mechanisms instructive or permissive? There is little evidence for instructive mechanisms in which the regulatory molecules provide genetic information or

**TABLE 3**

*Growth and Differentiation Factors with Hormone-Like Activity*

| Compound | Target Cells |
|---|---|
| Nerve Growth Factor | Sympathetic Nerves, Dorsal Ganglia |
| Epithelial Growth Factor | Epidermal Cells |
| Fibroblast Growth Factor | Mesodermal cells; fibroblasts, chondrocytes, glial, endothelial, smooth muscle |
| Ovarian Growth Factor | Ovary |
| Myoblast Growth Factor | Myoblasts |
| Erythropoietin | Erythroid |
| Thrombopoietin | Thrombocytes |
| Colony-Stimulating Factor | Granulocytes |
| Thymopoietin | Thymocytes |
| Thymosin | Thymocytes |
| Melanocyte Stimulating | Melanocyte |
| Somatomedins | Cartilage |

operate directly on the genome or transcription apparatus. The transfusion of specific differentiative information including genetic information seems feasible, though not necessary if each zygote and subsequent precursor cells retain all the genetic information. Most of the evidence suggests permissive mechanisms in which the regulatory molecules simply trigger differentiative transitions. Thus, the mechanisms of the transitions might be similar. It is even possible that all the differentiative switches might be accomplished by the same signal. Specificity then is determined by the receptor and the state of differentiation (competence to respond), but not necessarily by the signal itself. The nature of the signals has been a subject of intense interest since the time of Spemann. Hormone-like molecules which have little solubility in fluids may be transmitted by contact rather than by humoral means. This mechanism has the advantage that identical or similar mediating molecules can be involved in different systems: since the effector molecule is produced in juxtaposition to its receptor, the specificity of regulation is distributed topologically. Other more complicated mechanisms involving cell contact can be envisioned. For example, the basal lamina encasing epithelial cells may function by acting as a selective "filter" for required molecules, thus, selective loss of the basal lamina may regulate cell proliferation.

On the other hand, direct cellular contact may be involved (see Fig. 1). The inductive process is both biologically and chemically complex, since cell proliferation, morphogenesis, and differentiation are usually involved (16-18). Early experiments involving transplantation of inducing tissues into distant sites or of interposition of barriers between tissues strongly suggested that some kind of proximal regulation was involved. Grobstein and his colleagues observed induction of salivary gland and pancreatic epithelia by mesenchyme juxtaposed across a special thin Millipore filter (19,20). This suggested that a slowly diffusible substance mediated the biological response and that cellular contact was not required. Recently, however, Saxen and his colleagues (21) have analyzed the tissue interactions in kidney development. These workers have demonstrated contiguous cellular processes running through the filter between the two tissues, spinal cord and kidney mesenchyme. The degree of differentiation was roughly proportional to the degree of cell contact. These results have raised questions concerning the validity of the interpretation of transfilter experiments, and suggest that direct cell-cell contact may be required. In the pancreas system, however, a soluble factor has been isolated from mesenchymal rich embryonic tissues which can replace the mesenchymal tissue (22,23). In primary embryological induction as well, factors that induce the formation of mesodermal and endodermal structures have been detected and partially purified by using a sandwich culture assay technique for assay that restricts the diffusion of proteins or ribonucleoprotein (24). In these cases, therefore, the mode of regulation appears to involve the transmission of molecules between cells (although perhaps not by free diffusion).

The mechanism of the regulation could involve any molecule or structure required for function of the receptor cell. It appears, however, that a few strategems are favored. (a) Regulation of the intake of crucial molecules including ions or metabolic intermediates. The regulation of the intake of calcium, which may play a crucial role in secretion also influences the formation of junctional complexes. Control of the uptake of calcium may be an important developmental mechanism (25). In addition many metabolic processes may be regulated via the uptake of required nutrients (26,27). (b) The regulatory molecules or a receptor complex may interact with the chromosomes, to produce some change in the transcription of certain genes. The steroid hormones and the thyroid hormones appear to act in this fashion (28). (c) The molecules may act with a component of the cell membrane to stimulate a catalytic system which produces an internal regulatory molecule. The protein hormones may act on the adenylcyclase system to generate cyclic nucleotides such as cyclic AMP or cyclic GMP which are in turn mediators for intracellular processes, e.g., protein phosphorylation. It has been proposed that many developmental mechanisms are controlled via cyclic AMP (29-33). (d) The direct action of a protein/peptide molecule on a specific intracellular component is feasible. It is sometimes assumed dogmatically that all effector proteins interact with target cells via a surface receptor. This is certainly not the case. Certain toxins such as diptheria toxin (34,35), cholera toxin (36), abrin (37) and ricin (38) contain two components, an A peptide responsible for catalyzing intracellular damage, and a B peptide concerned with binding to cell membrane receptors. The specific entry of the A peptide can effectively control a critical aspect of macromolecular synthesis and, therefore, the viability of the cell. Nerve growth factor also enters the cell and is transported axonally to the cell body, where it induces the level of acetylocholinesterase in the Golgi (39). (e) Many of the functional activities of cells are mediated by membrane components and modification of these components may exert remarkable effects. For example, a specific phospholipase of snake venom exerts a selective effect on synaptic function (40), or selective destruction of certain components in the cell membrane by plasmin or by thrombin stimulates proliferation of certain cells (41-43).

This symposium focuses on factors which influence the differentiative process of specific cells. The challenge before us is not only to define the nature of the signals but the mechanisms by which they effectuate developmental and morphogenetic transitions.

## REFERENCES


1. Lowenstein, W. R. (1973). *Fed. Proc.* **32,** 60–64.
2. Furshpan, D. and Potter, M. (1968). *Curr. Top. Dev. Biol.* **3,** 95–127.
3. Hill, M. and Hillova, J. (1971). *Nature New Biol.* **231,** 261–265.
4. Mirrie, C. R., Geier, M. R. and Petricciani, J. C. (1971). *Nature* **233**, 398–400.

5. Willecke, K. and Ruddle, R. H. (1975). *Proc. Nat. Acad. Sci. U.S.A.* **72**, 1792–1796.
6. Burch, J. W. and McBride, O. W. (1975). *Proc. Nat. Acad. Sci. U.S.A.* **72**, 1797–1801.
7. De Haen, C., Neurath, H. and Teller, D. C. (1975). *J. Mol. Biol.* **92**, 225–259.
8. Ohta, T. and Kimura, M. (1971). *Nature* **233**, 118–119.
9. Sulston, J. E. and Brenner, S. (1974). *Genetics* **77**, 95–104.
10. Brenner, S. (1974). *Genetics* **77**,71–94.
11. Levine, S. and Mullins, R. F. Jr. (1966). *Science* **152**, 1585–1592.
12. Giulian, D., McEwen, B. S. and Pohorecky, L. A. (1974). *Proc. Nat. Acad. Sci. U.S.A.* **71**, 4106–4110.
13. Vonderhaar, B. K., Owens, I. S. and Topper, Y. S. (1973). *J. Biol. Chem.* **248**, 467–471.
14. Sato, G. This volume.
15. Fambrough, D. M. (1976) *In* "Neuronal Recognition" (S. H. Barondes, ed.), pp. 25–67. Plenum Press, New York.
16. Pictet, R. L. and Rutter, W. J. (1972). *In* "Handbook of Physiology, Section 7: Endocrinology" (D. F. Steiner and N. Freinkel, eds.), Vol. 1, pp. 25–66. Williams and Wilkins, Baltimore.
17. Pictet, R., Clark, W. R., Williams, R. H. and Rutter, W. J. (1972). *Develop. Biol.* **29**, 436–467.
18. Rutter, W. J., Wessells, N. K. and Grobstein, C. (1964). *J. Natl. Cancer Inst., Monogr.* **13**, 51–65.
19. Grobstein, C. (1967). *J. Natl. Cancer Inst. Monogr.* **26**, 279–282.
20. Kallman, F. and Grobstein, C. (1964). *J. Cell Biol.* **20**, 399–413.
21. Lehtonen, E., Wartiovaara, J., Nordling, S. and Saxen, L. (1975). *J. Embryol. exp. Morph.* **33**, 187–203.
22. Ronzio, R. A. and Rutter, W. J. (1973). *Develop. Biol.* **30**, 307–320.
23. Levine, S., Pictet, R. and Rutter, W. J. (1973). *Nature New Biol.* **246**, 49–52.
24. Born, J., Tiedemann, H. and Tiedemann, H. (1972). *Biochim. Biophys. Acta* **279**, 175–183.
25. Whitfield, J. F., McManus, J. P., Dixon, R. H., Boynton, A. L., Youdale, J. and Swierenga, S. (1976). *In Vitro* **12**, 1–18.
26. Holley, R. W. (1972). *Proc. Nat. Acad. Sci. U.S.A.* **69**, 2840–2841.
27. Holley, R. W. and Kiernan, J. A. (1974). *Proc. Nat. Acad. Sci. U.S.A.* **71**, 3976–3978.
28. Alberts, B. M. and Yamamoto, K. R. (1976). *Ann. Rev. Biochem.* **45**, 721–746.
29. McMahon, D. (1974). *Science* **185**, 1012–1021.
30. Rutter, W. J., Pictet, R. L. and Morris, P. W. (1973). *Ann. Rev. Biochem.* **42**, 601–646.
31. Wahn, H. L., Taylor, J. D. and Tchen, T. T. (1976). *Develop. Biol.* **49**, 470–478.
32. Desphande, A. R. and Siddiqui, M. A. Q. (1976). *Nature* **263**, 588–591.
33. Filosa, S., Pictet, R. and Rutter, W. J. (1975). *Nature* **257** 702–705.
34. Pappenheimer, A. M. Jr. and Gill, D. M. (1973). *Science* **182**, 353–358.
35. Kandel, J. L., Collier, R. J. and Chung, D. W. (1974). *J. Biol. Chem.* **249**, 2088–2097.
36. Gill, D. M. and King, C. A. (1975). *J. Biol. Chem.* **250**, 6424–6432.
37. Alsnes, S. and Pihl, A. (1973). *Eur. J. Biochem.* **35**, 179–185.
38. Alsnes, S. and Pihl, A. (1973). *Biochemistry* **12**, 3121–3126.
39. Thoenen, H., this symposium.
40. Strong, P. N., Goerke, J., Oberg, S. and Kelly, R. B. (1976). *Proc. Nat. Acad. Sci. U.S.A.* **73**, 178–181.
41. Burger, M. M. (1970). *Nature* **227**, 170–171.
42. Unkeless, J., Dano, K., Kelleman, G. M. and Reich, E. (1974). *J. Biol. Chem.* **249**, 4295–4305.
43. Chen, L. B. and Buchanan, J. M. (1975). *Proc. Nat. Acad. Sci. U.S.A.* **72**, 131–135.

# II. Growth Factors

# Biological and Molecular Studies of the Mitogenic Effects of Human Epidermal Growth Factor

Graham Carpenter and Stanley Cohen

*Department of Biochemistry*
*Vanderbilt University*
*Nashville, Tennessee 37232*

## I. INTRODUCTION

Epidermal growth factor isolated from mouse submaxillary glands (mEGF) is a single chain polypeptide (MW 6045) that contains 53 amino acid residues and three disulfide bonds. *In vivo* and in organ culture systems, mEGF stimulates the proliferation and keratinization of epidermal tissue. The effects of EGF, however, are not limited to epidermal cells. Mouse EGF is a potent mitogen when added to cell cultures of mouse and human fibroblasts (Armelin, 1973, Hollenberg and Cuatrecasas, 1973, Cohen *et al.*, 1975), human glia cells (Westermark, 1976), and rabbit lens epithelial cells (Hollenberg, 1976). The biological and chemical properties of mEGF have been reviewed (Cohen and Taylor, 1974; Cohen and Savage, 1974; Cohen *et al.*, 1975).

Although mEGF promotes the proliferation of cells derived from a number of species (mouse, human, rabbit, chick), the presence of EGF-like molecules in species other than the rodent has not, until recently, been demonstrated. In this report we summarize our data regarding the isolation of human EGF (hEGF) and its interactions with human fibroblasts.

A new aspect of the biology of EGF recently has emerged with the publication of the amino acid sequence and disulfide linkages of urogastrone, the gastric antisecretory hormone isolated from human urine (Gregory, 1975). Of the 53 amino acid residues comprising the urogastrone and mouse EGF molecules, 37 are common to both peptides, and the three disulfide bonds are formed in the same relative positions. Furthermore, mEGF has been shown to possess gastric antisecretory activity and urogastrone to possess the biological activity of EGF as judged by its ability to induce precocious eye opening in newborn mice. The results suggest that urogastrone may be identical, or very closely related, in both structure and function to human EGF.

## II. ISOLATION AND CHEMICAL PROPERTIES OF HUMAN EGF

The assay for the isolation of hEGF was based on the ability of both mouse and human EGF to compete with $^{125}$I-labeled mEGF for binding sites on human foreskin fibroblasts (Cohen and Carpenter, 1975). Under the assay conditions used, 2-20 ng of EGF were readily measurable. The starting material for the isolation was an acetone powder obtained from the urine of pregnant women. Ten grams of this preparation, obtained from 15 liters of urine, contained approximately 700 μg equivalents of hEGF; 100-150 μg of pure hEGF were isolated by the procedures outlined below. Although we have made use of urine from pregnant women, hEGF is not unique to pregnancy; it is also present in the urine of adult males.

The details of the isolation procedure have been published (Cohen and Carpenter, 1975). The main purification steps involved (a) gel filtration on Bio-Gel P-10, (b) gel filtration of Sephadex G-50, (c) ion exchange chromatography on CM-52 cellulose, and (d) ion exchange chromatography on DE-52 cellulose. Human EGF has also been partially purified from urine with the aid of an affinity column containing Sepharose-linked rabbit antibodies to mouse EGF (Starkey *et al.,* 1975).

The purity of the final preparation was examined by polyacrylamide disc gel electrophoresis under acid and alkaline conditions (Fig. 1). It may be seen that in each instance, hEGF migrates as a single band. Human EGF and mouse EGF migrate at approximately the same rate at pH 2.3. However, under alkaline conditions, hEGF migrates much more rapidly, suggesting that the net charge of hEGF at pH 9.5 is more negative than its mouse counterpart.

To establish whether the competitive binding activity of hEGF was associated with the stained band observed in the gel, we performed the following experiment. The alkaline gel (gel B, Fig. 1) was sliced into 1 mm sections, and each segment was fragmented into 400 μl of 10% $NaHCO_3$ containing albumin and incubated overnight at 5°. The extract from each slice

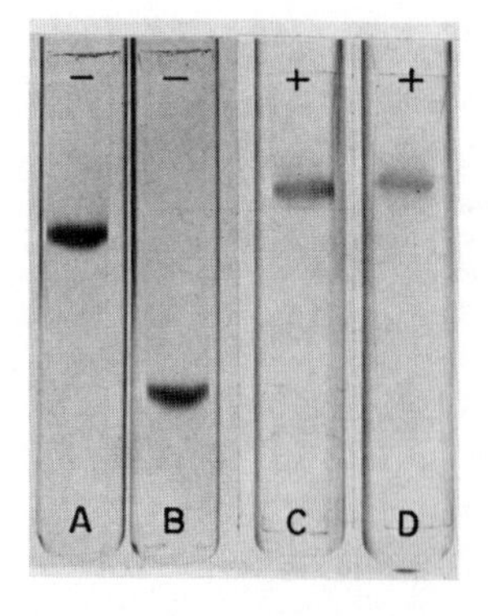


Fig. 1. Disc gel electrophoresis of hEGF and mEGF. Tubes A and C contain mEGF; tubes B and D contain hEGF. The pH of the gels in tubes A and B was 9.5 and in tubes C and D, 2.3. Samples of 10-20 μg of protein were applied. (from Cohen and Carpenter, 1975)

was assayed by competition with $^{125}$I-labeled in mEGF for binding to fibroblasts. Competitive binding activity was associated only with those fractions corresponding to the stained area of the gel.

The molecular weight of hEGF, as estimated by gel filtration on a column of Bio-Gel P-10, was approximately 5700. The sedimentation equilibrium of hEGF was examined by Dr. Leslie Holladay at Vanderbilt University. Linear plots of ln concentration against $r^2$ revealed no significant heterogeneity, and a weight average molecular weight of 5291 was calculated.

The results of preliminary amino acid analyses of 20-43 μg samples of hEGF indicated that mouse and human EGF exhibited both differences and similarities with respect to their amino acid compositions. For example, both molecules contained six half-cystines and each lacked phenylalanine; however, whereas mouse EGF was missing alanine and lysine, human EGF was missing threonine. Despite these differences the fibroblast binding assay suggested that the receptor binding sites of the two molecules were similar; the small, but definite, cross-reactivity of hEGF with antibodies to the mouse polypeptide (Starkey *et al.*, 1975) suggested that at least some of the antigenic sites on the two molecules also were related.

In view of the remarkable structural and biological similarities between mouse EGF and the human urinary hormone urogastrone (Gregory, 1975), it is probable, as suggested by Gregory, that urogastrone and human EGF are either identical or very closely related polypeptides. A comparison of the amino acid sequences of human urogastrone and mouse EGF is shown in Fig. 2. Although there are some differences in the reported amino acid composition of human EGF as compared to human urogastrone, these may be due to the analytical difficulties encountered when analyzing very small amounts of protein. It is of interest to note that during the isolation of both human EGF and urogastrone, the presence of at least two biologically active, related polypeptides was detected.

## III. BIOLOGICAL PROPERTIES OF hEGF

The biological effects of mouse-derived EGF include (i) induction of precocious eyelid opening when injected into newborn mice, (ii) hypertrophy and hyperplasia of corneal and skin epithelial cells in organ cultures, and (iii) stimulation of the proliferation of human fibroblasts and glia in cell culture. All these effects have been duplicated, at least qualitatively, with human EGF.

### *A. Biological Activity of hEGF in Newborn Mice*

Human EGF was assayed for precocious eyelid-opening activity by the daily subcutaneous injection into a newborn mouse (Cohen, 1962). Control mice opened their eyes at 14 days; mEGF (1 μg/g per day or 0.25 μg/g per

| | | | | | | | | | 10 | | |
|---|---|---|---|---|---|---|---|---|---|---|---|
| Asn | Ser | Tyr | Pro | Gly | Cys | Pro | Ser | Ser | Tyr | Asp | Gly |
| Asn | Ser | Asp | Ser | Glu | Cys | Pro | Leu | Ser | His | Asp | Gly |
| | | | | | | | 20 | | | | |
| Tyr | Cys | Leu | Asn | Gly | Gly | Val | Cys | Met | His | Ile | Glu |
| Tyr | Cys | Leu | His | Asp | Gly | Val | Cys | Met | Tyr | Ile | Glu |
| | | | | | 30 | | | | | | |
| Ser | Leu | Asp | Ser | Tyr | Thr | Cys | Asn | Cys | Val | Ile | Gly |
| Ala | Leu | Asp | Lys | Tyr | Ala | Cys | Asn | Cys | Val | Val | Gly |
| | | | 40 | | | | | | | | |
| Tyr | Ser | Gly | Asp | Arg | Cys | Gln | Thr | Arg | Asp | Leu | Arg |
| Tyr | Ile | Gly | Glu | Arg | Cys | Gln | Tyr | Arg | Asp | Leu | Lys |
| | 50 | | | | | | | | | | |
| Trp | Trp | Glu | Leu | Arg | ——— | EGF (mouse) | | | | | |
| Trp | Trp | Glu | Leu | Arg | ——— | UROGASTRONE | | | | | |

Fig. 2. Comparison of the amino acid sequences of human urogastrone and mouse epidermal growth factor with the changed residues enclosed. (from Gregory, 1975)

day) resulted in eyelid opening on days 9 or 11, respectively; hEGF (0.4 μg/g per day) produced eyelid opening on day 11. Human EGF thus appeared to be active in this *in vivo* assay.

### *B. Biological Activity of hEGF in Organ Culture*

Mouse EGF and hEGF, when assayed with organ cultures of chick embryo cornea (Savage and Cohen, 1973), were equally effective in causing proliferation of the corneal epithelium. Histological results (not shown) were identical to those previously described for mEGF (Cohen and Savage, 1974) and recently described for hEGF partially purified by affinity chromatography (Starkey *et al.,* 1975). The biological effects of mouse and human EGF on the corneal epithelium were completely and specifically inhibited by the addition of gamma globulin prepared from a rabbit antiserum against mEGF.

### *C. Biological Activity of hEGF in Cell Culture*

The effect of hEGF on the growth of human foreskin fibroblasts (HF cells) *in vitro* was studied by measuring cell numbers, incorporation of labeled thymidine, and autoradiography.

*1. Effect of hEGF on the Proliferation of Human Fibroblasts.* The data in Fig. 3 illustrate the effect of picomolar quantities (4 ng/ml) of hEGF on the proliferation of human fibroblasts plated in Dulbecco's Modified Eagle

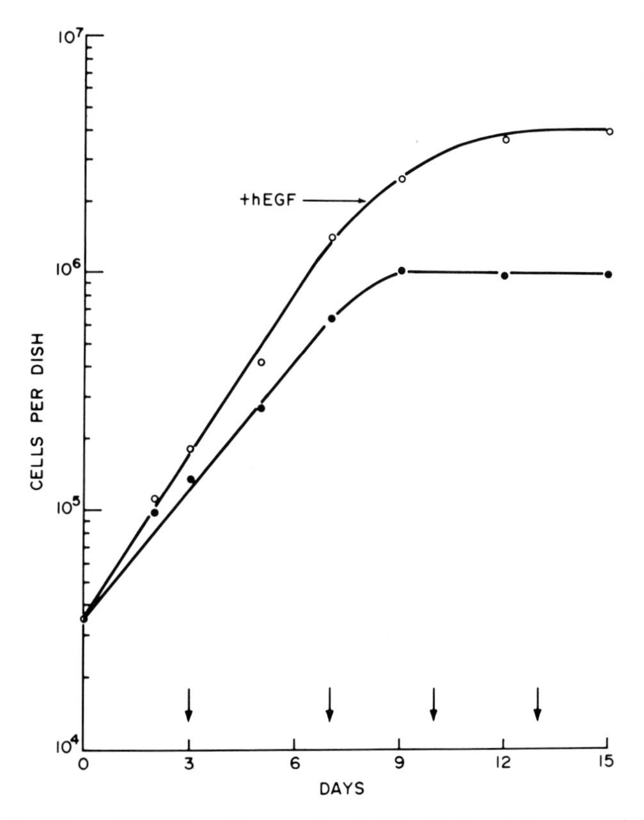


Fig. 3. Effect of hEGF on the growth of HF cells in medium containing 10% calf serum. HF cells were plated at approximately $3 \times 10^4$ cells per dish into 60 mm Falcon culture dishes containing DM plus 10% calf serum. The cells were incubated overnight, and hEGF (4 ng/ml) was added to one-half the dishes (day 0). At indicated times thereafter duplicate dishes from cultures growing in the presence (○—○) or absence (●—●) of hEGF were removed and the cell numbers determined. At the times indicated by the arrows, the medium in each set of cultures was removed, and fresh DM plus 10% calf serum was added. Human EGF was also added to the appropriate dishes at these times. (from Carpenter and Cohen, 1976a)

Medium (DM) and 10% calf serum. In the presence of hEGF the cells achieved a saturation density of $2 \times 10^5$ cells per $cm^2$, approximately four-fold higher than the density of $5 \times 10^4$ cells per $cm^2$ reached by cells grown in the absence of hEGF. The saturation density reached by the control cultures was not limited by the depletion of nutrients from the medium since no net increase in cell number was observed when fresh growth medium was added.

The morphological appearance of cultures of HF cells grown to saturation densities in the presence and absence of hEGF, as described above, is shown in Fig. 4. The cells grown in the absence of hEGF formed a tightly packed monolayer with typical fibroblastic morphology and orientation. These cultures showed only occasional areas of nuclear overlap. In contrast, the cells cultured in the presence of hEGF, while retaining their fibroblast appearance and orientation, grew in multiple cell layers and exhibited extensive areas of nuclear overlap.

The proliferation of fibroblasts in cell culture is characterized by various growth-regulating mechanisms. The behavior of cultures of HF cells grown to high cell density in the absence of hEGF is typical of a growth-limiting mechanism referred to as density dependent inhibition of growth (also as contact inhibition or topoinhibition). The data in Figs. 3 and 4 show that treatment of cultures of HF cells with hEGF results in cell proliferation that is not restricted in the usual manner by density dependent inhibition of growth.

Other growth-regulating mechanisms are defined by the serum requirement of fibroblasts. In order to proliferate, normal fibroblasts require a serum concentration of approximately 10% and also require complete serum. These cells do not grow if the serum concentration is lowered to 1% or if gamma globulin-free serum or plasma is present. To assess the influence of hEGF on the serum requirement of HF cells, cultures were placed in media containing serum alterations as described in Table I. The results presented in this table show that HF cells grew poorly in media containing 1% serum, gamma globulin-free serum or plasma. In the presence of hEGF, however, significant cell proliferation occurred in serum deficient media. The final cell densities achieved in serum deficient media containing hEGF were greater than those reached by cells grown in the presence of 10% serum.

The data described above show that the treatement of cells *in vitro* with hEGF results in cell proliferation which is not responsive to mechanisms which would otherwise limit cell growth. Interestingly, transformation of

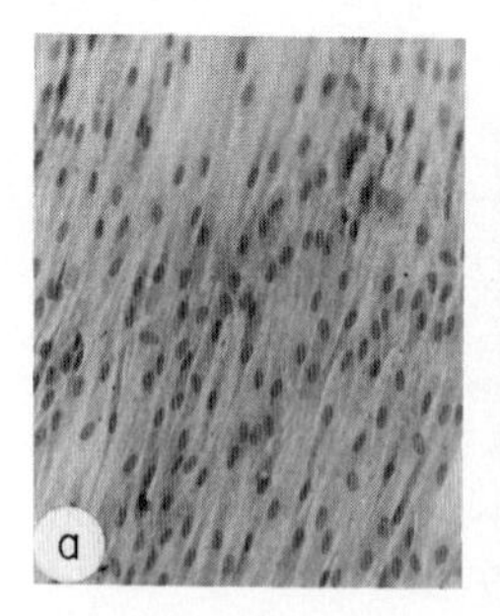


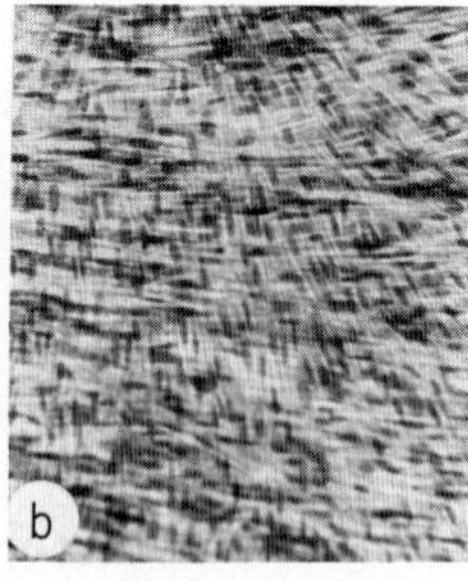


Fig. 4. Effect of hEGF on the morphological appearance of HF cells grown to saturation density in medium containing 10% calf serum. HF cells were grown to maximal saturation densities in DM containing 10% calf serum with (a) or without (b) hEGF. (from Carpenter and Cohen, 1976a)

**TABLE 1**

*Effect of hEGF on Proliferation of HF Cells in Serum-limited Media*

| Additions to Growth Medium | Day 0 | Cell Density (cells $\times 10^{-2}$ per $cm^2$) Day 3 | Day 6 | Day 9 |
|---|---|---|---|---|
| 10% Calf Serum | 64 | 290 | 624 | 681 |
| 10% Calf Serum + hEGF | | 410 | 1486 | 1866 |
| 1% Calf Serum | | 119 | 230 | 245 |
| 1% Calf Serum + hEGF | 63 | 247 | 571 | 848 |
| 10% γ-Globulin Free Calf Serum | | 162 | 335 | 389 |
| 10% γ-Globulin Free Calf Serum + hEGF | 95 | 263 | 1028 | 1437 |
| 10% Bovine Plasma | | 108 | 152 | 161 |
| 10% Bovine Plasma + hEGF | 96 | 217 | 780 | 1242 |

HF cells were plated in 60 mm Falcon culture dishes containing DM plus indicated serum components. The cells were incubated overnight, the cell density was determined (day 0); and hEGF (4 ng/ml) was added to the indicated dishes. At indicated times thereafter duplicate cultures of cells growing in the presence or absence of hEGF were removed and cell numbers determined. On day 3 and day 6 the medium in each group of cultures was removed and fresh media with appropriate serum additions were added. Human EGF was also added to the indicated dishes at these times.

fibroblasts by agents such as the oncogenic virus SV40 also renders cells insensitive to the growth limiting mechanisms described above. A further discussion of the growth of hEGF-treated cells and transformed cells has been presented elsewhere (Carpenter and Cohen, 1976a).

*2. Stimulation of DNA Synthesis by hEGF.* The following experiments were performed to characterize some of the parameters which may be involved in the hEGF-mediated stimulation of proliferation of cells subject to density dependent inhibition of growth. In these experiments HF cells were grown in DM plus 10% calf serum until a confluent monolayer was formed. Fresh medium containing 1% calf serum was added and 48 hours later the confluent, quiescent monolayer of cells was stimulated by the addition of hEGF.

The time course of DNA synthesis, as judged by the incorporation of labeled thymidine, following the addition of hEGF or fresh serum to confluent, quiescent HF cells is represented in Fig. 5. Under these conditions, an increased rate of DNA synthesis was detectable after 12 hours of incubation in the presence of either hEGF or serum; maximal stimulation occurred at approximately 24 hours. The incorporation of labeled thymidine was maximal in the presence of 2 ng/ml ($3.7 \times 10^{-10}$ M) hEGF; the response was half-maximal at a concentration of approximately 0.25 ng/ml (4.6 ×

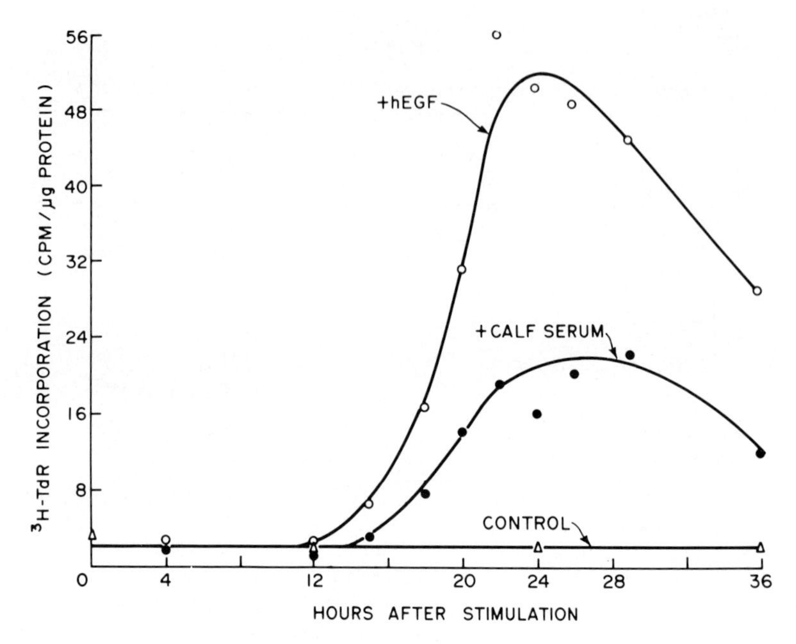


Fig. 5. Time course of [$^3$H]thymidine incorporation following stimulation of HF cells. Confluent, quiescent cultures of HF cells were stimulated by the addition of hEGF, 4 ng/ml (○——○) or fresh calf serum (●——●). Control cultures received no additions (△——△). At indicated times duplicate cultures were selected and labeled for one hour, 1 $\mu$C per ml and 0.15 $\mu$M with [$^3$H] thymidine. (from Carpenter and Cohen, 1976a)

$10^{-11}$ M) hEGF. These are concentrations comparable to the levels of mEGF in mouse plasma, $2.5 \times 10^{-10}$ M (Byyny *et al.,* 1974), and hEGF in human urine, $3.3 \times 10^{-8}$ M (unpublished results). The growth factor activity in human urine described by Holley and Kiernan (1968) might be attributable to hEGF.

The stimulation of DNA synthesis by hEGF in confluent, quiescent HF cells was significantly affected by the amount of serum in the medium and by the addition of ascorbic acid. The response of human fibroblasts to hEGF was greatest at serum concentrations above 2% and at approximately 2.5 $\mu$g/ml ascorbic acid. A similar serum requirement has been reported for the biological activities of mEGF (Cohen *et al.,* 1975) and fibroblast growth factor (Gospodarowicz and Moran, 1975).

The percentage of cells stimulated to synthesize DNA by hEGF under the conditions described above, was determined by autoradiography of cells incubated in the presence of labeled thymidine. The results showed that following the addition of hEGF to confluent, quiescent cultures maintained in medium containing 1% or 5% serum, 21% or 41% of the nuclei, respectively contained radioactivity. In the presence of ascorbic acid and hEGF an additional 13 to 15% of nuclei were labeled. In the absence of hEGF no more than 3.7% of the nuclei were labeled.

To analyze the stimulation of cells *in vitro* by a mitogenic agent it is important to determine whether the initial interaction (binding) of the mitogen to the cell surface is sufficient to ensure a sequence of intracellular events which will commit the cell to DNA synthesis and mitosis. Such a scheme has been referred to as a 'trigger' mechanism and been proposed to explain the metabolic effects of many polypeptide hormones. The stimulation of DNA synthesis in quiescent cells by serum (Ellem and Mironescu, 1972; Rubin and Steiner, 1975), multiplication stimulating factors (Smith and Temin, 1974) or Concanavalin A (Gunther *et al.*, 1974) has been reported to require the presence of the mitogen for several hours. Similar results are reported here for the stimulation of HF cells by hEGF. The data in Fig. 6 indicate that the addition of antibody prepared against hEGF, at any time during the first 3 hrs of exposure of fibroblasts to hEGF, blocked the stimulation of DNA synthesis. Addition of antibody at 7.5 hours after incubation of the

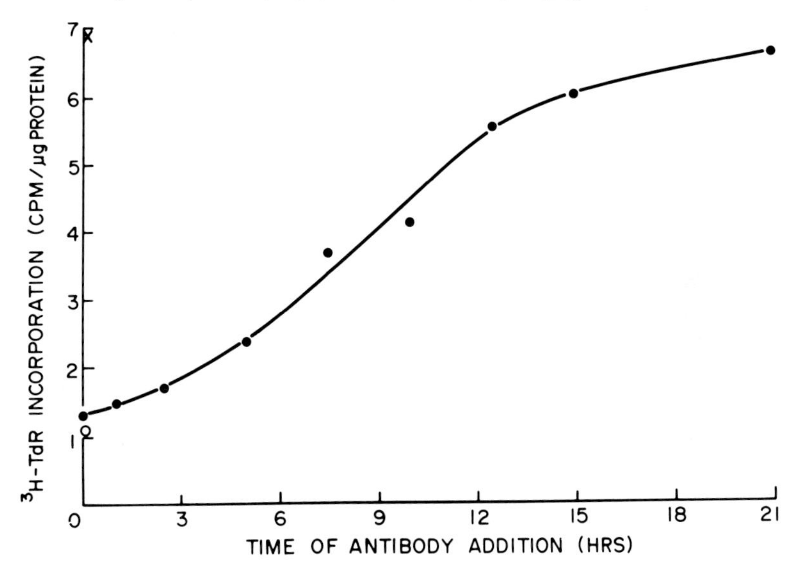


Fig. 6. Effect of the addition of antibody to hEGF on the stimulation of [$^3$H] thymidine incorporation into HF cells by hEGF. Human EGF (4 ng/ml) was added to quiescent, confluent cultures of HF cells and at various times thereafter 150 µg of DEAE purified antibody to hEGF were added to duplicate cultures. Twenty-one hours after the addition of hEGF to the cultures, [$^3$H] thymidine was added and the cells labeled for 4 hrs. Control cultures received hEGF but no antibody (X) or no additions (O). (from Carpenter and Cohen, 1976a)

cells with hEGF resulted in a stimulation of thymidine incorporation that was 50% of that observed in the absence of antibody. Since the binding of hEGF to human fibroblasts is a comparatively rapid event [maximal binding is achieved within 40 min. (see section IV B)] it does not appear that the stimulation of DNA synthesis by hEGF occurs by a 'trigger'-type mechanism.

## IV. INTERACTION OF $^{125}$ LABELED hEGF WITH HUMAN FIBROBLASTS

The data presented in previous sections demonstrate that hEGF is a potent mitogen for human fibroblasts *in vitro*, stimulating both the synthesis of DNA and cell proliferation. Although the ability of mitogens to stimulate the transport of nutrients, the synthesis of macromolecules, and cell division is well documented, little is known about the biochemical steps involved in the initial interaction of these agents with the cell surface. Since the binding of macromolecular mitogens to the cell surface membrane is reasoned to be the first step necessary for the expression of their biological activity, it is important to define the biochemical events involved in the binding of the hormone to the cell surface and to determine the metabolic fate of the bound hormone. Events which occur subsequent to binding are important because, as mentioned above, it does not appear that hEGF exerts its stimulatory effect by a 'trigger' type mechanism.

### A. $^{125}$*I-hEGF Binding Assay*

Confluent monolayer cultures of HF cells in 60 mm Falcon dishes were washed twice with Hanks' solution to remove growth medium, and 1.4 ml of binding medium were added to each dish. The binding medium consisted of DM medium plus bovine serum albumin (0.1%) and Gentamicin (50 $\mu$g/ml). Labeled hEGF or other components were added to a final volume of 1.5 ml. After incubation at 37° for indicated periods of time, the unbound radioactivity was removed by washing each monolayer with cold Hanks' solution containing bovine serum albumin (0.1%). The monolayers were then solubilized by the addition of sodium hydroxide and incubation at 37°. The solubilized material was transferred to counting vials and the amount of radioactivity determined with a gamma-spectrometer. Human EGF was iodinated by the chloramine T procedure (Hunter and Greenwood, 1962). The level of non-specific binding amounted to less than 2% of the total bound radioactivity. Additional details of the binding assay and iodination procedure have been reported (Carpenter *et al.,* 1975; Carpenter and Cohen, 1976b).

### B. *Binding of* $^{125}$*I-hEGF to Human Fibroblasts*

*1. Time course of Binding.* The time course of binding of $^{125}$I-hEGF to confluent monolayers of HF cells at 37° and 0° is shown in Fig. 7. Maximal binding was reached after incubations of 30-40 min at 37° or approximately

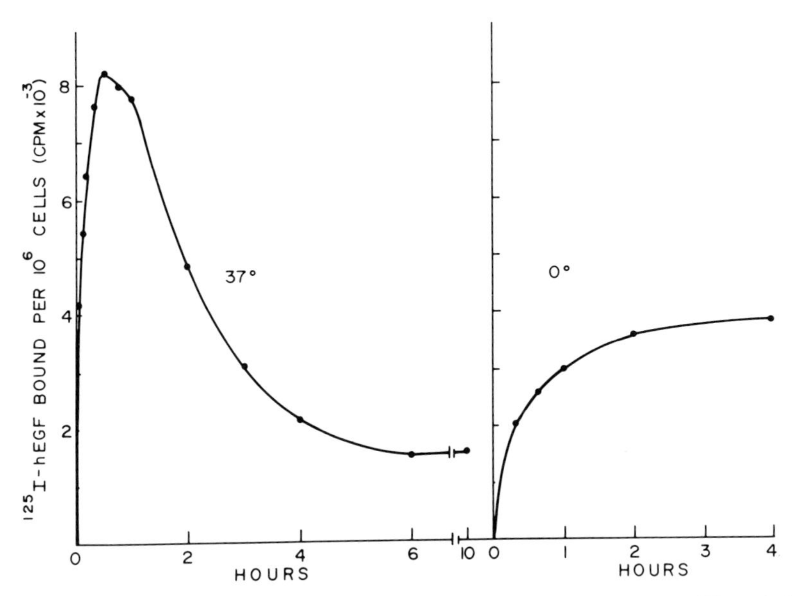


Fig. 7. Time course of $^{125}$I-hEGF binding to human fibroblasts at 37° and 0°. $^{125}$I-hEGF (final concentration 4 ng/ml, 24,400 cpm/ng) was added to each culture dish containing the standard binding medium (1.5 ml for the 37° experiments and 2.0 ml for the 0° experiments). At the indicated time intervals, duplicate dishes were selected and the cell-bound radioactivity was determined. (from Carpenter and Cohen, 1976b)

2.5 hours at 0°. On continued incubation at 37° the amount of cell-bound radioactivity decreased until a constant level of 15-20% of the initial maximal amount of cell-bound isotope remained associated with the cells. Continued incubation at 0°, however, did not result in a net decrease of cell-bound radioactivity suggesting the involvement of a temperature dependent mechanism for the loss of cell-bound radioactivity observed at 37°. The following experiments were conducted to determine the basis for the decrease in binding at 37°. Labeled hormone was incubated with cells for 6 hours at 37°, the medium was removed and added to a monolayer of 'fresh' cells. The results showed that the hormone remaining in the medium was fully active in binding to the monolayer of 'fresh' cells. This indicates that the loss of binding observed at 37° is not due to extensive degradation or inactivation of the free hormone in the medium. That the loss of binding which occurs at 37° is due to a cellular function is indicated by the demonstration that cells which had been incubated with the hormone for 6 hours at 37° and washed to remove unbound radioactivity were unable to bind 'fresh' hormone. Control experiments showed only a slight decrease (10%) in hEGF binding capacity following a 6 hour incubation in binding medium in the absence of hEGF. The loss of cell-associated radioactivity at 37°, therefore, is due to a hormone-dependent cell function. The simplest explanation for this phenomenon is the loss or inactivation of hEGF-specific cell surface receptors subsequent to hormone binding.

2. *Effect of hEGF Concentration on Binding.* The effect of increasing concentrations of hEGF on the binding indicates that the binding reaction is a saturable process with maximal binding at 8 ng/ml ($1.5 \times 10^{-9}$ M) and half maximal binding at 1.8 ng/ml ($3.3 \times 10^{-10}$M). These results and the data previously mentioned in section III C-2 show that the concentration of hEGF required to achieve maximal binding is approximately four-fold higher than the concentration of growth factor required for maximal stimulation of DNA synthesis. The results indicate that only a fraction (less than 25%) of the binding sites needs to be occupied for cells arrested in the $G_1$ phase of the cell cycle to enter the S phase. These results and conclusions are in agreement with the report of Hollenberg and Cuatrecasas (1975).

3. *Autoradiography of Cell Bound* $^{125}$*I-hEGF*. The binding of $^{125}$I-hEGF to HF cells in the presence and absence of an excess of unlabeled mEGF is shown in Fig. 8. The binding reaction was specific as evidenced by the absence of radioactive grains in the cells incubated with unlabeled hormone (Fig. 8a). The pattern of grains presented in Fig 8b was typical of all cells incubated with labeled hEGF.

4. *Specificity of Binding.* Previously it was shown (Carpenter *et al.*, 1975) that neither insulin, prolactin, FSH, TSH, growth hormone, nor glucagon competed with $^{125}$I-mEGF in the fibroblast receptor assay. Additional experiments have demonstrated that neither fibroblast growth factor, multipli-

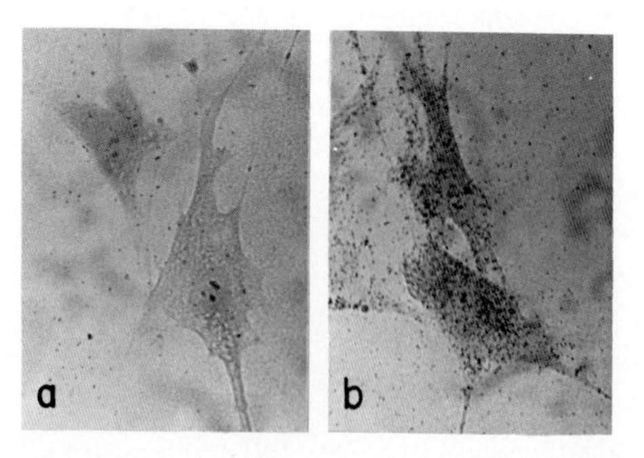


Fig. 8. Autoradiography of cell-bound $^{125}$I-hEGF. Labeled hEGF (3.5 ng/ml, 145,000 cpm/ng) was added to sparse cultures of human fibroblasts in the standard binding medium (a) in the presence and (b) in the absence of unlabeled mEGF (10 μg/ml). After a 40 min binding period at 37°, the cells were washed, fixed with glutaraldehyde, covered with a layer of NTB-2 emulsion, and exposed for approximately 6 weeks. (from Carpenter and Cohen, 1976b)

cation stimulating activity, somatomedins B and C, nor non-suppressible insulin like activity competes with $^{125}$I-hEGF in the binding assay. Both hEGF and mEGF effectively compete with $^{125}$I-hEGF in the binding assay.

*5. Nature of the Cell-Bound $^{125}$I-hEGF.* The cell-bound $^{125}$I-hEGF was extracted with 0.1 N HCl, characterized by gel filtration, and tested for its ability to rebind to fibroblasts.

HF cells were incubated with labeled hEGF for 40 min at 37° and extracted with acid. Eighty percent of the radioactivity was extracted by this procedure and, of this amount, 98% was of the same molecular weight as native hEGF as determined by gel filtration chromatography (see Carpenter *et al.,* 1975, and Carpenter and Cohen, 1976b, for details of procedure). Low molecular weight products of $^{125}$I-hEGF degradation do not appear to accumulate within the cell.

The ability of cell-bound $^{125}$I-hEGF to rebind to 'fresh' cells was examined. Labeled hEGF was incubated with HF cells for 15 min at 37° prior to extraction with acid. The extracted material was lyophilized, rehydrated in binding medium, and was tested for binding activity. The results showed that the extracted material was able to rebind to fibroblasts with an affinity almost identical to that of the native hormone. This result and that described above suggest that the structure of the hEGF is not grossly altered during the initial binding reaction.

### *C. Degradation of Cell-Bound $^{125}$I-hEGF*

*1. Kinetics of Degradation.* The rate of release of cell-bound $^{125}$I-hEGF was measured after incubating monolayers of HF cells with the labeled hormone for 40 min at 37°. The monolayers were washed to remove unbound hormone, fresh binding medium (without hEGF) was added, and the amount of cell-bound radioactivity was determined at various times thereafter. The results of this experiment showed that the amount of cell-bound radioactivity decreased rapidly with a half-life of approximately 20 min. After 2 hours incubation, over 85-95% of the initial cell-bound radioactivity was no longer associated with the cells. Upon addition of 'fresh' $^{125}$I-hEGF at this time only 10-20% of the original binding capacity could be detected. The rapid loss of cell-bound radioactivity under these conditions was largely temperature dependent. When the labeled hormone was bound to the cells at 37°, less than 10% of the cell-bound radioactivity was released during a subsequent 6 hour incubation at 0°. The loss of cell-bound radioactivity was also measured after incubation of HF cells with labeled hEGF for 2 hours at 0°, washing, and re-incubation at 0°. Under these conditions the amount of cell-bound radioactivity slowly decreased until approximately 55% of the cell-bound material was released from the cells after 4-6 hours. Our interpretation of these data is discussed in section D.

*2. Nature of Radioactive Material Released.* The nature of the radioactive material released into the medium at 37° and 0° was determined by gel filtration. The data in Table II show that two peaks of radioactivity were detected: a high molecular weight fraction the elution of which corresponded to that of intact hEGF and a low molecular weight fraction with an elution volume corresponding to that of monoiodotyrosine. Essentially all of the radioactive material released at 0° had the same elution volume as intact $^{125}$I-hEGF; however at 37°, most of the radioactivity was released as low molecular weight material. This low molecular weight material was identified as 80-85% [$^{125}$I]-monoiodotyrosine and 5-10% [$^{125}$I]-diiodotyrosine on the basis of co-chromatography on paper with authentic mono- and diiodotyrosine. These data indicate that whereas at 0° cell bound $^{125}$I-hEGF dissociates from the cell as an intact molecule, at 37° most of the cell-bound hormone is rapidly and extensively degraded prior to release from the cell.

*3. Inhibition of Degradation of Cell-Bound $^{125}$I-hEGF.* To investigate the mechanism involved in the rapid degradation of cell-bound $^{125}$I-hEGF we attempted to inhibit this process. Previous studies (Carpenter, *et al.,* 1975) demonstrated that the trypsin inhibitors tosyl-L-lysine chloromethyl ketone and the benzyl ester of guanidobenzoic acid inhibited the degradation of $^{125}$I-mEGF by HF cells. At concentrations of 50-100 μg/ml these protease inhibitors also blocked the degradation of $^{125}$I-hEGF by approximately 55%.

**TABLE 2**

*Nature of Cell-bound $^{125}$I-hEGF Released into Medium*

| Temperature | Time interval | Radioactivity released | |
|---|---|---|---|
| | | High MW | Low MW |
| | hrs | % | % |
| 37° | 0.0 – 0.5 | 32 | 68 |
| | 0.5 – 2.0 | 10 | 90 |
| 0° | 0.0 – 2.0 | >99 | <1 |
| | 2.0 – 8.0 | >99 | <1 |

$^{125}$I-hEGF (1.6 ng/ml, 24,700 cpm/ng) was allowed to bind to confluent cultures of HF fibroblasts at 37° for 40 min or at 0° for 60 min; approximately 5000 or 1000 cpm of $^{125}$I-hEGF were bound, respectively. The culture dishes were then washed free of unbound hEGF using Hanks' solution, 2 ml of the standard binding medium were re-added, and the cultures were incubated at the indicated temperatures. The medium, containing the released radioactivity, was collected at the intervals indicated and lyophilized. The distribution of the radioactivity in each sample was examined by gel filtration (Carpenter *et al.,* 1976b), using aliquots containing 400 – 5000 cpm of $^{125}$I.

A number of inhibitors of metabolic energy production were tested for their ability to influence the degradation of cell-bound $^{125}$I-hEGF. In a medium lacking glucose and amino acids, dinitrophenol (lmM), sodium azide (5mM), or sodium cyanide (5mM) inhibited the degradation of cell-bound radioactivity by 73%, 52% or 48%, respectively. The results indicate that the degradation of $^{125}$I-hEGF is dependent upon the generation of metabolic energy.

Several other potent and interesting inhibitors of the degradation process were found. The following compounds inhibited the degradation of cell-bound $^{125}$I-hEGF by the indicated percentages: 0.1 mM chloroquine, 72%; 1 mM lidocaine, 66%; 1 mM cocaine, 78%; 1 mM procaine, 64%; and 10 mM ammonium chloride, 98%. At these concentrations none of the inhibitors altered the cell morphology as judged by light microscopy nor did these agents, including those mentioned above, inhibit the initial binding of $^{125}$I-hEGF to HF cells. The inhibitory effect of chloroquine suggests that the degradation might be mediated by lysosomal enzymes. The lysosomal action of chloroquine has been reported by various investigators (DeDuve *et al.*, 1974; Lie and Schofield, 1973; Goldstein *et al.*, 1975; and Wibo and Poole, 1974). Although the mechanism of action of the local anesthetics and ammonium ion are not known, it is possible that their activity may be similar to that of chloroquine as all these agents are lipid soluble and permeable to cell membranes. Recently, Nicholson *et al.* (1976) have found that local anesthetics induce a disruption of submembraneous systems of microfilaments and microtubules.

### D. *Internalization of Cell-Bound $^{125}$I-hEGF*

Experiments were performed to provide evidence, albeit indirect, that cell-bound $^{125}$I-hEGF is internalized prior to degradation. The rationale for these experiments is that if internalization (endocytosis) of membrane-bound hormone is minimized by lowering the temperature to 0°, the accessibility of the bound hormone to reagents in the medium, such as trypsin or antibodies to hEGF, will be increased.

The experiment reported in Fig. 9 demonstrates that $^{125}$I-labeled antibodies to hEGF were able to bind to HF cells which had been previously incubated with hEGF at 0°. The capacity of the hEGF treated cells to bind antibody was rapidly lost upon warming for 1-8 min at 37° prior to the addition of the labeled antibody. The half-life for the loss of antibody binding capacity was 1-2 min. Control experiments utilizing $^{125}$I-hEGF demonstrated that under these conditions the labeled hormone did not dissociate from the surface into the medium.

The sensitivity of cell-bound $^{125}$I-hEGF to trypsin is markedly dependent on the temperature at which the binding of labeled hormone to the cells was

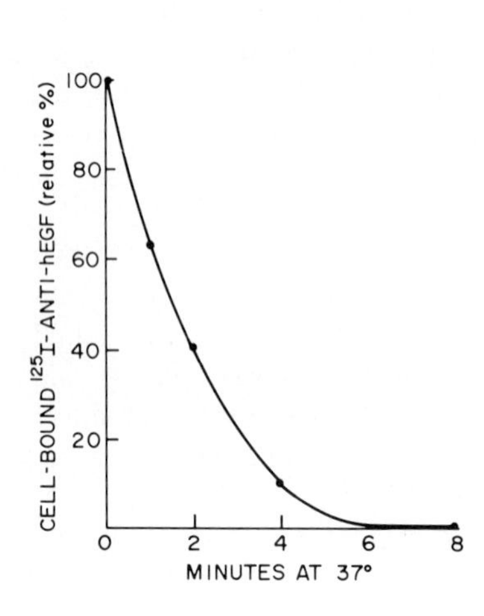


Fig. 9. Binding of $^{125}$I-labeled rabbit hEGF-antiserum to fibroblasts pre-incubated with hEGF. Unlabeled hEGF (6 ng/ml) was added to the culture dishes containing 2 ml of the binding medium at 0°. After a 2 hr incubation, the cells were washed free of unbound hEGF, and 2 ml of binding medium were re-added. $^{125}$I-labeled antiserum to hEGF (0.75 $\mu$g, 3 × $10^6$ cpm/$\mu$g) was added and the cultures were re-incubated for 90 min at 0°. Then the cultures were washed and the cell-bound radioactivity determined; the amount of radioactivity present (corrected for "non-specific" binding) was taken at 100%. Replicate cultures, taken just prior to the addition of antiserum, were incubated at 37° for the indicated time intervals, re-chilled, and then tested for their ability to bind the $^{125}$I-labeled antiserum. The radioactivity which was cell-bound in the absence of added hEGF was considered to be the "non-specific blank" and was subtracted from the experimental values obtained. In separate experiments this "blank" value of approximately 3,000 cpm/dish amounted to 25-50% of the value obtained in the presence of hEGF. (from Carpenter and Cohen, 1976b)

carried out. $^{125}$I-hEGF was incubated with confluent monolayers of HF cells for 40 min at 37° or 60 min at 0°. All the cultures were then placed at 0° and washed free of unbound hormone. To each dish 1 ml of 0.25% trypsin was added, and the trypsinization was allowed to proceed for 30 min at 0°. Serum-containing medium was added to stop the trypsin activity, the detached cells were collected by centrifugation, and the radioactivity present in the cell pellets and the supernatants was determined. The results showed that approximately 50% of the radioactivity bound to the cells at 0° was liberated by trypsin; however only 10% of the material bound to the cells at 37° was released by this procedure.

As previously described (Section IV,C-1) when the labeled hormone was bound to cells at 37°, the expected release of the cell-associated radioactivity did not occur when the temperature was lowered to 0°. These results and those described above are consistent with a mechanism in which $^{125}$I-hEGF if bound to the cell surface and subsequently internalized (most probably by endocytosis) prior to degradation within lysosomes. The inhibition of degradation by inhibitors such as chloroquine is also consistent with this hypothesis.

*E. hEGF-Mediated Loss and Recovery of Receptor Activity.*

The results of previous experiments (see Section IV, B-1 and C-1) indicated that following the binding and degradation of hEGF, human fibroblasts were capable of rebinding only a small quantity of 'fresh' growth factor (approximately 10-20% of the initial value). The following experiment (Fig. 10) was performed to examine the kinetics and the metabolic requirements for the recovery of hEGF binding capacity by these cells. The binding sites on replicate cultures of HF cells were saturated with unlabeled EGF, unbound hormone was removed, and the cell-bound material was allowed to degrade. The time course of recovery of hEGF binding capacity by these cells then was examined. The results indicate that: 1) approximately 10 hours of

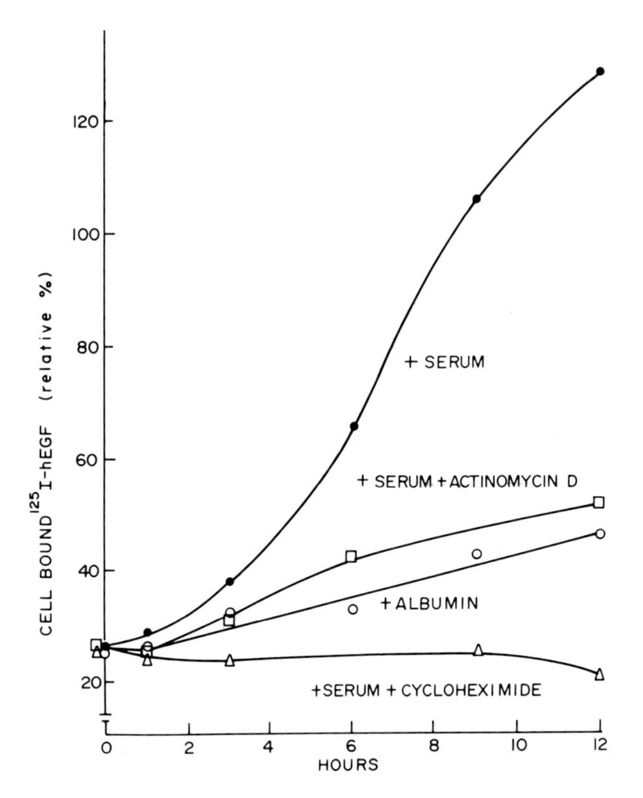


Fig. 10. Recovery of hEGF binding capacity in fibroblasts. An excess of unlabeled mEGF (1 μg/ml) was added to replicate culture dishes containing 1.5 ml of the standard binding medium at 37° . After a 1 hr period, the cells were washed to remove unbound mEGF, 1.5 ml of the binding medium were re-added and the cultures were incubated for 4 hrs at 37° to permit degradation of the bound hormone. Then the medium was replaced with one of the following: (a) DM plus 0.1% albumin (○——○), (b) DM plus 10% calf serum (●——●), (c) DM plus 10% calf serum plus cycloheximide (20 μg/ml) (△——△), or (d) DM plus 10% calf serum plus actinomycin D (1 μg/ml) (□——□). At the indicated times duplicate dishes from each group were selected, washed and assayed for their binding capacity for $^{125}$I-hEGF (6 ng/ml) in the standard binding medium (40 min at 37°). The extent of $^{125}$I-hEGF binding in replicate cultures, *not* exposed to mEGF, was taken as 100% and the results are expressed as the relative percentage of cell-bound radioactivity. (from Carpenter and Cohen, 1976b)

incubation in serum-containing medium were required for the complete recovery of binding capacity and 2) the synthesis of RNA and protein was necessary for the reappearance of hEGF binding in these cells.

The decreased ability of cells to bind $^{125}$I-hEGF following the binding and degradation of the hormone suggests that either the entire hormone-receptor complex is removed from the cell surface by endocytosis or that the receptor has been masked or otherwise inactivated. The return of binding capacity in the presence of serum and the inhibition of this recovery by cycloheximide or actinomycin D suggests that the receptor is resynthesized following endocytosis of the hormone-receptor complex. The possibility that macromolecular synthesis is required for the recycling of 'old' receptors or for the insertion into the membrane of receptors from an intracellular pool has not been excluded. The regulation of the concentration of membrane receptors by other polypeptides has been reported (Gavin *et al.*, 1974; Hinkle and Tashjian, 1975; Devreotes and Fambrough, 1975; Oliver *et al.*, 1974).

## V. CONCLUDING REMARKS

The evolutionary conservation of similar EGF polypeptides in man and mouse and the presence of receptors for these molecules in a wide variety of cells and species argue for a normal and essential function for EGF in cell control. *In vivo* effects of the hormone on such diverse processes as epidermal growth and gastric secretion are readily demonstrable, as are its mitogenic properties in cell and organ culture. However, the exact physiological significance of EGF in normal development and homeostasis remains to be established.

### ACKNOWLEDGMENTS

These studies were supported by U.S. Public Health Services Grant HD-00700. G.C. is a postdoctoral fellow of the USPHS, 5 F22 AMO1176.

### REFERENCES

Armelin, H. A. (1973). *Proc. Nat. Acad. Sci. U.S.A.* **70**, 2702–2706.

Byyny, R. L., Orth, D. N., Cohen, S. and Doyne, E. S. (1974). *Endocrinology* **95**, 776–782.

Carpenter, G. and Cohen, S. (1976a). *J. Cell. Physiol.* **88**, 227–238.

Carpenter, G. and Cohen, S. (1976b). *J. Cell. Biol.* **71**, 159–171.

Carpenter, G., Lemb]ch, K. J., Morrison, M. and Cohen, S. (1975). *J. Biol. Chem.* **250**, 4297–4304.

Cohen, S. (1962). *J. Biol. Chem.* **237**, 1555–1562.

Cohen, S. and Carpenter, G. (1975). *Proc. Nat. Acad. Sci. U.S.A.* **72**, 1317–1321.

Cohen, S. and Savage, C. R., Jr. (1974). *Rec. Prog. Horm. Res.* **30**, 551–574.

Cohen, S. and Taylor, J. M. (1974). *Rec. Prog. Horm. Res.* **30**, 533–550.

Cohen, S., Carpenter, G. and Lembach, K. J. (1975). *Adv. Metab. Dis.* **8**, 265–284.
DeDuve, C., DeBarsy, T., Poole, B., Trouet, A., Tulkens, P. and Van Hoof, F. (1974). *Biochem. Pharmacol.* **23**, 2495–2531.
Devreotes, P. N. and Fambrough, D. M. (1975). *J. Cell Biol.* **65**, 335–358.
Ellem, K. A. O. and Mironescu, S. (1972) *J. Cell. Physiol.* **79**, 389–406.
Gavin, J. R., III., Roth, J., Neville, D. M., Jr., De Meyts, P. and Beull, D. N. (1974). *Proc. Nat. Acad. Sci. U.S.A.* **71**, 84–88.
Goldstein, J. L., Brunschede, G. Y. and Brown, M. S. (1975). *J. Biol. Chem.* **250**, 7854–7862.
Gospodarowicz, D. and Moran, J. S. (1975). *J. Cell Biol.* **66**, 451–457.
Gregory, H. (1975). *Nature* **257**, 325–327.
Gunther, G. R., Wang, J. L. and Edelman, G. M. (1974). *J. Cell Biol.* **66**, 451–457.
Hinkle, P. M. and Tashjian, A., Jr. (1975). *Endocrinology* **14**, 3845–3851.
Hollenberg, M. D. (1976). *Arch. Biochem. Biophys.* **171**, 371–377.
Hollenberg, M. D. and Cuatrecasas, P. (1973). *Proc. Nat. Acad. Sci. U.S.A.* **70**, 2964–2968.
Hollenberg, M. D. and Cuatrecasas, P. (1975). *J. Biol. Chem.* **250**, 3845–3853.
Holley, R. W. and Kiernan, J. A. (1968). *Proc. Nat. Acad. Sci. U.S.A.* **60**, 300–304.
Hunter, W. M. and Greenwood, F. C. (1962). *Nature* **194**, 495–496.
Lie, S. O. and Schofield, B. (1973). *Biochem. Pharmacol.* **22**, 3109–3114.
Nicholson, G. L., Smith, J. R. and Poste, G. (1976). *J. Cell Biol.* **68**, 395–402.
Oliver, J. M., Ukena, T. E. and Berlin, R. D. (1974). *Proc. Nat. Acad. Sci. U.S.A.* **71**, 394–398.
Rubin, H. and Steiner, R. (1975). *J. Cell. Physiol.* **85**, 261–270.
Savage, C. R., Jr. and Cohen, S. (1973). *Exp. Eye Res.* **15**, 361–366.
Smith, G. L. and Temin, H. (1974). *J. Cell. Physiol.* **84**, 181–192.
Starkey, R. H., Cohen, S. and Orth, D. N. (1975). *Science* **189**, 800–802.
Westermark, B. (1976). *Biochem. Biophys. Res. Comm.* **69**, 304–310.
Wibo, M. and Poole, B. (1974). *J. Cell Biol.* **63**, 430–440.

# Cellular Specificities of Fibroblast Growth Factor and Epidermal Growth Factor

Denis Gospodarowicz, John S. Moran and Anthony L. Mescher

*Cancer Research Institute*
*School of Medicine*
*University of California*
*San Francisco, California 94143*

## I. INTRODUCTION

The proliferation of animal cells is strictly regulated by mechansims that are largely unknown. It has been known for many years that hormones such as ACTH, growth hormone, and estrogen are important for the proliferation and differentiation of their target cells *in vivo* but these hormones do not stimulate the proliferation of their target cells *in vitro* (Mueller, 1953; Salmon and Daughaday, 1975; Masui and Garren, 1971). This lack of effect of classical hormones has led to the concept that the stimulation of cell proliferation by hormones could be indirect and that growth control may be mediated by a recently discovered class of polypeptides – the mitogenic growth factors. Such a class of biological agents is by no means new to biologists; a polypeptide involved in the control of erythroblast proliferation and differentiation was identified as early as 1906 (Carnot and Deflandre, 1906). In recent years, much attention has been given to two purified, mitogenic growth factors: epidermal growth factor (EGF) and fibroblast growth factor (FGF).

EGF, first isolated by Stanley Cohen from the submaxillary glands of adult, male mice (Cohen, 1962), is an acidic polypeptide with a molecular weight of 6045 (Savage *et al.,* 1972). It was first shown to stimulate the proliferation of epidermal cells both *in vivo* (Cohen, 1962) and *in vitro* (Cohen, 1965). It has recently been reported to be a potent mitogen for human diploid fibroblasts (Hollenberg and Cuatrecasas, 1973; Cohen and Carpenter, 1975; Lembach, 1976). The structure of mouse EGF is similar to that of human urogastrone, a polypeptide that inhibits gastric acid secretion (Gregory, 1975).

FGF has been purified in our laboratory from the pituitary and brain of mammals and is a basic polypeptide of 13,400 molecular weight

(Gospodarowicz, 1975). Named on the basis of its mitogenic potency for Balb/c 3T3 cells (Gospodarowicz, 1974), it has recently been shown to stimulate the division of a wide variety of cells (Table I).

In the research described here we compared the effects of FGF and EGF on a variety of mesoderm-derived cells *in vitro* in an effort to elucidate possible roles of these growth factors during development. We also investigated the possibility that FGF may be a "neurotrophic" growth factor by testing its effect *in vivo* on amphibian limb regeneration, an example of nerve-dependent cell proliferation.

## II. RESULTS

### *A. 3T3 Cells*

3T3 cells are permanent lines of cells derived from minced, 14-17 day-old mouse embryos. They have been much studied as prototype nontumorigenic "fibroblast" lines. These cells proliferate rapidly in 10% serum until they form a confluent monolayer at which time cell division ceases (Todaro and Green, 1963).

When sparse populations of Balb/c 3T3 cells are maintained in low serum (<1%) cell division ceases with over 90% of the cells coming to rest in the $G_1$ phase of the cell cycle. The addition of as little as 0.01 ng/ml ($10^{-12}$M) of either brain or pituitary FGF to such cells stimulates the initiation of DNA synthesis is some cells and 2.5 ng/ml cause a maximal stimulation which is 30-50% of that observed after the addition of an optimal concentration of serum (Gospodarowicz, 1974). In the presence of glucocorticoid, FGF has a stimlatory effect on both DNA synthesis and cell division which is comparable to the effect of serum (Gospodarowicz, 1974, 1975; Gospodarowicz and Moran, 1974a). The effect is more pronounced when cells are maintained in the presence of platelet-poor plasma serum than when blood serum is used (Gospodarowicz *et al.*, 1975a). Of special interest is the fact that FGF also causes a marked overgrowth of "contact-inhibited" cells maintained in 10% serum (Fig. 1).

In addition to Balb/c 3T3 cells, FGF has also been shown to be mitogenic for Swiss 3T3 cells (Rudland *et al.*, 1974; Holley and Kiernan, 1974) and for thermosensitive mutants of polyoma virus-transformed Balb/c 3T3 cells maintained at the unpermissive, but not at the permissive, tmperature (Rudland *et al.*, 1974b). Armelin (1973) and Rose *et al.* (1975) have shown that 3T3 cells can also be stimulated to proliferate by low concentrations ($10^{-10}$M) of EGF.

### *B. Diploid Fibroblasts*

Since Todaro and Green (1963) describe the establishment of 3T3 cell lines as a process of selection for mutant, usually heteroploid, cells adapted to *in vitro* culture, it cannot be assumed that growth control in these cells is the

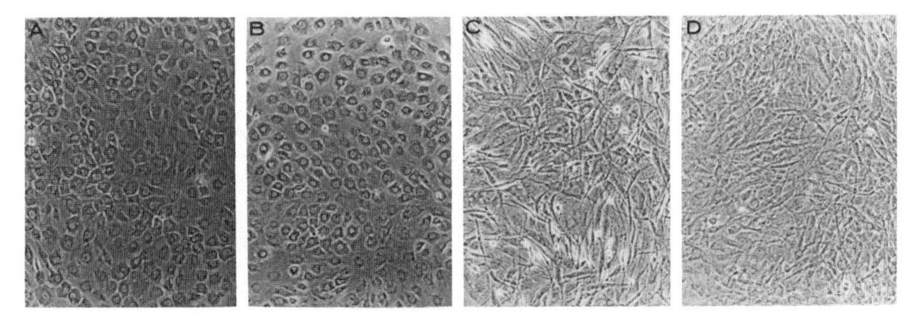


Fig. 1. Effect of FGF and glucocorticoid on Balb/c 3T3 cell "density-inhibition" in the presence of 10% serum. Balb/c 3T3 cells were grown in Dulbecco's modified Eagle's medium with 10% calf serum. Two days after reaching confluency (A), either 10% serum (B), 25 ng/ml FGF (C), or 25 ng/ml FGF plus 1 $\mu$g/ml dexamethasone (D) were added to the cells. FGF or FGF plus dexamethasone were added daily; serum was added only once. Four days later photographs were taken at a magnification of 40X with phase contrast optics. While the addition of serum did not change the morphology of the cells, the addition of FGF alone to cultures grown in 10% serum caused the cells to become elongated, density inhibition was overcome, and the cells lost their orientation and grew in multiple layers. When FGF plus dexamethasone was added, the cells were flatter but grew to an even higher density than with FGF alone. When the cells were trypsinized, replated, and grown in the presence of serum, they resumed their usual morphology, thus demonstrating that the transformation observed in the presence of FGF is reversible.

same as that in the populations from which they originate. Therefore, in order to further characterize the role of FGF in the control of cell division, we examined the mitogenic effect of FGF on early passage cells derived from human foreskin.

When subconfluent populations of foreskin fibroblasts had become quiescent in 5% serum (3% labelled nuclei after a 12 hour incubation with [$^3$H]thymidine), the addition of 10% fresh serum resulted in the initiation of DNA synthesis in 14% of the cells. However, the addition of FGF alone (50 ng/ml) caused the initiation of DNA synthesis in 55% of the cells and with FGF plus fresh serum more than 85% of the cells were labelled (Gospodarowicz and Moran, 1975). The additive effect of FGF over serum was also seen when population growth was measured. In the presence of FGF and 10% serum the cells grew twice as fast and grew to a higher density than did cells in 10% serum alone (Fig. 2).

Cohen and Carpenter (1975) have shown that EGF also is mitogenic for human foreskin fibroblasts. It stimulates DNA synthesis and it increases the final cell density of cultures grown in either 1 or 10% serum. Lembach (1976) has shown that the response of cultured fibroblasts to EGF depends on the presence of small amounts of serum in the medium. In serum-free medium, the response to EGF is negligible. Likewise, in serum-free medium the response of these cells to FGF is slight (Gospodarowicz and Moran, 1975).

**TABLE 1**

*Comparison of the Mitogenic Effects of FGF and EGF*

| Cell Type | Species | FGF Sensitivity | EGF Sensitivity | References |
|---|---|---|---|---|
| Balb/c 3T3 | mouse | +++ | + | Gospodarowicz, 1974; Gospodarowicz & Moran, 1974a, b; Armelin, 1973 |
| Swiss 3T3 | mouse | ++ | + | Rudland *et al.*, 1974a |
| 3T3 thermosensitive mutant | permissive temperature | – | NT* | Rudland *et al.*, 1974b |
| | unpermissive temperature | +++ | NT | |
| foreskin fibroblasts | human | ++ | ++ | Cohen & Carpenter, 1975; Gospodarowicz & Moran, 1975 |
| glial cells | human | + | ++ | Westermark & Wasteson, 1975, 1976 |
| kidney fibroblasts | cow | + | NT | Gospodarowicz, unpublished observations |
| amniotic cells (fibroblasts) | human, cow | +++ | + | Gospodarowicz *et al.*, 1977f |
| chondrocytes | rabbit | ++ | +++ | Gospodarowicz *et al.*, 1976a, Gospodarowicz Mescher 1977b,c |
| myoblasts | cow | ++ | – | Gospodarowicz *et al.*, 1976c, Gospodarowicz and Mescher 1977c |
| vascular endothelial cells | cow umbilical cord | ++ | – | Gospodarowicz, 1976; Gospodarowicz *et al.*, 1976b, 1977e |
| vascular smooth muscle cells | cow-human | ++ | + | Gospodarowicz *et al.*, 1976b, 1977c; Ross & Glomset, 1976 |

| | | | | |
|---|---|---|---|---|
| vascular endothelial cells | cow aortic arch | ++ | – | Gospodarowicz *et al.*, 1976b, 1977e |
| cornea endothelial cells | cow | ++ | +++ | Gospodarowicz, *et al.*, 1977d |
| cornea epithelial cells | rabbit | NT | +++ | Cohen & Savage, 1974 |
| Y1 adrenal cortex cells | mouse | ++ | – | Gospodarowicz & Handley, 1975b |
| adrenal cortex cells | cow | ++ | – | Gospodarowicz *et al.*, 1976a, 1977a |
| granulosa cells | cow | ++ | +++ | Gospodarowicz *et al.*, 1977a |
| liver cells | rat (endoderm) | – | – | Gospodarowicz, 1976 |
| anterior pituitary cell | rat (ectoderm) | – | NT | Gospodarowicz, 1976 |
| thyroid cells | cow (endoderm) | – | – | Gospodarowicz, unpublished observations |
| epidermal cells | human, rabbit chick (ectoderm) | – | + | Gospodarowicz, 1976; Cohen & Savage, 1974 |
| pancreatic cells | rat (endoderm) | –– | NT | Gospodarowicz, 1976 |
| blastemal cells | frog | ++ | NT | Gospodarowicz *et al.*, 1975 |
| blastemal cells | *Triturus viridescens* | ++ | – | Mescher & Gospodarowicz, in preparation |

*NT = not tested; – = not active; ++ = active; +++ = very active.

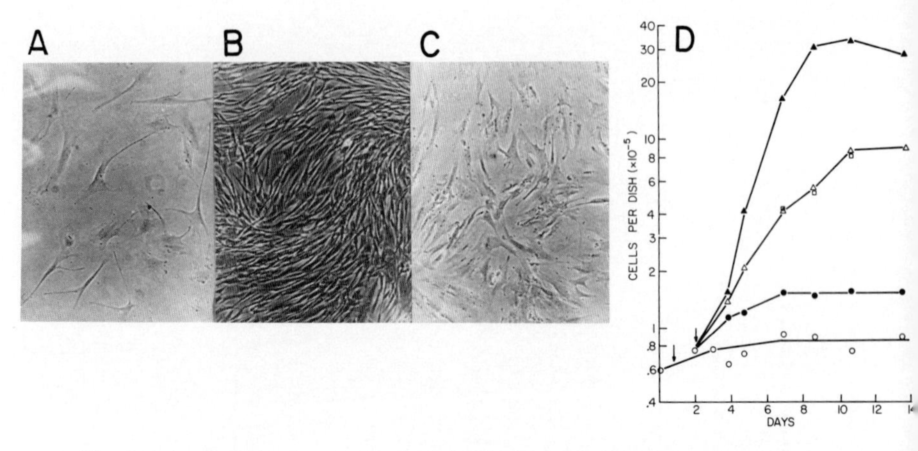


Fig. 2 A, B, C. Appearance of human foreskin fibroblast cells maintained in 0.2% calf serum, 10% calf serum, or 10% calf serum plus FGF. Cells were plated and grown as described for the growth curve (Fig. 2D). Photos were taken at 100X magnification under phase contrast after 6 days of growth and 2 days before the cultures reached their maximal density. (A) 0.2% calf serum; (B) 10% calf serum plus FGF (25 ng/ml added daily); (C) 10% calf serum. As for the 3T3 cells, these cells were morphologically transformed when grown in the presence of FGF; this transformation was reversed when FGF was removed from the medium.

Fig. 2 D. Growth of human foreskin fibroblasts in the presence and in the absence of FGF. $6 \times 10^4$ cells were plated per 6 cm dish in 2.5% calf serum (day 0). On day 1 (first arrow), the medium was changed to 0.2% calf serum. On day 2 (second arrow), either 20% fetal calf serum (□——□), 10% calf serum (△——△), FGF (●——●), or 10% calf serum plus FGF (▲——▲) were added. FGF was added daily (25 ng/ml). Duplicate cultures were counted in a Coulter counter after trypsinization. Control (○——○). Cells maintained in the presence of 0.2% serum went through only one cycle of division when FGF was added; however, in the presence of high serum concentrations (10 or 20%), the addition of FGF reduced the division cycle from 2 days to 1 day, and final cell density was increased 4-fold.

## *C. Amniotic Fluid-derived Cells*

Since FGF increases the growth rate of diploid, human fibroblasts maintained with an optimal serum concentration, the ability of FGF to accelerate the growth in culture of cells derived from amniotic fluid was tested (Gospodarowicz et al., 1977f). The addition of FGF (50 ng/ml) greatly increased the rate of proliferation of these cells. EGF at the same concentration had a smaller effect (Fig. 3).

## *D. Limb Regeneration*

The observation that FGF was most abundant in the brain raised the possibility that it was a "neurotrophic" growth factor similar to the neural agent that stimulates blastema formation during regeneration of amphibian

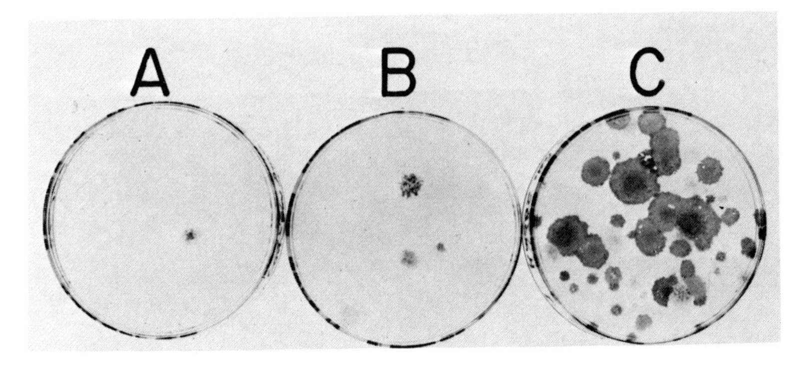


Fig. 3. Bovine amniotic fluid-derived fibroblast cells grown for 15 days in the presence of Dulbecco's modified Eagle's medium containing 20% fetal calf serum without addition of growth factors (A), or with 50 ng/ml EGF (B), or 50 ng/ml brain FGF (C). Cells were plated at an initial concentration of 100 cells/6 cm dish. On day 1, FGF or EGF was added, and the medium was renewed 2 days later and every other day thereafter. Dishes were fixed with 10% formalin, pH 7.2, on day 15 and stained with 1% crystal violet.

limbs. This possibility was first tested by implanting pellets of agar containing 100 μg FGF into the stumps of amputated forelimbs of adult *Rana pipiens,* since it has been suggested that the lack of regeneration ability in post-metamorphic frog limbs results from the lack of a sufficient quantity of nerves to promote the cell proliferation leading to blastema formation (Singer, 1954). The release of FGF from the implanted agar, supplemented by twice-weekly injections of 10 μg FGF in physiological saline into the limb, resulted in increased proliferation of the cells in the distal, dedifferentiated tissues of the limb stump so that instead of the normal, rapid healing of the amputation surface with connective tissue (Forsyth, 1946), a sequence of events more closely resembling blastema formation was initiated (Gospodarowicz *et al.,* 1975d). By 6 or 7 weeks after amputation, when growth had ceased, the FGF-treated limbs displayed regenerates averaging 9mm in length (Fig. 4A). The new limb parts consisted of well-differentiated cartilage and skeletal muscle proximally with much loose connective tissue in distal regions and, like the regenerates produced by surgical augmentation of the nerve supply (Singer, 1954), were invariably heteromorphic, showing little or no evidence of digit formation. Although it is well-known that traumatization alone or other nonspecific treatments can elicit limited regeneration in amputated frog limbs (Carlson, 1974), control animals receiving untreated agar pellets and saline injections in our experiments showed no evidence of regeneration following the initial healing of the amputation wound (Fig. 4B).

The newt, *Notophthalmus (Triturus) viridescens,* has a well-developed regenerative ability that has been shown to require a polypeptide factor

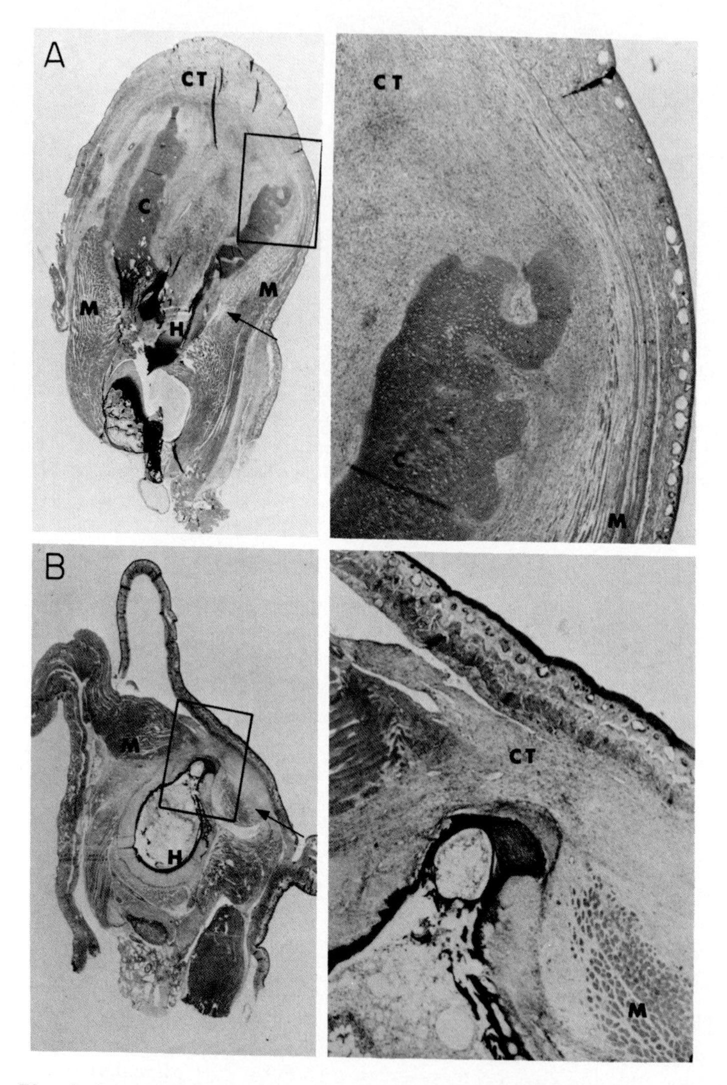


Fig. 4. Longitudinal sections through adult frog forelimb stumps 7 weeks after implantation with agar pellets containing 100 μg FGF and repeated FGF injections (A) or with agar alone and repeated saline injections (B). Arrows indicate approximate level of amputation through the humerus (H). The saline-treated limb shows healing typical of anuran limb stumps (Forsyth, 1946), with a thickened pad of connective tissue (CT) at the site of amputation covering the bone stump. In contrast, the FGF-treated limb formed a distal accumulation of mesenchymal cells which has differentiated proximally into muscle (M) and cartilage (C). A mass of connective tissue has formed near the tip of the "regenerate" and morphogenesis of distal limb elements has not occurred.

provided by nerves (Singer, 1974). If the regenerating forelimb of a newt is denervated prior to the completion of blastema formation, cell proliferation quickly slows, no further growth occurs, and the early blastema is resorbed (Singer, 1952). We have found that injection of physiological saline containing 50 μg FGF into early regenerates 3 days after denervation, when the mitotic index of the denervated limb is only 10-20% that of the contralateral undenervated limb, causes a resumption of mitotic activity in the distal, dedifferentiated cells (Fig. 5). During the 24 hours following the injection, mitotic activity in the denervated, FGF-injected limbs increases to approximately 70% of the levels shown by the contralateral controls (Fig. 6). Control injections of saline containing 50 μg of bovine serum albumin, cytochrome c, or growth hormone have little or no effect on mitotic activity, indicating that the effect seen in the denervated, FGF-injected blastemas is a result of the mitogenic properties of FGF rather than a non-specific response to injected protein. The mitotic response to FGF is dose-dependent, falling off rapidly at doses below 50 μg. The need for a large injection dose is probably due to a number of factors, including proteolysis by the abundant histolytic enzymes in the dedifferentiating tissues of the limb stump and loss of material to the aquatic environment through the loosely organized wound epidermis.

We have also examined the mitogenic effects of EGF in denervated blastemas and, significantly, 50 μg doses of this growth factor are not effective in stimulating mitoses in the dedifferentiated blastemal cells (Fig. 6). Since EGF is not produced by neurons (Turkington *et al.*, 1971), this observation is consistent with earlier data reviewed by Singer (1952,1974) which suggest that formation of a regeneration blastema is dependent on specific, nerve-derived growth factors.

It has not yet been possible to induce or maintain macroscopically visible evidence of regeneration of newt limbs following their denervation. Although single injections of FGF stimulate mitotic activity, their effects are short-lived. Repeated injections lead to trauma that only accelerates the process of regression in the denervated limb stump. The possibility that FGF may be similar to the neurotrophic agent which controls the early phase of amphibian limb regeneration requires further investigation using radioimmunological and immunofluorescent methods.

### *E. Chondrocytes*

Since the cells of regeneration blastemas are derived from cartilage and muscle, among other tissues (Hay, 1966), we investigated the effect of FGF on the rate of proliferation and differentiation of primary cultures of chondrocytes and myoblasts. We also compared the effect of FGF in these experiments to the effect of EGF (Gospodarowicz and Mescher, 1977b).

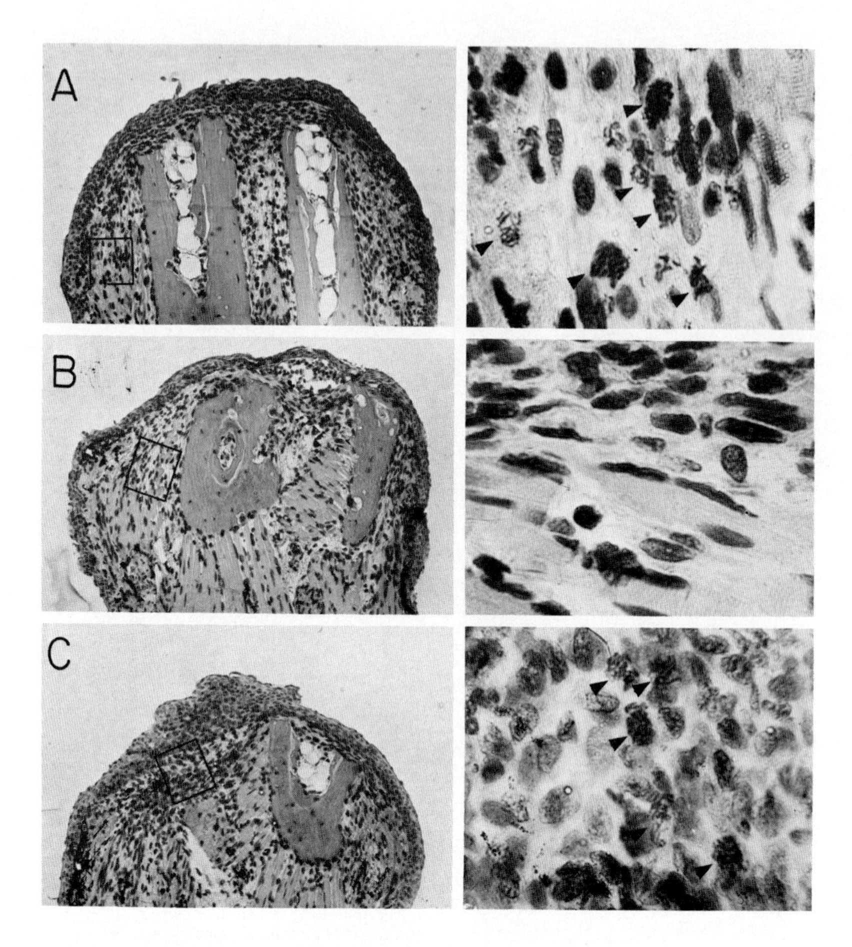


Fig. 5. Longitudinal sections through distal areas of regenerating and denervated forelimbs of adult newts, 14 days after amputation through the radius and ulna and 1 day after an intraperitoneal injection of colchicine, 0.5 mg/g body weight. (A) Control regenerate. (B) Limb stump, denervated 10 days after amputation and injected with 5 $\mu$l saline 24 hours before fixation. (C) Limb stump denervated 10 days after amputation and injected with 5 $\mu$l saline containing 50 $\mu$g FGF. The higher magnifications of the enclosed areas show numerous colchicine-collected mitotic figures in the innervated (A) and FGF-injected (C) limb stumps, but none in the saline-injected limb stump (B). These areas are typical of the regions sampled, at several planes through the limb, for the determination of mitotic indices.

With cultured cartilage cells, one can study not only proliferation, but also differentiation, quite easily. Colonies of rabbit ear chondrocytes can be detected with the naked eye after long-term culture by their three-dimensional, nearly hemispherical appearance (Fig. 7).

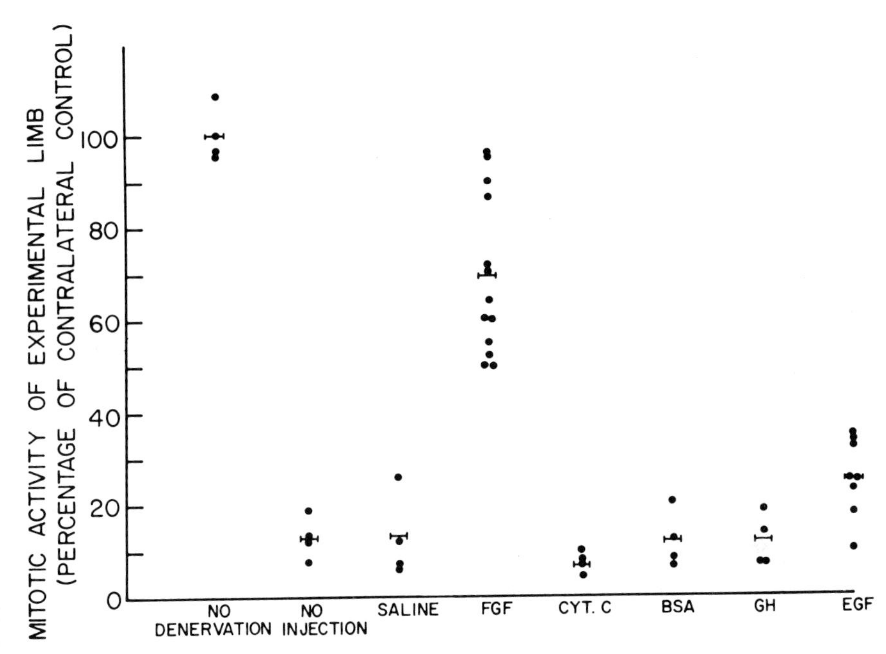


Fig. 6. Mitotic activity of denervated limb stumps following injection with various purified proteins. Ten days after bilateral amputation one forelimb of each newt was denervated. The denervated limb stumps were injected in 5 μl saline in every case, and each newt was injected intraperitoneally with colchicine, 0.5 mg/g body weight. Both forelimbs were fixed 1 day later and mitotic indices were determined. Mitotic activity in the denervated limb was 10-20% that of the contralateral regenerating limb and was unaffected by injection of cytochrome c, bovine serum albumin, ovine growth hormone, or epidermal growth factor. Injection of FGF stimulated mitotic activity in the denervated limb stumps to levels averaging 70% that of the contralateral controls. Each point indicates a comparison of the forelimbs of one animal, the cross bar indicating the mean of the values.

Differential staining with Alcian green and metanil yellow can also be used to demonstrate colonies with cartilage-like matrix in 1-2 week cultures (Fig. 8). Our results indicate that both FGF and EGF are mitogenic for chondrocytes (Fig. 8,9). Addition of either mitogen speeds proliferation and delays differentiation, which is consistent with the postulate that differentiation and proliferation are mutually exclusive. EGF is a more potent mitogen for chondrocytes than is FGF, however, since 0.1 pg/ml caused a significant increase in cell number while 0.1 ng/ml FGF was needed to produce an effect (Fig. 9).

*F. Myoblasts*

Very different results were obtained with myoblasts. FGF is mitogenic but EGF has no effect. The addition of FGF to primary cultures of fetal

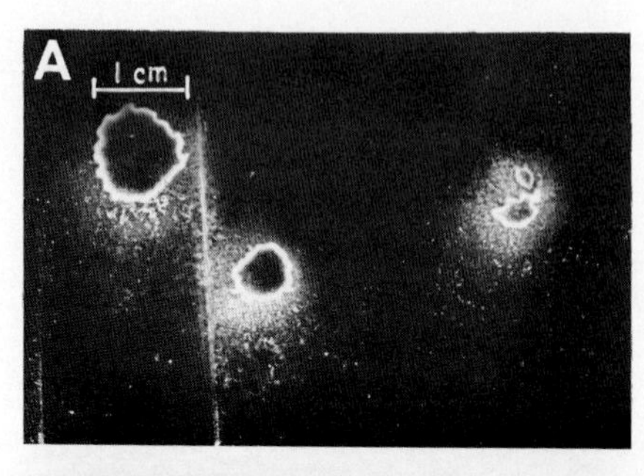


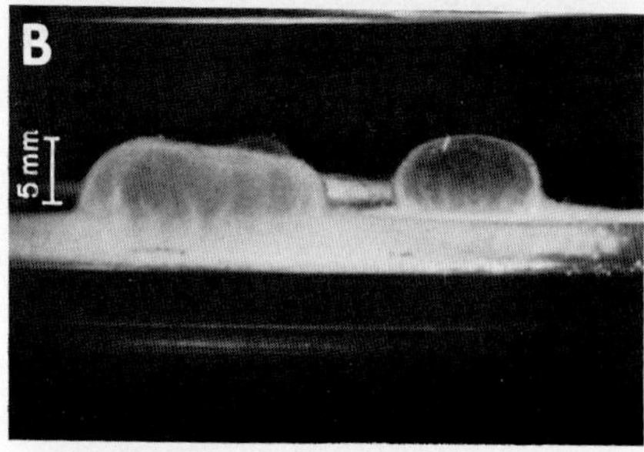


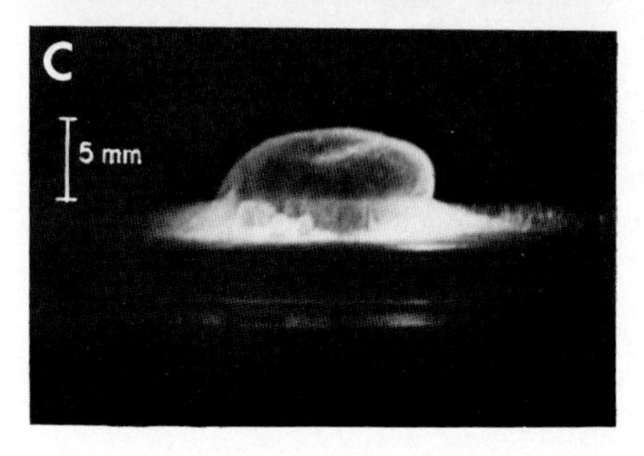


Fig. 7. Appearance of chondrocyte colonies after long-term culture. Chondrocytes obtained from rabbit ear by collagenase dissociation were plated at a density of 5 cells per 15 cm dish in the presence of 5% fetal calf serum and 100 ng/ml FGF in F12 medium. Cultures were fixed in 10% formalin and photographs were taken after 3 weeks of culture. (A) Top view of three colonies. The dark centers of the colonies are cartilage; the light perimeters are chondrocytes. (B) Side view of the two colonies shown to the left in (A). Cartilage projects 5 mm from the surface of the culture dish. (C) Side view of another colony.

bovine myoblasts induces rapid proliferation and significantly delays both fusion and the appearance of acetylcholine receptors (Fig. 10) (Gospodarowicz *et al.,* 1976c). EGF has no effect (Fig. 11). The observation that FGF, but not EGF, stimulates the proliferation of cultured myoblasts is consistent with the hypothesis that FGF may be a neurotrophic agent since the promotion of growth and differentiation of muscle cells in tissue culture by ganglion explants (Crain and Peterson, 1974) and brain extracts (Oh, 1975 has provided strong evidence for the existence of neurotrophic agents and has been shown to correlate well with the large number of in vivo observations (Guttman, 1976).

The effect of FGF on myoblasts plated at high density should not be confused with that of another agent identified in brain crude extract and

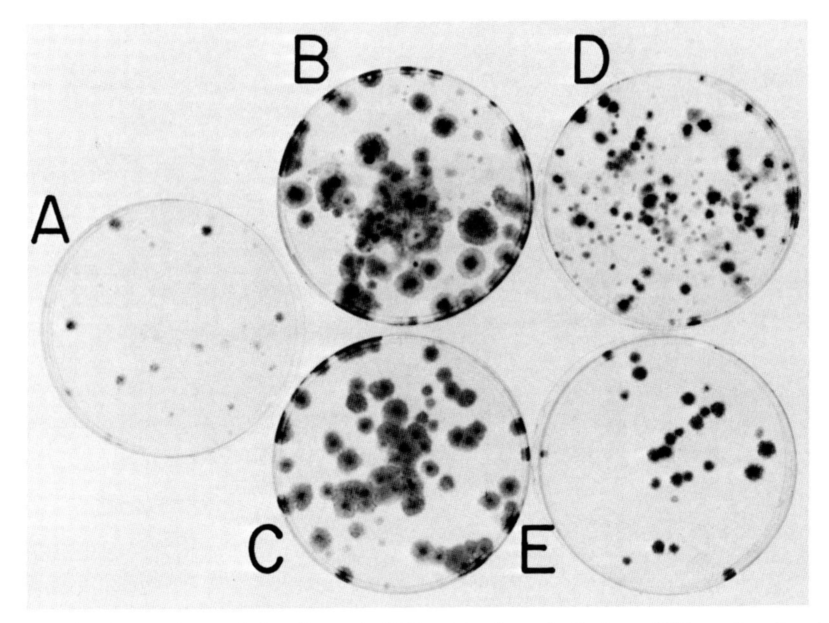


Fig. 8. Comparison of the effect of EGF and brain and pituitary FGF on the clonal growth of chondrocytes plated at low density. Chondrocytes obtained by collagenase dissociation of rabbit ear cartilage were plated at a concentration of 50 cells/6 cm dish in the presence of 5 ml F12 medium with 2.5% fetal calf serum. The following additions were made: (A) none; (B) pituitary FGF (100 ng/ ml); (C) brain FGF (100 ng/ml); (D) EGF (100 ng/ml). Media were renewed once every 5 days. FGF was added every other day and crude extracts were added with each media change. After 12 days the plates were washed and fixed with 10% formalin, pH 7.2, and the cells were stained with Alcian green to stain the cartilage green and counterstained with metanil yellow which stains the chondrocytes yellow – differentiated colonies appear as yellow with a green spot in the center (Ham, 1972). Over 80% of the colonies produced cartilage as indicated by green staining. On the figure, some of the most dense concentrations of cartilage appear as dark spots in the center of gray colonies. However, in most colonies the green staining was light. This could be due to the fact that since the cells divided actively they did not make cartilage. To demonstrate that all clones could differentiate in cartilage-forming colonies in the absence of mitogens, cells were grown in the presence of FGF for 4 days (E). The media were then renewed without FGF and the cells were further incubated for 3 more days. The plates were then fixed and stained as already described. All the clones showed heavy cartilage production when stained with Alcian green.

named by us, "myoblast growth factor" (Gospodarowicz *et al.*, 1975b). While the effect of FGF is best seen with myoblasts seeded at high density and will delay fusion, the effect of myoblast growth factor is best seen on low density cultures of myoblasts ($3 \times 10^{-2}$ cells/cm$^2$) and will accelerate fusion. Also, while FGF is best solubilized from brain tissue at acid pH's (4.5), the myoblast growth factor activity is solubilized at basic pH's (8.5) indicating that it is a different entity from FGF.

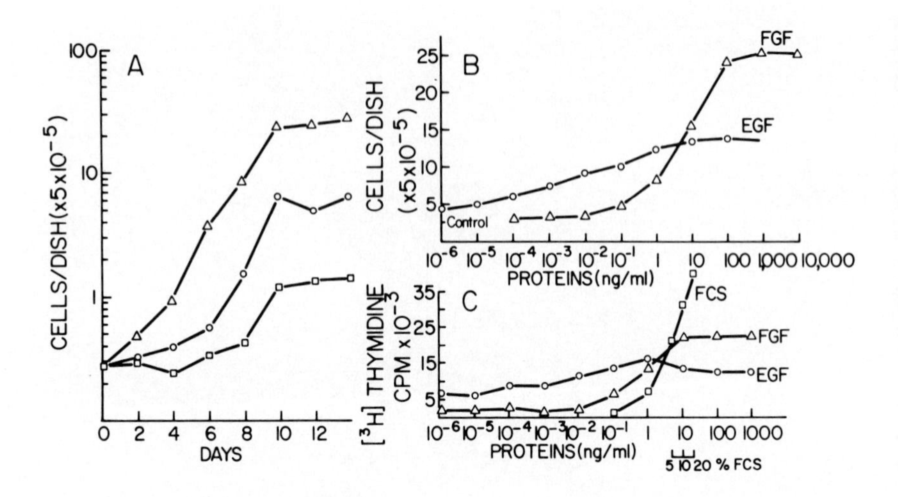


Fig. 9. Effects of FGF and EGF on cell proliferation and initiation of DNA synthesis in cultures of chondrocytes. (A) Chondrocytes were plated at 5000 cells per 6 cm dish in 5 ml of F 12 medium with 10% calf serum; 16 hours later, the medium was renewed with fresh F 12 containing 2.5% calf serum without (□——□) or with 100 ng/ml of EGF (○——○) or FGF (△——△). Every other day cells in triplicate dishes were counted. Standard deviation did not exceed 10% of the mean. (B) Effect of increasing concentrations of EGF (○——○) and FGF (△——△) on the proliferation of chondrocytes. The cells were plated as described in A, and maintained in the presence of increasing concentrations of EGF and FGF. At day 10, the cells in triplicate dishes were trypsinized and counted. Controls had 48,000 cells. (C) Chondrocytes were plated at 50,000 cells per 3.5 cm dish in F 12 with 10% fetal calf serum. Sixteen hours later, the medium was removed and replaced with F 12 and with 0.5% fetal calf serum. The cells were left for 72 hours before sample additions. FGF (△——△), EGF (○——○) and fetal calf serum (△——△) were added to the plates in various concentrations and 12 hours later, the cells were pulsed for 16 hours with ($^3$H)-thymidine. Each point was done in triplicate and standard deviation did not exceed 10% of the mean.

## G. *Adrenal Cortex Cells*

The cells of the adrenal cortex produce steroids (glucocorticoids and mineralocorticoids) that help to maintain homeostasis. ACTH, the trophic hormone for these cells, stimulates steroidogenesis both *in vivo* and *in vitro* (Ramachadran and Suyama, 1975).

Since FGF (Gospodarowicz *et al.,* 1976b) and EGF (Byyny *et al.,* 1974) are both present in serum, we compared the effects of FGF and EGF on bovine adrenal cortex cells *in vitro* (Figs. 12,13).

As little as 1 ng/ml FGF in the incubation medium is sufficient to stimulate the initiation of DNA synthesis and 100 ng/ml FGF is saturating. In contrast, EGF has no effect. Physiological concentrations of ACTH block the effect of FGF. Measurements of cell number corroborate the measurements of DNA synthesis (Gospodarowicz *et al.,* 1977a).

With the highly differentiated Yl adrenal tumor cell line which originates from mouse adrenal cortex, similar results have been obtained. FGF is mitogenic and ACTH at physiological concentrations blocks the FGF effect (Gospodarowicz and Handley, 1975b) (Fig. 14). As with the normal adrenal cells, EGF has no mitotic activity (Gospodarowicz *et al.*, 1977a). The observation that EGF is not mitogenic for adrenal cells provides us with another cellular system in which the action of the two mitogens, EGF and FGF, differ. Others have reported that the Yl cells, which are transformed adrenal cells, do not have receptor sites for EGF (Carpenter *et al.*, 1975), and our studies indicate that EGF does not affect Y1 cell growth (Gospodarowicz *et al.*, 1977a). While the Y1 cells' lack of response to EGF could have been due to their being transformed cells, our observation that normal adrenal cells do not respond to EGF demonstrates that the Yl cell resembles the normal bovine adrenal cortical cell in its response to these growth factors. The presence of receptors for FGF and the absence of receptors for EGF in the Yl cell is then not so much a characteristic of transformation as it is of adrenal cells in general. The observation that EGF is not mitogenic for adrenal cells is in contrast with the observation that EGF and FGF are both mitogenic for granulosa cells, with EGF potent at a concentration as low as $3 \times 10^{-14}$ M (Gospodarowicz *et al.*, 1977). Since these two cells are derived from the same embryonic anlage and contain similar steroidogenic pathways, it is clear that the sensitivity of cells to growth factors is not dictated only by their embryological origin or function.

*H. Granulosa Cells*

The cells surrounding the oocyte, the granulosa cells, are the major cell type involved in the formation of the primary follicle. After ovulation, granulosa cells give rise to the corpus luteum. In the early stages of follicle development the granulosa cells are in intimate contact with capillaries; in the later stage they are in an avascular medium, isolated from the richly vascularized thecae interna and externa by a basement membrane. Factors regulating the transformation of primordial follicles (oocytes surrounded by a single layer of granulosa cells) into growing follicles (oocytes surrounded by multiple layers of granulosa cells) are unknown but the consensus is that gonadotrophins are not involved (Greenwald, 1974). Because the most dramatic event observed during the transformation of primordial follicles into growing follicles is the proliferation of granulosa cells, we investigated the effects of FGF and EGF on bovine granulosa cell proliferation *in vitro* in an effort to ascertain whether these growth factors could be involved in follicle development.

Both EGF and FGF were found to be very potent stimulators of the initiation of DNA synthesis by granulosa cells (Fig. 15). A half-maximal

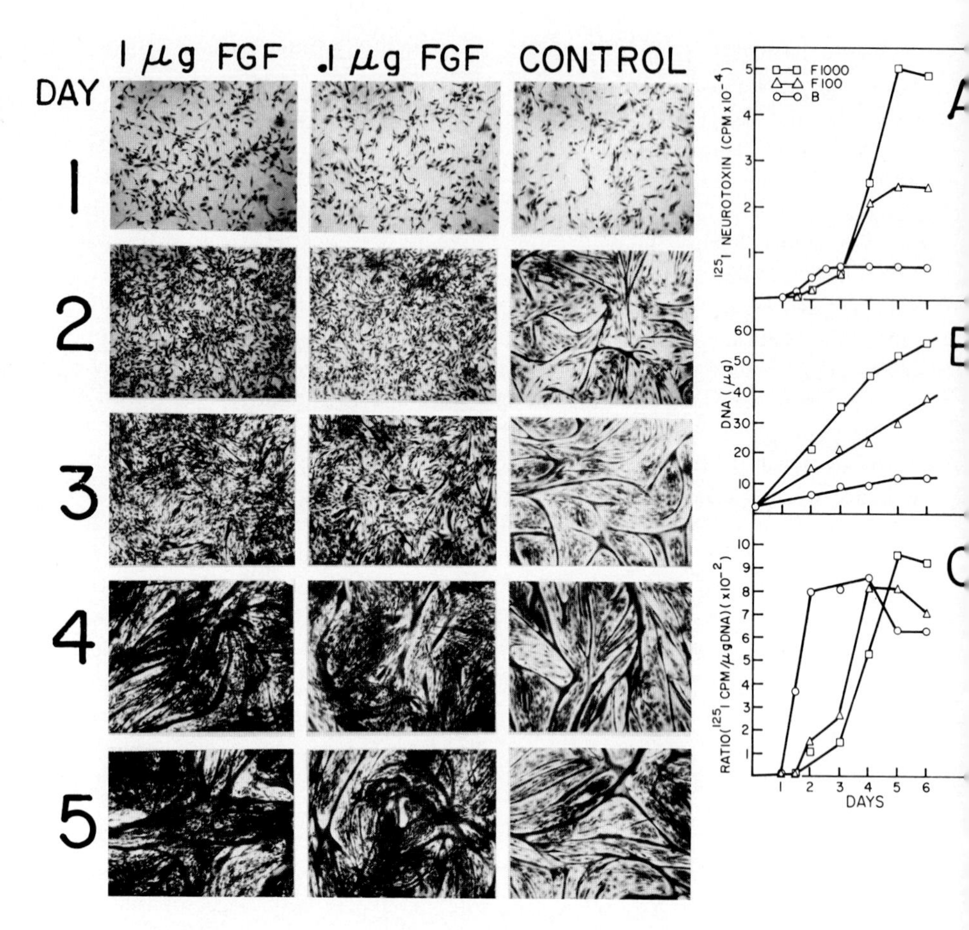


Fig. 10. *(Left Panel)*. Appearance of bovine myoblast cultures maintained in the presence or absence of FGF for 5 days. Myoblasts were obtained by trypsinization of hind leg muscles of 3-month bovine fetuses. They were seeded on gelatinized dishes at a density of 8000 cells/cm$^2$ in Medium 199, Dulbecco's modified Eagle's medium, and horse serum (8:2:1). FGF, at a final concentration of 1 μg/ml (left column) or 0.1 μg/ml (middle column), was added every other day. Controls (right column) received no additions. Every day a set of plates were washed and fixed for 1 hour with 10% formalin and stained with 1% crystal violet. Pictures represent plates at daily intervals from days 1 to 5.

Fig. 10. *(Right Panel)*. Comparison of the binding of neurotoxin and the synthesis of DNA as a function of time in cultures of bovine myoblasts. (A) Determination of the binding of $^{125}$I-neurotoxin as a function of time in myoblast cultures maintained in the presence of FGF [0.1 μg [△—△), 1 μg (□—□)], or in its absence (○—○). (B) Determination of DNA content as a function of time in myoblast cultures maintained in either the presence of FGF [0.1 μg/ml (△—△) and 1 μg/ml (□—□)] or in its absence (○—○). (C) Ratio of the binding of neurotoxin to the DNA content of the myoblast cultures maintained in either the presence of FGF [0.1 μg (△—△) and 1 μg (□—□)] or in its absence (○—○). In controls, the neurotoxin specific binding was noticeable at 24 hours and increased until 60 hours when it reached a plateau. With 0.1 μg or 1 μg of brain or

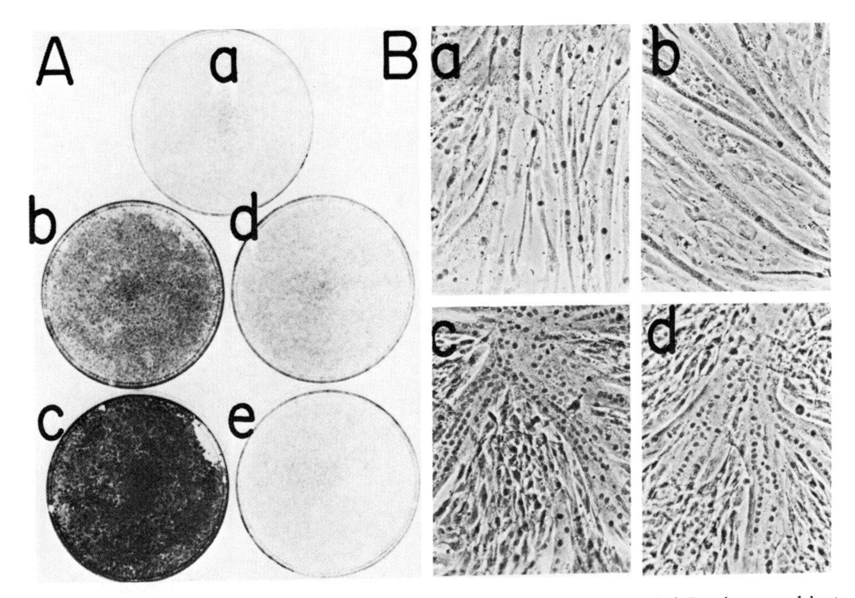


Fig. 11. Comparison of the EGF and FGF effects on myoblasts. (A) Bovine myoblasts were distributed as described in Fig. 10 and maintained for 5 days without mitogens (a) or with 0.1 μg/ml FGF (b), 1 μg/ml FGF (c), 0.1 μg/ml EGF (d) or 1 μg/ml EGF (e). On day 5 the plates were washed and fixed with 10% formalin for 1 hour and stained with 0.1% crystal violet. (B) Morphological appearances of myoblast cultures maintained for 5 days in the absence of mitogens (a) or in the presence of 1 μg/ml EGF (b), 0.1 μg/ml FGF (c), or 1 μgml FGF (d). (Phase contrast 150 X.) The observation that *in vitro* FGF, but not EGF, is mitogenic for myoblasts eliminates the remote possibility that FGF could increase the fusion of myoblasts through an indirect effect by stimulating contaminating fibroblasts to divide. Since EGF is a mitogenic agent for fibroblasts and since no increased fusion is seen with EGF, it is unlikely that the increased fusion seen with FGF could be due to a density of fibroblasts.

pituitary FGF/ml, the amount of binding during the first 72 hours was lower than in controls, reflecting a lower rate of fusion in the presence of FGF. At 72 hours, the cultures in the presence of 0.1 μg FGF bound as much neurotoxin as did controls. Binding increased to a maximum between days 4 and 5. Maximal binding to cultures maintained in the presence of 0.1 μg/ml of FGF was 5-fold higher than in controls; with 1 μg/ml, it was 10-fold higher.

The final specific binding per culture under each condition was proportional to the final DNA content per culture (Fig. 10 B,C). As shown in Fig. 10 C, the number of binding sites per cell (as determined by cpm of toxin bound/μg DNA) was similar at the end of the experiment. This demonstrates that the increased cell proliferation seen with FGF was due to an increased proliferation of myoblasts, and that the percentage of non-myoblasts, if any, was similar under all experimental conditions. Although the amounts of toxin bound per cell were nearly the same at the end of the experiment for all cultures, the time course of appearance of toxin binding sites was quite different. Whereas the ratio of toxin bound per cell reached a maximum on day 2 in controls, it reached a maximum on day 4 with 0.1 μg/ml FGF and on day 5 with 1 μg/ml FGF. In all cases, the maximum toxin binding per cell reached a maximum shortly after the rate of DNA synthesis per culture (determined by [$^3$H] thymidine pulse labelling) dropped from its maximum.

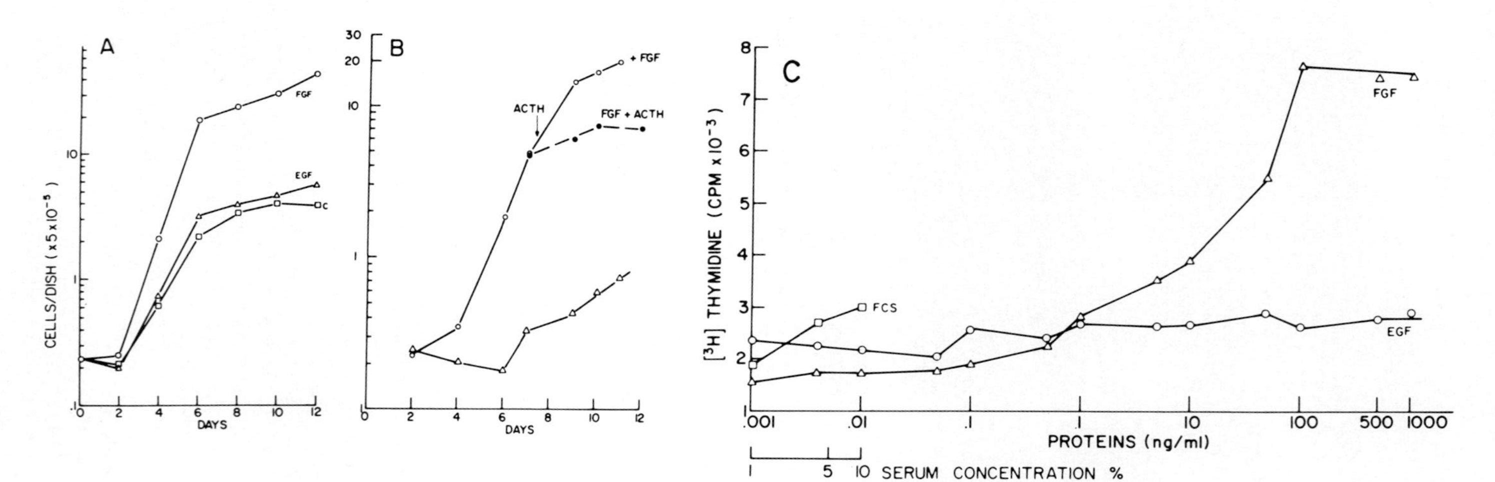


Fig. 12. Effect of FGF on the proliferation of bovine adrenal cortex cells. (A) Cloned adrenal cortex cells were plated at 3000 cells/6 cm dish. Cells were maintained in F12 medium with 12.5% horse serum and 2.5% fetal calf serum without (□) or with 100 ng/ml of EGF (△) or FGF (○). Each point was done in triplicate. Standard deviation was 10% that of the mean. (B) Primary culture of adrenal cortex cells (3000 cells/6 cm dish). On day 6, $3 \times 10^{-7}$M ACTH were added to half of the plates containing FGF. All cells were maintained in the presence of 12.5% horse serum and 2.5% fetal calf serum. Each point was done in triplicate. Standard deviation was 10% that of the mean. (Δ), control cells grown in 12.5% horse serum and 2.5% fetal calf serum; (○), plus FGF, 100 ng/ml; (●), plus FGF, 100 ng/ml and ACTH, 1 μg/ml. (C) Initiation of DNA synthesis in subcultures of adrenal cortex cells in response to increasing concentrations of EGF, FGF and serum. Adrenal cortex cells were plated at 35,000 cells/3.5 cm dish in F12 with 12.5% horse serum and 2.5% fetal calf serum. Sixteen hours later the medium was removed and replaced with F12 with 2.5% fetal calf serum. FGF, EGF and fetal calf serum were added to the plates in various concentrations and 12 hours later the cells were pulsed for 48 hours with $[^3H]$thymidine. Determinations of $[^3H]$thymidine incorporated into DNA were done. The control values were 1200 ± 230 cpm. Each point was done in triplicate. Standard deviation was 10% that of the mean.

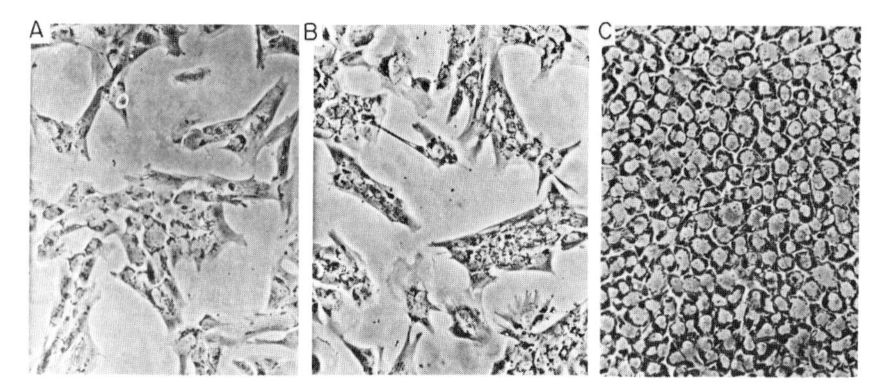


Fig. 13. Morphological appearance of bovine adrenal cortex cells maintained in the absence and presence of FGF or EGF. (A) Adrenal cortex cells maintained in the absence of EGF and FGF. Cells were plated as described in Fig. 12 and maintained in the presence of 12.5% horse serum and 2.5% fetal calf serum for 7 days. (Phase contrast optics 143 X.) (B) Adrenal cortex cells maintained in the presence of EGF (100 ng/ml) for 7 days. (Phase contrast optics 143 X.) (C) Same as (B), but the cells were maintained in the presence of FGF (100 ng/ml) for 7 days. (Phase contrast optics 143 X.)

response to EGF was seen at 1 pg/ml ($2 \times 10^{-13}$M) and a half-maximal response to FGF was seen at 500 pg/ml ($4 \times 10^{-11}$M). Increases in cell number in response to EGF and FGF correlated with the observed amounts of DNA synthesis (Figs. 16,17). In contrast to EGF and FGF, neither FSH nor highly purified LH had any mitogenic activity. High concentrations of a partially purified LH preparation (NIH-LH-B9) were mitogenic, but this preparation has been shown to contain FGF as a contaminant (Gospodarowicz, 1974).

The mitogenic effect of EGF on granulosa cells differs from its effect on both human fibroblasts and lens epithelial cells in that the molar concentration of EGF needed to induce a half-maximal initiation of DNA synthesis in granulosa cells ($2 \times 10^{-13}$M) is 300-fold lower than for human fibroblasts ($7 \times 10^{-11}$) (Hollenberg and Cuatrecasas, 1975) and 3000-fold lower than for lens epithelium ($6 \times 10^{-10}$M) (Hollenberg, 1975).

The response of granulosa cells to EGF also differed from the response of human fibroblasts and lens epithelium to EGF in other ways. With human fibroblasts and lens epithelium, the response to EGF depends on the cell density, the concentration of serum in the medium, and the length of time the cells are left without renewal of the medium. Under ideal conditions EGF has as great a mitogenic effect as serum, but slight variations in this assay procedure can result in a diminution of the EGF effect to a level lower than that of serum. In contrast, the response of sparse cultures (100 cells per cm$^2$)

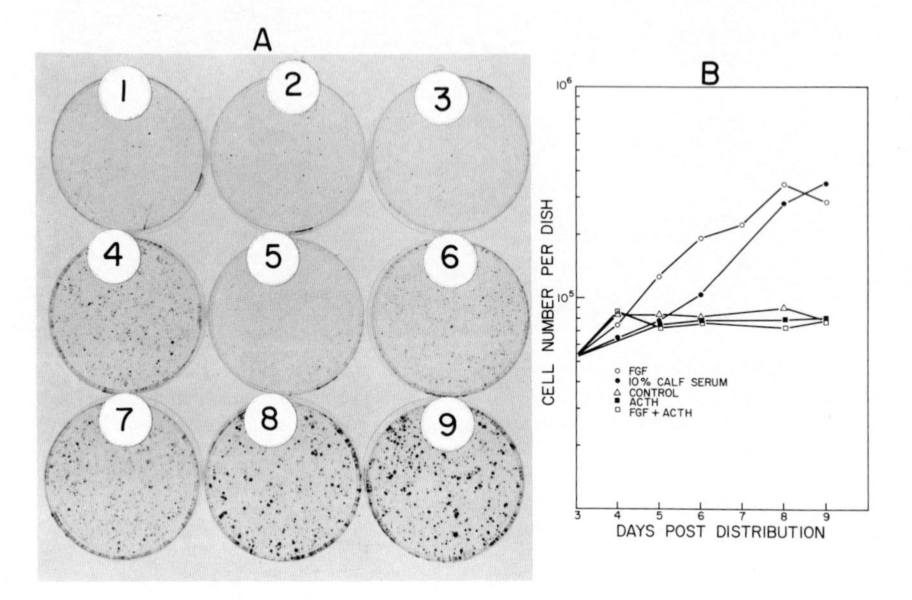


Fig. 14. Effect of FGF on the growth of Yl adrenal cells. (A) Effect of various agents on the final cell density of Yl adrenal cells. Two × $10^4$ cells in 5 ml of Ham's F10 medium with 5% calf serum were seeded in 6 cm dishes. One day later the serum concentration was changed to 0.2% calf serum. On day 3 the following additions were made: (1) control, (2) 1 μg/ml dexamethasone, (3) 500 ng/ml insulin, (4) 50 ng/ml FGF, (5) 50 ng/ml FGF plus 1 μg/ml dexamethasone, (6) 50 ng/ml FGF plus 1 μg/ml dexamethasone plus 500 ng/ml insulin, (7) 50 ng/ml FGF plus 500 ng/ml insulin, (8) 10% calf serum, and (9) 12.5% horse serum plus 2.5% fetal calf serum. All agents were added every other day for 9 days, except serum which was added only on day 3. At day 7, the plates were fixed with ethanol-methanol (1:1) and stained with 1% crystal violet. Of all the agents added, only FGF was mitogenic. Insulin had no effect and dexamethasone was slightly inhibitory. With the Yl, FGF replaced serum in contrast to other cell lines where FGF has an additive effect. (B) Effect of ACTH on the growth rate of Yl cells maintained in the presence of serum or FGF. Cells were cultured as described in (A). FGF was added at a final concentration of 5 ng/ml and ACTH at 0.75 IU/ml. While with FGF the cells proliferated at the same rate as in the presence of 10% calf serum, as soon as ACTH was added to cultures maintained in the presence of FGF the cells ceased proliferating.

of granulosa cells to EGF is not dependent on the length of time the cells are mitogen as serum (Gospodarowicz *et al.*, 1977). With granulosa cells as with human fibroblasts and lens epithelium, the magnitude of the response to EGF depends on the serum concentration in the medium. The differences between the response of granulosa cells and human fibroblasts to FGF are generally the same as the differences between the responses of the two cell types to EGF.

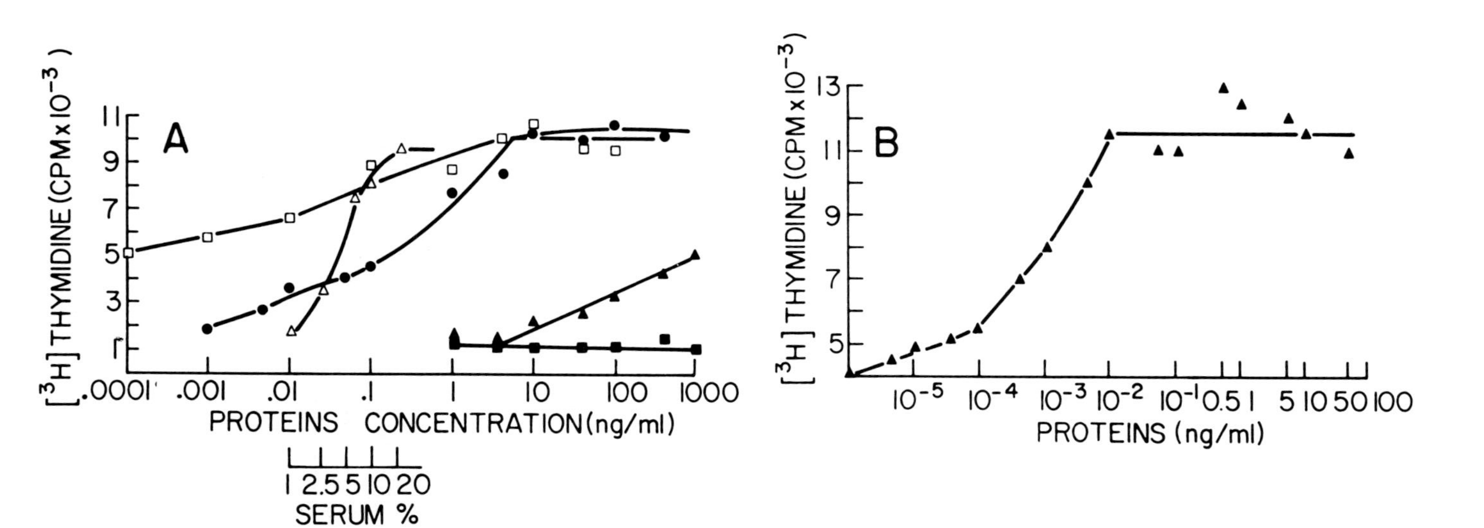


Fig. 15. Initiation of DNA synthesis in subcultures of bovine granulosa cells in response to increasing concentrations of EGF, FGF and serum. (A) Granulosa cells were plated at 30,000 cells/3.5 cm dish in F12 with 10% calf serum. Twenty-four hours later the medium was removed and replaced with F12 with 1% calf serum. The cells were left for 3 days before sample additions. FGF (●——●), EGF (□——□), highly purified LH (■——■), NIH-LH-B9 (▲——▲), and calf serum (△——△) were added to the plates in various concentrations and 12 hours later the cells were pulsed for 16 hours with [$^3$H] thymidine. Determinations of [$^3$H] thymidine incorporated into DNA were done as described (Gospodarowicz, 1974). The control values were 1200 ± 130 cpm. NIH-FSH-S9 and insulin from 1 ng to 1 μg/ml gave 1200 ± 250 cpm. Every point was done in triplicate. Standard deviation was 10% that of the mean. (B) Same as (A) but the EGF (▲——▲) was tested in a different experiment since, due to its potency, it was necessary to go to lower concentrations than with FGF. Control values were 3800 ± 280 cpm. Every point was done in triplicate. Standard deviation was 10% that of the mean.

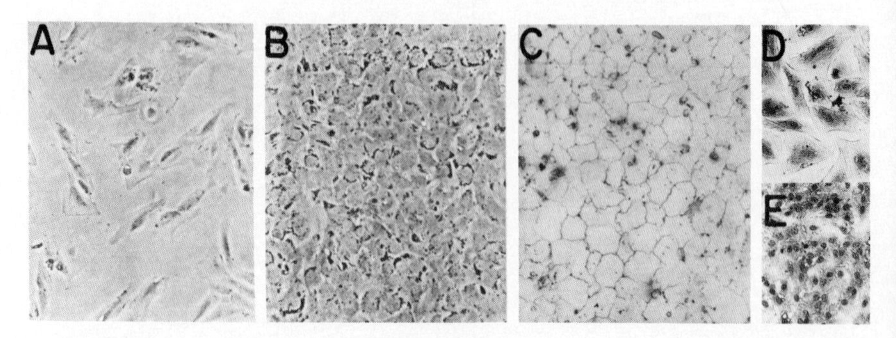


Fig. 16. Morphological appearance of bovine granulosa cells maintained in the absence and in the presence of FGF or EGF. (A) Granulosa cells maintained in the absence of EGF and FGF. Cells were plated at 100 cells/cm² and maintained in F12 medium, 1% serum for 7 days. The cells are very large and contain a single nucleus with 1-3 nucleoli and a granular perinuclear region surrounded by a broad expanse of cytoplasm with longitudinal ridges. (Phase contrast optics 143 X.) (B) Granulosa cells maintained in the presence of FGF (100 ng/ml) for 7 days. The cells are much smaller than the control and are tightly packed with discrete granules within the cytoplasm. The cells did not overgrow each other in confluency but remained in a monolayer. (Phase contrast optics 143 X.) Similar results were obtained with EGF (100 ng/ml). (C) Same as (B) but the cells were stained with silver nitrate to show the cell border. The cells are clearly density-inhabited. (Phase contrast optics 143 X.) (D) Same as (A) but the cells were fixed with 10% formalin and stained with 0.1% Giemsa (57 X). (E) Same as (B) but the cells were fixed with 10% formalin and stained with 0.1% Giemsa (57 X).

The extraordinary potencies of EGF and FGF on the proliferation of granulosa cells *in vitro* led us to investigate their effects on these cells *in vivo*. During normal ovarian development, the growth of primary follicles is characterized by the proliferation of granulosa cells (10 through 17 days after birth in the rat). The injection of EGF for 5 successive days into 15-day-old rats led to greater development of the follicles than in controls. A similar, but less pronounced, effect was produced by FGF. The increased number of small follicles observed after EGF injection caused the experimental ovaries to be visibly larger than control ovaries (Fig. 18). In view of the *in vitro* and *in vivo* data, it appears that EGF and/or FGF may be involved in controlling the early phase of follicular development in the ovary, a period during which the granulosa cells are in contact with mitogenic factors in the bloodstream.

*J. Vascular Endothelial Cells*

Endothelial cells constitute the inner lining of the blood vascular system. Formation of new capillaries during tissue regeneration and neoplastic growth requires the proliferation of vascular endothelial cells. Because of their location at the interface between blood and tissue, they are the chief elements involved in the permeability of blood vessels. Abnormalities of endothelial cell

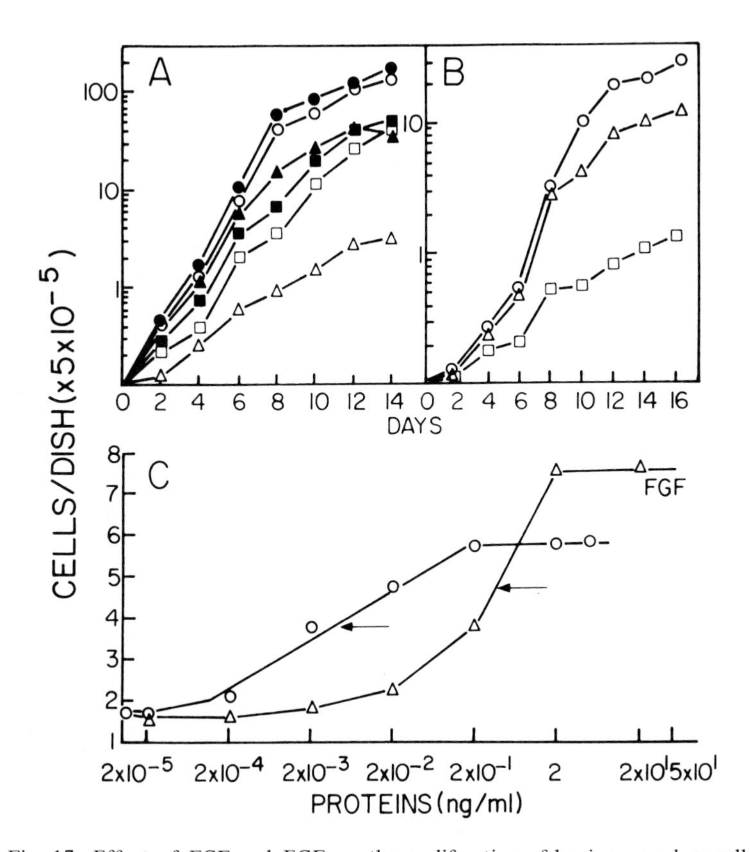


Fig. 17. Effect of FGF and EGF on the proliferation of bovine granulosa cells. (A) Bovine granulosa cells were plated at 3,000 cells/6 cm dish. They were then maintained in the presence of 1% calf serum without (△——△) or with 100 ng/ml of EGF (□——□) or FGF (■——■). The proliferation effect of FGF and EGF was compared to that of 10% calf serum (▲——▲) to which 100 ng/ml EGF (○——○) or 100 ng/ml FGF (●——●) was added. Every point was done in triplicate. Standard deviation was 10% that of the mean. (B) Similar to (A) but the cells were maintained in 1% calf serum without (□——□) or with 100 ng/ml of EGF (△——△) or FGF (○——○). Every point was done in triplicate. Standard deviation was 10% that of the mean. (C) Effect of increasing concentrations of EGF (○——○) and FGF (△——△) on the proliferation of bovine granulosa cells. The cells were plated (100 cells/cm$^2$) and maintained in the presence of increasing concentrations of EGF and FGF. At day 7 the cells were trypsinized and counted. Control gave 30,000 ± 1800 cells. Every point was done in triplicate.

structure and function are prominent in the pathology of a number of diseases of blood vessel walls such as thromboangitis and microangiopathy.

Since the continuity of the vascular endothelium is essential for the growth and survival of the organism, the elucidation of the factors involved in endothelial cell survival and proliferation is important. This can be examined

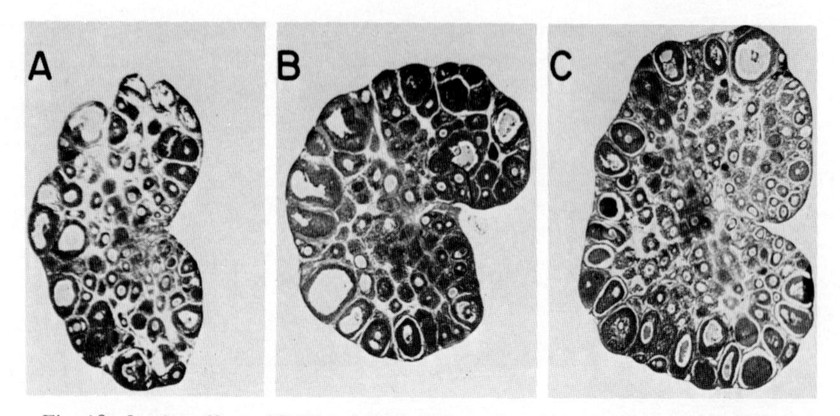


Fig. 18. *In vivo* effect of EGF and FGF. Fifteen-day-old rats were injected twice every day with 10 μg FGF or EGF for 5 days and killed on day 6. (A) Control (BSA). (B) FGF. (C) EGF. Sections were stained with hematoxylin and photographed at the same magnification (36 X).

most easily in tissue culture. These cells have been cultured in a number of laboratories but with limited success because it has proven to be difficult to maintain pure cultures of vigorously growing endothelial cells for the long term necessary for complete studies of factors influencing growth. We have examined the effects of FGF and EGF on the survival and proliferation of these cells *in vitro* (Gospodarowicz, 1976) in an attempt to determine whether such mitogens could promote the maintenance and growth of capillaries *in vivo.* The effects of FGF on these cells were of special interest since Balb/c 3T3 cells, for which FGF was originally shown to be an extremely potent mitogen, are probably derived from vascular endothelial cells (Boone *et al.,* 1976).

When primary cultures of cells from bovine aortic endothelium are started at low density (30 cells/$cm^2$), formation of a monolayer depends on whether or not FGF is added to the culture medium (Gospodarowicz *et al.,* 1976b). In 10% calf serum without FGF, small colonies developed from cell aggregates during the first few days, but the cells looked unhealthy and soon became vacuolated. In contrast, if FGF was present the cells proliferated vigorously and formed a monolayer (Fig. 19). As little as 1 ng/ml FGF stimulated the initiation of DNA synthesis in endothelial cells, with a maximal effect at 25-50 ng/ml. EGF did not stimulate DNA synthesis in these cells even at concentrations as high as 1 μg/ml, nor did insulin or glucagon (Fig. 20B). The addition of as little as 0.1 ng/ml of FGF also increased the growth rate of cultures maintained in 10% calf serum; 10-100 ng/ml reduced the population doubling time to as little as 24 hours (Fig. 20A). As predicted from the

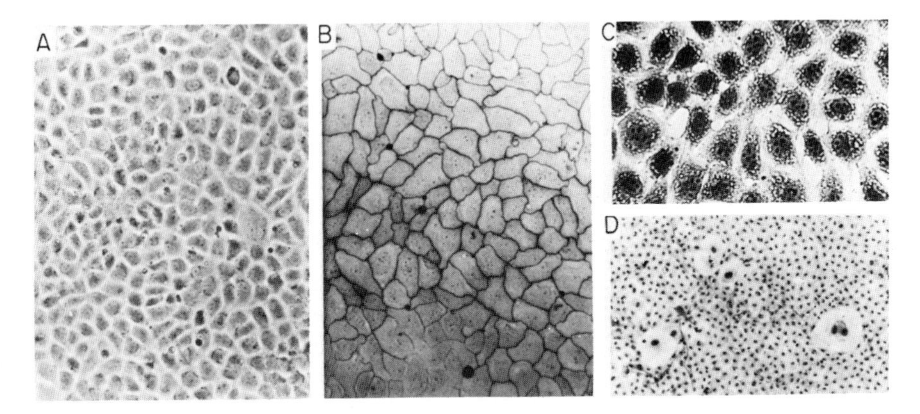


Fig. 19. Appearance of cloned adult bovine aortic endothelial cells maintained in tissue culture in the presence of FGF. (A) Monolayer of endothelial cells (3 passages, 27 generations) maintained in the presence of 10% calf serum and 100 ng/ml FGF. The cells are polygonal, closely apposed, and with an indistinct border (phase contrast, 150 X). (B) Same monolayer as in (A) but stained with silver nitrate to show the cell borders. The cells showed the same organization as did preparations of endothelium stained *in situ* (220 X). (C) Same monolayer as in (A) but fixed with 10% formalin and stained with 0.1% Giemsa. The cells have central nuclei with 4-6 nucleoli and a small rim of cytoplasm with perinuclear vacuoles (400 X). (D) Monolayer of endothelial cells after 42 generations (6 passages). A few large mono- and binucleated cells are visible as well as fibrous material reflecting the intensive secretory activity of these cells (cells were fixed and stained as described for [C]) (100 X).

observations on the initiation of DNA synthesis, EGF (100 ng/ml) did not increase the growth rate of the endothelial cells (Gospodarowicz *et al.*, 1977e).

Having demonstrated the mitogenic activity of FGF for endothelial cells *in vitro*, we then tested the ability of FGF to induce the proliferation of endothelial cells *in vivo*. Endothelial cell proliferation is perhaps most easily observed when capillaries proliferate to vascularize a normal avascular space. Using the technique described by Gimbrone *et al.* (1974), we found that FGF was a potent inducer of neovascularization in the rabbit cornea (Fig. 21).

## III. DISCUSSION

Our results indicate that EGF and FGF are not mitogenic for all cultured cells. For example, EGF stimulates the division of granulosa cells and skin fibroblasts but not adrenal cortex cells or vascular endothelial cells. FGF stimulates myoblasts and chondrocytes but not liver cells or thyroid cells, and that EGF and FGF act on specific cell types, and not on cultured cells in general, suggests that these growth factors have some physiological role in

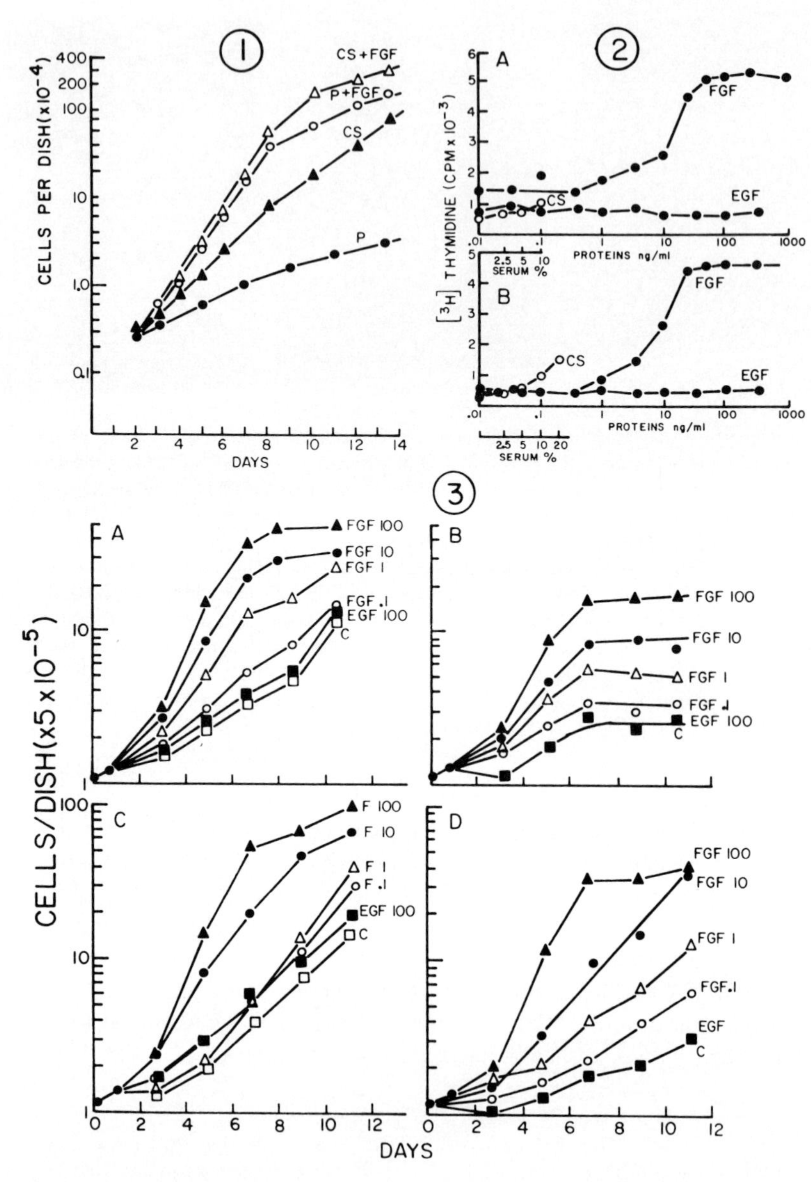


Fig. 20. Mitogenic effect of FGF with endothelial cells. (A) Growth rates of primary cultures of fetal bovine vascular endothelial cells maintained in the presence of plasma serum or blood serum with or without FGF. Endothelial cells were obtained from fetal bovine aortic arches as described by Gospodarowicz *et al.* (1976b). Cell aggregates were dissociated by treatment with 0.25% trypsin and seeded at 1000 cells/6 cm dish in the presence of Dulbecco's modified Eagle's medium (DME) with 10% calf serum. After 2 days,

regulating growth and are not merely factors that promote the growth of cells under the artificial conditions of tissue culture.

FGF has been proven to be mitogenic for all mesoderm-derived cells tested to date. Although this suggests that FGF is rather unspecific (there being a great variety of cell types derived from that germ layer), one should consider not only specificity of target, but also specificity of function. FGF, like insulin, acts on a wide variety of cells, but its action is specific: it induces mitosis.

A key question that must be answered before the physiological role of growth factors can be understood is: what are the factors that control cell's

---

the medium was removed and replaced with either 10% calf serum (CS) or 10% bovine plasma serum (P) with or without 100 ng/ml FGF. The medium was changed every other day. In the absence of FGF and in the presence of plasma serum the cells went through 3-4 generations in 12 days. In the presence of calf serum they went through 8 generations, while with FGF added to plasma serum or serum they went through 10-12 generations in 12 days. Similar results were obtained with primary cultures of adult bovine aortic endothelial cells. (B) Initiation of DNA synthesis in cloned vascular endothelial cells in response to various concentrations of FGF, EGF and serum. Endothelial cells were plated at 20,000 cells/6 cm dish in DME with 10% calf serum. Twenty-four hours later the medium was removed and replaced with DME with 2.5% bovine plasma serum. Twenty-four hours later various concentrations of FGF, EGF, glucagon, insulin and serum were added to the cells. $[^3H]$thymidine was added 1 day later. The increase in cell number in the presence of FGF was monitored every 12 hours. When the number of cells in cultures with FGF had doubled (approximately 48 hours after the addition of mitogens), the media were removed and the cells lysed with 0.5 N NaOH. $[^3H]$thymidine incorporation into DNA was measured as described by Gospodarowicz (1974). ($B_1$) shows the incorporation of $[^3H]$thymidine observed with adult bovine aortic endothelial cells (5 passages, 37 generations since cloning). Controls were 600 cpm/dish. Insulin and glucagon from 0.1-10 $\mu$g/ml gave values of 650 ± 70 cpm/dish. ($B_2$) shows fetal bovine aortic endothelial cells (3 passages, 27 generations). Controls and insulin or glucagon from 0.1-10 $\mu$g/ml gave values of 500 cpm ± 100 cpm/dish. (C) Effect of various concentrations of FGF on the proliferation of cloned vascular endothelial cells maintained in the presence of calf serum or plasma serum. Endothelial cells were plated as described in (B). Twenty-four hours later the medium was removed and replaced with DME with 2.5% calf serum or 2.5% bovine plasma serum. FGF (0.1, 2, 10, or 100 ng/ml) or EGF (100 ng/ml) was added. The additions of FGF or EGF were repeated every other day. Triplicate cultures were counted every other day. ($C_1$) shows adult bovine aortic endothelial cells (6 passages, 42 generations) maintained in 2.5% calf serum. Without FGF and with or without EGF the cells have gone through 2 generations in 8 days. With 100 ng/ml FGF they have gone through 5.5 generations. ($C_2$) is the same as ($C_1$) but the cells were maintained in the presence of 2.5% plasma serum. Control and EGF have gone through 1 generation in 7 days. In the presence of FGF (100 ng/ml) they have gone through 4 generations in 7 days. ($C_3$) shows fetal bovine aortic endothelial cells (4 passages, 32 generations) maintained in 2.5% calf serum with or without FGF. In the absence of FGF and in the presence of EGF the cells went through 2 generations by day 7. With 100 ng/ml FGF they have gone through 5.5 generations. ($C_4$) is the same as (C) but the cells were maintained with 2.5% bovine plasma serum. The control and EGF by day 7 have gone through 1.7 generations while with 100 ng/ml FGF they have gone through 5 generations.

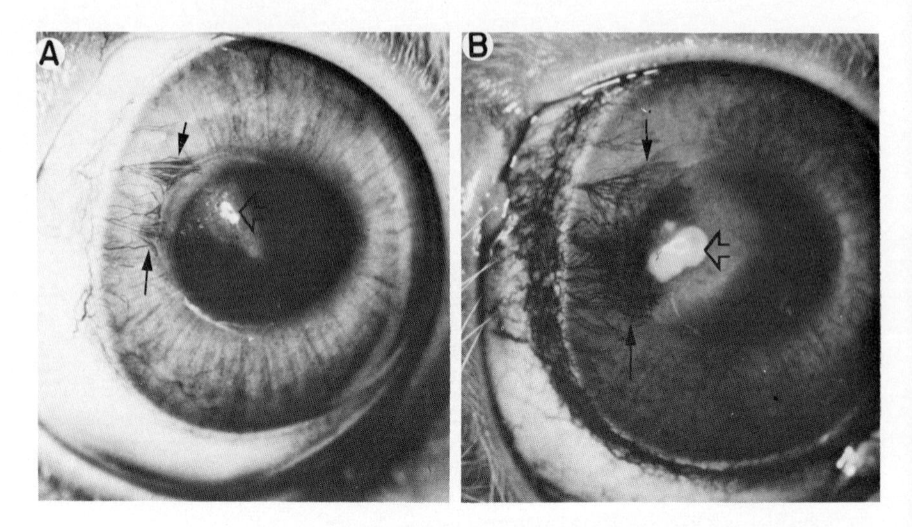


Fig. 21. Rabbit corneal vascularization induced *in vivo* by FGF (A) FGF was administered to the cornea in an acrylamide pellet implant (2 μl acrylamide, 2 μg FGF). Growth of capillaries (arrows) took place in an eccentric position in the direction of the acrylamide implant. In the contralateral eye an implant of acrylamide containing 0.2% formic acid was done to control for an inflammatory response. A photo of the FGF implanted eye was taken 7 days later. No sign of inflammation was present in the cornea, which received the FGF containing acrylamide. Seven out of ten of the animals implanted with acrylamide pellets containing FGF showed corneal capillary proliferation. (B) Similar to A but with 20 μl acrylamide implanted (20 μg FGF).

sensitivity to growth factors? In the adult animal, some FGF-sensitive cells (e.g., adrenal cortex) may proliferate while others do not, despite the fact that many other cell types are just as sensitive to FGF *in vitro* (e.g., endothelial cells, myoblasts, etc.). If a mitogen like FGF (or EGF) is responsible for stimulating cell proliferation *in vivo*, either its local concentration must somehow be controlled or the sensitivity of each cell type must be controlled.

The local concentration of mitogens could be controlled by sequestration and transport of mitogens to specific sites. For example, there is evidence that mitogens are stored in platelets and released during platelet lysis at sites of injury (Ross and Glomset, 1976). In the case of FGF, which is found in neural tissue and which can partially replace the neurotrophic factor responsible for limb regeneration in amphibians (Gospodarowicz *et al.*, 1975d), there could be a direct transport of the mitogen through he peripheral nervous system to target organs.

The sensitivity of target cells to mitogens could be controlled by the levels of circulating hormones or other agents. Examples of agents that control the response of target cells to mitogens are found in tissue culture: for 3T3 cells, the response to FGF is greatly potentiated by glucocorticoid (Gospodarowicz, 1974); for human fibroblasts, the responses to both FGF (Gospodarowicz and Moran, 1975) and EGF (Lembach, 1976) depend on the presence of some serum component. *In vivo,* an example of agents controlling a mitogenic response are estrogens. Estrogens promote cell proliferation in the endometrium *in vivo,* yet have no mitogenic effect on the same cells *in vitro* (Mueller, 1953). One explanation for these observations is that estrogens are not mitogenic by themselves, and therefore have no mitogenic activity by themselves *in vitro*, but they do cause target cells *in vivo* to become responsive to mitogens already present.

Another question that arises from the data presented here is: what is the significance of the observation that one cell type can be responsive to more than one mitogen? For example, granulosa cells respond to both EGF and FGF. This situation can be compared to the case of fat cells that respond to both glucagon and epinephrine; both hormones stimulate lipolysis. In each case at least two different agents can cause the same response, and the relative importance of each agent in vivo is presumably dependent on the relative sensitivity of each target cell to each agent (i.e., the relative binding affinities of each agent to each target cell).

In addition to having direct, mitogenic effects on target cells, some growth factors could promote growth indirectly as well. Factors, like FGF, that stimulate vascular endothelial cell proliferation could indirectly influence organogenesis, wound healing, and tumor growth by promoting vascularization. Cell proliferation in a three-dimensional structure depends on the cells' access to nutrients in the bloodstream. Therefore, factors affecting capillary density, such as endothelial cell mitogens, can regulate the structure's overall growth. This growth control mechanism is particularly evident during embryonic development and in tumor growth, when the active growth of an organ is preceded by capillary invasion (Folkman, 1975).

Classical studies of vertebrate developmental biology have been done using organ culture systems and unpurified tissue extracts and secretions. Advances in tissue culture and biochemistry have made possible experiments such as those described here using purified mitogens and homogeneous, primary cell cultures. These experiments should prove to be a rich source of information on the role of hormones and growth factors in animal growth, development and repair.

## ACKNOWLEDGMENTS

We thank Academic Press, the Rockefeller University Press and the Excerpta Medica Foundation for permission to reproduce figures. We are indebted to the Talone Meat Company for providing us with the fetuses and organs needed for our work. Since this paper was limited in length, we have presented mostly our work; contributions by others to the field of growth factors are acknowledged in the references of the publications to which the reader is referred.

This work was supported by the National Institutes of Health and the American Cancer Society.

## REFERENCES

Armelin, H. (1973). *Proc. Nat. Acad. Sci. U.S.A.* **70,** 2702–2706.

Boone, C. W., Takeichi, N., Paranjpe, M. and Gilden, R. (1976). *Cancer Res.* **36,** 1626–1633.

Bynny, R. L., Orth, D. N., Cohen, S. and Doyne, E. S. (1974). *Endocrinology.* **95,** 776–782.

Carlson, B. M. (1974). *In* "Neoplasia and Cell Differentiation" (G. C. Sherbet, ed.), p. 60. S. Karger, Basel.

Carnot, P. and Deflandre. (1906). *C. R. Acad. Sci.* **143,** 432–435.

Carpenter, G., Lembach, K. J., Morrison, M. M. and Cohen, S. (1975). *J. Biol. Chem.* **250,** 4297–4304.

Cohen, S. (1962). *J. Biol. Chem.* **237,** 1555–1562.

Cohen, S. (1965). *Develop. Biol.* **12,** 394–407.

Cohen, S. and Savage, C. R., Jr. (1974). *Rec. Prog. Hormone Res.* **30,** 551–574.

Cohen, S. and Carpenter, G. (1975). *Proc. Nat. Acad. Sci. U.S.A.* **72,** 1317–1321.

Crain, S. M. and Peterson, E. R. (1974). *Ann. N. Y. Acad. Sci.* **228,** 6–34.

Folkman, J. (1975). *In* "Cancer: A Comprehensive Treatise. Vol. 3: Biology of Tumors: Cellular Biology and Growth" (F. F. Becker, ed.), p. 355–388. Plenum Press, New York.

Forsyth, J. W. (1946). *J. Morph.* **79,** 287–321.

Gimbrone, M., Cotran, R., Leapman, S. and Folkman, J. (1974). *J. Nat. Cancer Inst.* **52,** 413–427.

Gospodarowicz, D. (1974). *Nature* **249,** 123–127.

Gospodarowicz, D. (1975). *J. Biol. Chem.* **250,** 2515.

Gospodarowicz, D. (1976). Progr. in Clin. and Biol. Res. Membranes and Neoplasia: New Approaches and Strategies. A. R. Liss, Inc., 1–19.

Gospodarowicz, D., Greene, G., and Moran, J. (1975a). *Biochem. Biophys. Res. Comm.* **62,** 779.

Gospodarowicz, D., and Handley, H. (1975b). *Endocrinology* **97,** 102–107.

Gospodarowicz, D., Ill, C. R., and Birdwell, C. R. (1977). *Endocrinology* **100,** 1108–1121.

Gospodarowicz, D., Ill, C. R., Hornsby, P. J., and Gill, G. N. (1977a). *Endocrinology* **100,** 1080–1091.

Gospodarowicz, D., and Mescher, A. L. (1977b). *J. Cell Physiol.,* in press.

Gospodarowicz, D., and Mescher, A. L. (1977c). *In* "Gene Expression and Regulation in Cultured Cells," *Nat. Cancer Inst.* Monographs, in press.

Gospodarowicz, D., Mescher, A. L., and Birdwell, C. R. (1977d). *Exp. Eye Res.,* in press.

Gospodarowicz, D., and Moran, J. (1974a). *Proc. Nat. Acad. Sci. U.S.A.* **71,** 4584–4588.

Gospodarowicz, D., and Moran, J. (1974b). *Proc. Nat. Acad. Sci. U.S.A.* **71,** 4648–4652.

Gospodarowicz, D., and Moran, J. S. (1975c). *J. Cell Biol.* **66,** 451.

Gospodarowicz, D. and Moran, J. (1975). *J. Cell. Biol.* **66,** 451–457.

Gospodarowicz, D., Moran, J., and Bialecki, H. (1976a). Third Int. Symp. on Growth Hormone and Related Peptides, Excerpta Medica, p. 141–157.

Gospodarowicz, D., Moran, J., and Braun, D. (1977e). *J. Cell Physiol.*, in press.

Gospodarowicz, D., Moran, J., Braun, D., and Birdwell, C. R. (1976b). *Proc. Nat. Acad. Sci. U.S.A.* **73,** 4120.

Gospodarowicz, D., Moran, J., and Owashi, N. (1977f). *J. Clin. Endocr. Met.* **44,** 651–659.
Gospodarowicz, D., Rudland, P., Lindstrom, J., and Benirschke, K. (1975d). *Adv. Metab. Disord.* **8,** 302–335.
Gospodarowicz, D., Weseman, J., and Moran, J. (1975e). *Nature* **256,** 216–219.
Gospodarowicz, D., Weseman, J., Moran, J., and Lindstrom, J. (1976c). *J. Cell. Biol.* **70,** 395–405.
Greenwald, G. S. (1974). *In* "Handbook of Physiology, Endocrinology V", Part 2, p. 293–323.
Gregory, H. (1975). *Nature* **257,** 325–327.
Gutmann, E. (1976). *Ann. Rev. Physiol.* **38,** 177–216.
Ham, R. G. (1972). *In* "Methods in Cell Physiology" (D. Prescott, ed.), Vol. V. Academic Press, New York. 37–74.
Hay, E. D. (1966). "Regeneration". Holt, Rinehart, and Winston, New York. 1–148.
Hollenberg, M. D. (1975). *Arch. Biochem. Biophys.* **171,** 371–377.
Hollenberg, M. D. and Cuatrecasas, P. (1975). *J. Biol. Chem.* **250,** 3845–3853.
Holley, R. W. and Kiernan, J. A. (1974). *Proc. Nat. Acad. Sci. U.S.A.* **71,** (8) Aug. 2942–2950.
Lembach, K. J. (1976). *Proc. Nat. Acad. Sci. U.S.A.* **73,** 183–187.
Masui, H. and Garren, L. D. (1971). *Proc. Nat. Acad. Sci. U.S.A.* **68,** 3206–3210.
Mueller, G. C. (1953). *J. Biol. Chem.* **204,** 77–90.
Oh, T. H. (1975). *Exp. Neurol.* **46,** 432–438.
Ramachadran, J. and Suyama, A. T. (1975). *Proc. Nat. Acad. Sci. U.S.A.* **72,** 113–117.
Rose, S. P., Pruss, R. M. and Herschman, H. R. (1975). *J. Cell. Physiol.* **86,** 593–598.
Ross, R. and Glomset, J. (1976). *J. Cell Biol.,* in press.
Rudland, P., Seifert, W. and Gospodarowicz, D. (1974a). *Proc. Nat. Acad. Sci. U.S.A.* **71,** 2600–2604.
Salmon, W. D. and Daughaday, W. H. (1957). *J. Lab. Clin. Med.* **49,** 825–836.
Savage, C. R., Jr., Inagami, T. and Cohen, S. (1972). *J. Biol. Chem.* **247,** 7612–7621.
Singer, M. (1952). *Quart. Rev. Biol.* **27,** 169–200.
Singer, M. (1954). *J. Exp. Zool.* **126,** 419–471.
Singer, M. (1974). *Ann. N. Y. Acad. Sci.* **228,** 308–322.
Todaro, G. J. and Green, H. (1963). *J. Cell Biol.* **17,** 299–313.
Turkington, R. W., Males, J. L. and Cohen, S. (1971). *Cancer Res.* **31,** 252–256.
Westermark, B. and Wasteson, A. (1975). *Adv. Metab. Disord.* **8,** 85–100.
Westermark, B. and Wasteson, A. (1976). *Biochem. Biophys. Res. Com.* **69,** 304–310.

# Growth Control by Serum Factors and Hormones in Mammalian Cells in Culture

Dieter Paul, Hans J. Ristow, H. T. Rupniak and Trudy O. Messmer*

*Cell Biology and *Molecular Biology Laboratories*
*The Salk Institute*
*Post Office Box 1809*
*San Diego, California 92112*

## I. INTRODUCTION

Cells in culture require serum for growth. Serum constituents are known to be involved in growth control in normal and transformed mammalian cells in culture. Therefore, we are interested in studying the interactions between purified serum growth factors and cells in order to learn how cell growth is controlled. During the last several years we have been involved in systematically fractionating serum in order to use purified factors as culture medium supplements in attempts to reconstitute the growth promoting activity of serum. We have identified several growth factors in serum and recently we have been able to replace most of the serum growth promoting activities by highly purified serum factors using 3T3 mouse cells and SV40 virus transformed 3T3 (SV3T3) cells.

## II. SERUM FACTORS REQUIRED BY 3T3 AND SV3T3 CELLS

Serum is the most active of the body fluids we have tested, although plasma has similar, perhaps slightly lower activity as assayed by DNA synthesis initiation in quiescent 3T3 cells (Table I). Lymph and cerebrospinal fluid are much less active (Table I).

Initial studies using 3T3 and SV3T3 cells, indicated that several growth factors are required simultaneously by cells (Paul *et al.*, 1971). Therefore, we kept 3T3 cells quiescent in low serum levels and assayed for activities that lead

Abbreviations: FGF, fibroblast growth factor; FRL, primary fetal rat liver cells; REF, secondary rat embryo fibroblasts; PI, phosphatidylinositol; ODC, ornithine decarboxylase; PRPP, phosphoribosyl pyrophosphate; SFI, serum factor 1; SFII, serum factor 2.

**TABLE 1**

*Growth Promoting Activity of Body Fluids*

| | 3T3 | SV3T3 |
|---|---|---|
| Serum (10%) | 100% | 100% |
| Plasma (10%) | 90% | 100% |
| Lymph (10%) | 25% | 70% |
| CSF (10%) | 10% | 25% |

Growth promoting activity of various body fluids in 3T3 and SV3T3 cells. Cells were cultured in Dulbecco's medium with the supplement as indicated: calf serum; human plasma prepared in the presence of citrate; rat lymph from the thoracic duct; human cerebrospinal fluid (CSF). Cells were counted 4 days after plating with a Coulter Counter and cell counts normalized.

to DNA synthesis initiation. In other assays factors were identified that are required for cell survival (Paul *et al.*, 1971), cell migration (Lipton *et al.*, 1971) and cell attachment to collagen (Klebe, 1974). Table II shows a summary of the serum factors required by 3T3 and SV3T3 cells that we have worked with. In 3T3 cells, the assay for DNA synthesis initiation in the presennce of low serum (0.3%) is convenient to identify the most limiting factor(s) that cells require for $G_1 \rightarrow S$ transition. Ideally, we intend to remove serum completely from the

**TABLE 2**

*Serum Factors*

| Function | Source | Molecular weight |
|---|---|---|
| 1. Survival factor[1] | Calf serum | 150,000 daltons |
| 2. Attachment factor[2] | Calf serum | 200,000 daltons |
| 3. Migration factor[3] | Calf serum | 100,000 daltons |
| 4. Growth Factors (DNA synthesis initiation) | | |
| $S_1$[4] | Fetal calf serum | 620,000 daltons |
| $S_2$[4] | Fetal calf serum | 26,000 daltons |
| SFII[5] | Calf serum | 500 daltons |

[1] Paul *et al.* (1971).
[2] Klebe (1974).
[3] Lipton *et al.* (1971).
[4] Hoffman *et al.* (1973).
[5] Paul, unpublished.

cultures. Therefore, the depleted medium of arrested cultures is removed, and the appropriate factors added in fresh serum-free medium. As can be seen in Table III most of the growth stimulating activity present in serum can be replaced by a mixture of several factors: $S_1$ (Hoffmann *et al.*, 1973) or FGF (Gospodarowicz, 1974), SFII, insulin and dexamethasone.

The attachment of 3T3 cells to plastic is independent of serum (Paul *et al.*, 1971; Lipton *et al.*, 1972). A serum factor is required to keep cells alive during prolonged incubation in serum-free medium, which we have called "survival factor" (Paul *et al.*, 1971). Therefore, even though the bulk of the serum in the experiment summarized in Table III has been removed, it is clear that the system is still undefined because traces of serum necessary to keep cells surviving during the 24 hour period after removal of the depleted medium are presumably still present in the cultures. This presumably causes the high variability of our data from experiment to experiment. In order to be able to study initiation of DNA synthesis in 3T3 cells in the absence of serum in response to growth factors and hormones, cells were cultured in serum-free medium in the presence of survival factor that has been obtained in highly purified form from calf serum using a complex purification procedure that included Con A-sepharose affinity chromatography, gel filtration, extraction with chloroform/methanol (D. Paul and M.Henahan, unpublished). Optimal results were obtained by plating 3T3 cells in serum-free medium in the presence of survival factor and SFII and allowing cells

**TABLE 3**

*Initiation of DNA Synthesis in 3T3 Cells by Serum Factors and Hormones*

| Compound added to culture | TCA precipitable cpm after $[^3H]$ thymidine pulse (18-24 hrs) % |
|---|---|
| 10% Calf serum | 100 |
| No addition | 1–4 |
| $S_1$ (10 $\mu$g/ml) | 10–20 |
| SFII (50 ng/ml) | 2–6 |
| Insulin (100 ng/ml) | 12–15 |
| Dexamethasone (500 ng/ml) | 1–4 |
| Insulin + dexamethasone | 20–25 |
| SFII + Ins + dx | 40–70 |
| $S_1$ + SFII + Ins + dx | 70–120 |

3T3 cells ($10^5$) were plated in 0.3% serum in 3cm dishes. Twenty-four hours later the medium was aspirated and replaced with Dulbecco's medium containing hypoxanthine (25 $\mu$g/ml), biotin (0.2 $\mu$g/ml) and proline (34 $\mu$g/ml). Additions were made as indicated and the cultures incubated for 18 hours. After addition of $[^3H]$thymidine, cultures were incubated for 5 additional hours and TCA precipitable radioactivity determined and related to serum control ≡ 100%.

to become arrested in $G_1$ and to eventually regenerate trypsin-labile hormone receptors. When cells are ready for stimulation, FGF is added together with insulin and dexamethasone or hydrocortisone. Initiation of DNA synthesis by two serum factors + FGF + insulin + dexamethasone proceeds with a rate similar to that observed with 10% serum (Table IV) (D. Paul and H. J. Ristow, manuscript in preparation). It is interesting that the growth factor mixture [FGF (or $S_1$) + insulin + dexamethasone] has little if any effect on cells that were plated in serum-free medium, although most cells are viable and can be stimulated by serum (Table IV). SFII and survival factor appear to be important for maintaining the cells receptive for the growth factors. Although the cells proceed through one cycle in response to the five factors (Table IV), cells deteriorate presumably because the nutritional conditions are not yet optimal for keeping cells in continuous culture. Presumably, additional serum constituents have to be present in serum-free cultures for long-term cell propagation, such as transferrin, lipid carriers, etc.

**TABLE 4**

*Initiation of DNA Synthesis in $G_1$-arrested 3T3 Cells by Growth Factors and Hormones in Serum-free Medium*

| Plate in (−24 hrs) | Additions (0-time) | TCA ppt. cpm after [$^3$H]thymidine (18-24 hrs) (cpm) |
|---|---|---|
| No | No | 130 |
| No | ins + dx | 153 |
| No | FGF + ins + dx | 160 |
| No | 10% serum | 930 |
| SFII | No | 163 |
| SFII | ins + dx | 462 |
| SFII | FGF + ins + dx | 855 |
| SFII + survival | No | 158 |
| SFII + survival | ins + dx | 213 |
| SFII + survival | FGF + ins + dx | 1012 |
| SFII + survival | 10% serum | 987 |

SFII: 50 ng/ml; survival factor: 5 μg/ml; insulin: 100 ng/ml; dexamethasone (dx): 500 ng/ml; FGF: 50 ng/ml.

3T3 cells were trypsinized from stock dishes, washed 3X with serum-free medium and plated into medium as described in Table III with or without serum factor addition as indicated in the left column. After cells have attached, growth factors and/or hormones were added 24 hrs later. Cultures were pulsed with [$^3$H]thymidine 18-24 hours thereafter and TCA precipitable radioactivity determined.

## III. SERUM FACTOR REQUIREMENTS OF OTHER CELL TYPES

We have studied the serum factor requirements of secondary rat embryo fibroblasts (REF) (Frank *et al.*, 1972) and primary fetal rat liver cells (FRL) (Leffert and Paul, 1972). Although culture conditions comparable to those described in Table IV for 3T3 cells have not yet been found for culturing REF or FRL cells, experiments similar to those presented in Table III using REF or FRL cells indicate that these cells require different serum factors for DNA synthesis initiation. For example, REF do not require dexamethasone but respond well to $S_2$, whereas 3T3 and FRL cells do not respond to this protein. FRL cells do not require survival factor or $S_2$, dexamethasone or FGF, but triiodothyronine ($T_3$) stimulates DNA synthesis initiated by insulin (Leffert, 1974). On the other hand, $T_3$ is not required by REF or 3T3 cells.

Therefore, it appears that each cell type requires a specific mixture of hormones plus growth factors for DNA synthesis initiation and cell growth (Hayashi and Sato, 1976). Such a complex regulatory system might represent a highly efficient means to control growth of a large number of different cell types independently from each other in an animal. Similarly, the control of the expression of differentiated functions is effected by synergistic effects of various hormones, e. g. in the mammary gland in response to insulin, hydrocortisone and prolactins (Mukherjee *et al.*, 1973). Since the control of the expression of differentiated functions is presumably independent of the control of cell growth, it appears that serum growth factors might represent a new class of molecules, different from classical hormones, that control growth of cells provided that certain specific hormonal requirements are fulfilled. Such a hierarchial system would very much simplify the complexity of regulatory networks controlling growth or specific differentiated functions independently, without loss of specificity or flexibility.

## IV. GROWTH CONTROL BY SERUM FACTORS, HORMONES AND CELL ANCHORAGE

In "normal" cells in culture growth is controlled by the rate at which cells inactivate the growth factor(s) required for initiation of DNA synthesis. In 3T3-4a cells the factor that is depleted first can be replaced by either FGF, $S_1$, or EGF and presumably other factors. In other 3T3-clones, the most limiting factor can be replaced by dexamethasone or cortisol (Thrash and Cunningham, 1973). Therefore, different cells not only require different serum factors but they differ by the rate at which they deplete various factors. "Controlled" growth in culture means (1) that cells deplete a growth factor fast enough, thus preventing that high cell densities are attained, and (2) that cells arrest in $G_1$ ($G_0$) when the level of the limiting factor(s) is low enough. In 3T3 cells these criteria are given because cells were selected for these properties (Todaro and Green, 1963).

Transformed cells could be cell types that do not fulfill one or both criteria. In fact, benzo(a)pyrene-transformed 3T3 cells (BP3T3) deplete the factors they require for growth much slower than 3T3 cells (Fig. 1), but they have retained the capacity to become arrested in $G_1$ (Holley *et al.*, 1976). SV3T3 cells not only deplete the growth factors which they require at slower rates than 3T3 cells, but these cells do not become arrested in $G_1$, even in the complete absence of serum (Table V). Therefore, it appears that SV3T3 cells initiate DNA synthesis constitutively, i. e. in the absence of specific serum factors or hormones. Indeed, SV3T3 cells can be cultured for a few days in medium supplemented with survival factor + SFII (Fig 2). Cells divide 2-3 times before they die, presumably due to nutritional and/or hormonal deficiencies in the culture medium. SV3T3 cells do not require the serum factors that initiate DNA synthesis in 3T3 cells (FGF or $S_1$, insulin, dexamethasone). We do not know whether these or equivalent factors are synthesized in SV3T3 cells replacing exogenous factors, or whether the functions which are induced by these factors in 3T3 cells have become constitutive after transformation with SV40 virus and are no longer subject to controls by external signals. Both the slower rate of depletion of growth factors and the lower serum factor requirement provide, at least in part, an explanation for the fact that transformed cells require "less serum" for growth. Also, these characteristics of SV3T3 cells can account for uncontrolled growth in culture and *in vivo*. Whether or not invasive growth *in vivo* is related to such aberrant growth patterns is not known.

Most transformed cells in monolayer, including HeLa cells, initiate DNA synthesis without specific serum factor or hormone requirements (D. Paul, submitted for publication). Even in the absence of serum, SV3T3 or HeLa cells

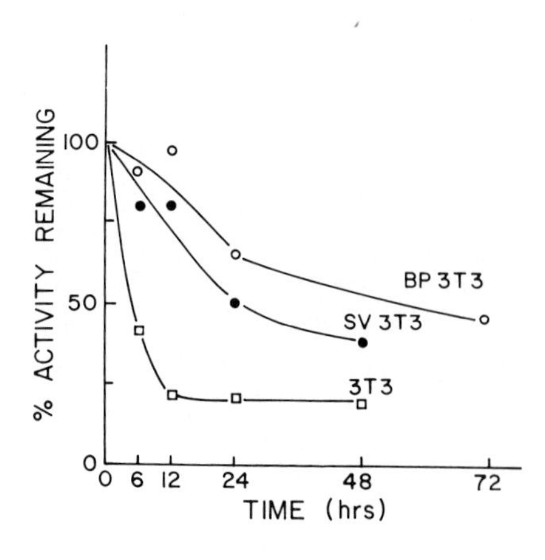


Fig. 1. Depletion of growth stimulating activity by untransformed and transformed 3T3 cells. Cells ($10^6$/5cm dish) were cultured in 10% serum for various times, after which the media were removed, filter-sterilized and kept frozen until all samples were collected. Sparse cultures of the three indicated cell types were prepared ($5 \times 10^4$ cells/dish) in 0.1% serum. After 24 hours the medium was removed and replaced with 3 ml of the depleted media collected previously. At day 6 cells were counted and the results normalized. Fresh, undepleted media containing 10% fresh serum ≡ 100%. BP3T3: benzpyrene-transformed balb/c-3T3 cells (obtained from Dr. J. diPaolo).

**TABLE 5**

*Labeling Index of SV3T3 and HeLa Cells in Different Serum Concentrations*

| Serum % | percent labeled cells SV3T3 | HeLa |
|---|---|---|
| 10 | 52 | 49 |
| 1 | 49 | 58 |
| 0.3 | 48 | 53 |
| 0.1 | 52 | 55 |
| 0.01 | 50 | 52 |
| No (0-time) | 53 | 60 |
| No (12 hrs) | 54 | 58 |
| No (24 hrs) | 58 | 50 |

Cells (1-2 × $10^5$ from growing monolayer stock cultures were plated in 1% (HeLa) or 0.3% (SV3T3) serum. After 8 hrs the medium was aspirated, cultures washed twice with Tris-saline buffer pH 7.4 and incubated for 24 hours in medium with the indicated serum concentrations. [$^3$H] thymidine was added for 5 hours (HeLa) or 15 minutes (SV3T3) except in the experiments performed in serum-free medium. In these experiments the label was added at the indicated time (0, 12, or 24 hours after medium change). After incubation the cultures were washed, stripped with Kodak AR10 autoradiographic film and exposed for 3-6 days and the fraction of labeled nuclei determined.

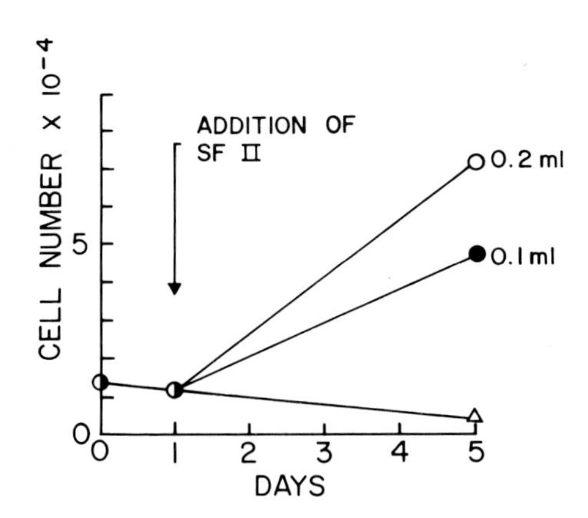


Fig. 2. SV3T3 cell growth in serum-free medium. Cells were trypsinized from stock cultures, washed 3× with serum-free medium and cultured in the presence of 5 μg/ml survival factor in medium as indicated in Table III. At day 1 SFII was added and cells counted in a Coulter Counter. ▲—▲ no addition; ●—● 50 ng/ml SFII; ○—○ 100 ng/ml SFII.

enter the cell cycle before they die and come off the dish as shown by autoradiographic studies (Table V). HeLa cells in suspension become arrested in $G_1$ due to serum factor limitation, i. e. DNA synthesis in suspended HeLa cells is under the control of serum factors (Fig. 3). Such quiescent $G_1$-arrested HeLa cells in suspension initiate the cycle after they attached to an appropriate surface, even in their depleted, unchanged spinner medium (Fig. 3). It appears that attachment *per se* provides a signal equivalent or similar to that of a serum factor required for DNA synthesis. Since such a signal is not depletable like a serum factor, cells grow constitutively and without restraint in monolayer.

Dense 3T3 cultures require more serum than sparse cultures. In crowded cultures the available area per cell for cell attachment is decreased. As a consequence, the anchorage-dependent signal required for growth (Clarke *et al.*,

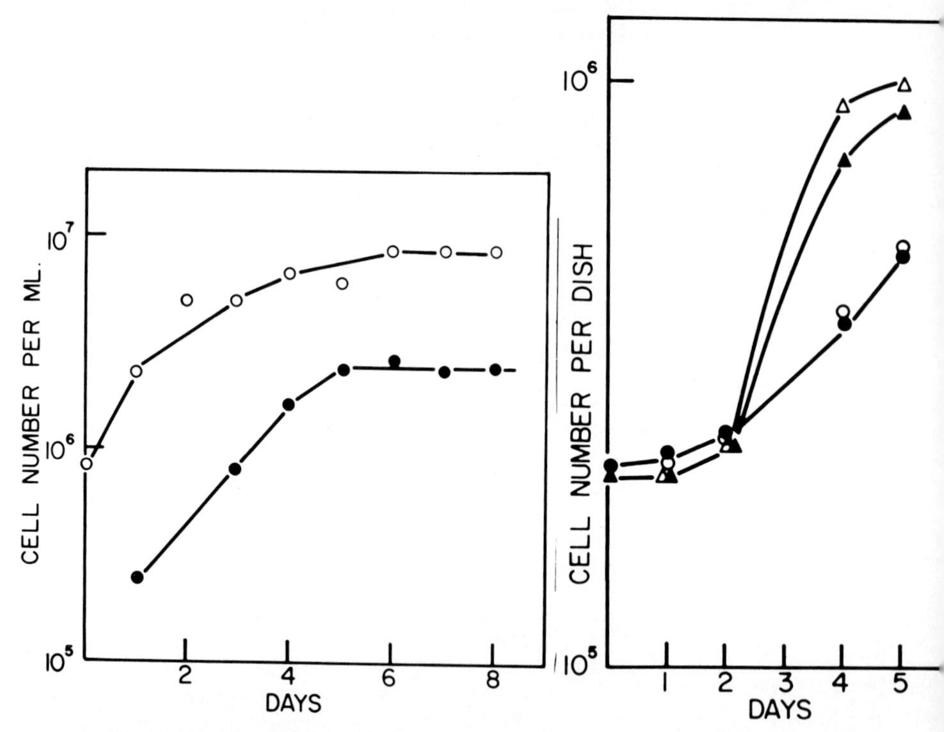


Fig. 3. Growth of HeLa cells in suspension and in monolayer. *Left Panel.* HeLa cells were cultured in spinner medium supplemented with 2% serum (●——●) or with 10% serum (○——○). Cells were centrifuged daily and the medium replaced with fresh medium and cell counted daily. *Right Panel.* $G_1$-arrested cells from suspension cultures in 2% serum were transferred into 3cm tissue culture dishes and cells counted after trypsinization. *Circles:* depleted spinner medium; *triangles:* fresh medium containing 2% serum; *open symbols:* 1.8 mM $Ca^{++}$; *closed symbols:* no calcium except that introduced by undialyzed serum.

1970) could become limiting and cells would stop growing. For example, in a wounded confluent 3T3 monolayer cells start spreading at the edge of the wound in response to a migration factor isolated from serum (Lipton *et al.*, 1971), i. e. the extent of interaction between cells at the wound edge with the substratum increases. It is possible that after cells spread out at the wound edge, the anchorage-dependent signal is no longer limiting and cells become responsive to serum growth factors, whereas cells in the monolayer remain quiescent.

Therefore, the difference between anchorage dependent growth (Clarke *et al.*, 1970) in 3T3 or other "normal" cells and HeLa cells is that in "normal" cells sufficient anchorage-dependent signal is required for growth *in addition* to the full serum factor supplement, while HeLa cells grow in suspension until the most limiting serum factor has been depleted; growth can be reinitiated by *either* addition of the limiting serum factor in suspension *or* by cell anchorage in monolayer (D. Paul, submitted for publication).

Cell attachment to collagen-coated dishes is dependent on the presence of a serum attachment factor (Klebe, 1974). Therefore, cell growth under these conditions is under the control of this additional serum factor that mediates cell anchorage via $Ca^{++}$-bridges.

## V. CELLULAR RESPONSES AFTER STIMULATION OF $G_1$-ARRESTED CELLS WITH SERUM FACTORS

Most of our studies concerning the mechanism of action of serum factors and hormones that initiate the growth cycle in $G_1$-arrested cells have focused on early cellular events after addition of growth stimulation to the cultures. One important cellular response after stimulation of $G_1$-arrested cells with serum is an increase in the rate of rRNA synthesis that takes place within 10 min (Green, 1974). Therefore, a mechanism depending on an increase in the synthesis of a specific protein, itself depending on an increase of its mRNA is presumably excluded since mRNA synthesis and its transport into the cytoplasm requires at least 15-20 minutes (Penman *et al.*, 1968). If a change of the rate of synthesis of a specific protein were involved in response to serum, this change would have to be affected at the translational level in order for the protein to be active on time. Alternatively, post-translational mechanisms involving activation of pre-existing proteins could be postulated. These considerations make it probable that critical changes in cellular metabolism that are responsible for triggering events that lead to the initiation of DNA synthesis occur very quickly after stimulation with serum, i. e. within 5-10 minutes. Therefore, some of our initial studies have focused on early effects in $G_1$-arrested cells in response to serum or serum factors.

In REF, DNA synthesis initiation is dependent on an increased $Ca^{++}$ influx in response to purified serum factors (Frank, 1973). In contrast, 3T3-4a cells enter the cell cycle in response to serum in the absence of $Ca^{++}$ in the medium, although complete removal of $Ca^{++}$ by chelating it to EGTA results in a complete inhibition of DNA synthesis initiation (H.-J. Ristow and D. Paul, manuscript in preparation). The fact that DNA synthesis is initiated by serum factors after $Ca^{++}$ has been washed out of the culture means that in contrast to REF, 3T3-4a cells keep intracellular $Ca^{++}$ levels high enough for many hours, so that intracellular $Ca^{++}$ levels do not become limiting. If serum is added 24 hrs after $Ca^{++}$ has been washed out, initiation of DNA synthesis is poor. In REF $Ca^{++}$ leaks out faster than in 3T3-4a cells. Therefore, these cells initiate DNA synthesis only in the presence of sufficient $Ca^{++}$ in the medium. It is concluded, that in 3T3-4a cells $Ca^{++}$ is not a signal in response to serum factors. Whether or not $K^{+}$ transport may play a role in the initiation of the cell cycle in 3T3 cells remains to be determined (Rozengurt and Heppel, 1975).

The metabolism of phosphatidyl inositol (PI) in response to various stimuli in many biological systems has been reviewed (Lapetina and Michell, 1975). In REF, PI metabolism (rates of synthesis and breakdown) is dramatically affected in response to serum factor addition to quiescent cultures (Hoffmann *et al.*, 1974). Similar observations were made in 3T3-4a cells in response to serum factors or serum (H.-J. Ristow and D. Paul, manuscript in preparation). It was shown that PI synthesis increases immediately after serum addition to $G_1$-arrested cells. This $Ca^{++}$-independent increase in the rate of PI synthesis is a necessary event required for the initiation of DNA synthesis since inhibition of PI synthesis with delthexane (δ-hexachloro cyclohexane, an analogue of myoinositol) suppresses DNA synthesis initiation. Inhibition of PI synthesis for 4 hours with delthexane (0-4 hours after serum factor addition to quiescent cultures) leads to a shift in the time course of DNA synthesis initiation (R. Hoffmann, unpublished; Fig. 4). It is possible that specific subclasses of PI characterized by specific fatty acid residues are preferentially synthesized or broken down in response to serum, therefore leading to a net alteration of the phospholipid bilayer structure of the cell membrane, and changing the hydrophilic/hydrophobic microenvironment in the lipid bilayer, that might lead to changes in the rate of nutrient transport, activity of membrane-bound enzymes, etc.

Cells can also be arrested in $G_1$ due to the limitation of certain amino acids or other nutrients. For example, it was found that primary liver cells become arrested in $G_1$ when arginine is limiting in the culture medium (Paul and Walter, 1975). Similarly, 3T3 cells and other mammalian cells become arrested in $G_1$ when some individual or all amino acids in the culture medium become limiting (Holley and Kiernan, 1974). Addition of the limiting amino acids leads to initiation of DNA synthesis.

The increased turnover of PI observed after serum stimulation of serum-arrested cells is not observed after stimulating amino acid-starved cells with the limiting amino acids (Table VI). This suggests that the restriction point in $G_1$ at

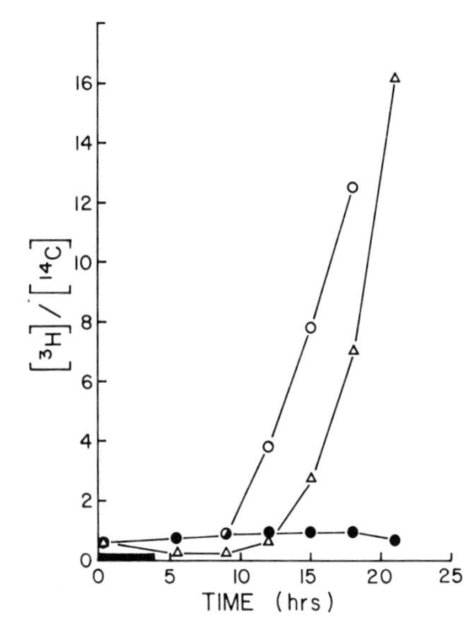


Fig. 4. Inhibition of phosphatidyl inositol (PI) synthesis in response to serum factors by delthexane in 3T3 cells. 3T3 cells ($10^5$/dish) prelabeled with $[^{14}C]$thymidine were plated in 0.3% serum. After 24 hours some cultures received 10% serum with or without 10 μg/ml delthexane (δ-hexachlorocyclohexane). The drug was washed out after 4 hours and 10% serum added back to the cultures. At the indicated times the cultures were pulsed with $[^3H]$thymidine for 1 hour and TCA precipitable radioactivity determined. The data are shown as $[^3H]/[^{14}C]$ ratios. ●—●, no addition; ○—○, 10% serum; △—△, 10% serum + 10 μg/ml delthaxane (0-4 hours, *black bar*).

which cells are arrested at low serum is different from that at which cells are arrested due to amino acid limitation. This brings up the question whether or not there is in fact one single restriction point at which all cells are arrested in low serum or whether perhaps different serum factors and/or hormones control different restriction points.

Elevated levels of polyamines and ornithine decarboxylase (ODC) activity are associated with a range of rapidly growing tissues and cell types. Therefore, we studied whether or not induction of ODC activity which is observed in response to serum growth factors (H. T. Rupniak, unpublished) is an essential requirement for DNA synthesis initiation. When polyamine synthesis is inhibited

**TABLE 6**

*Incorporation of $[^3H]$inositol into Phosphatidy inositol of Amino Acid Starved Cells Arrested in $G_1$*

| | 3T3 | | FRL | |
|---|---|---|---|---|
| | $[^3H]$inositol (0-60 min) | $[^3H]$thymidine (22-24 hrs) | $[^3H]$inositol (0-60 min) | $[^3H]$thymidine (22-24 hrs) |
| $G_1$ arrested | | | | |
| minus amino acids | 137 | 227 | 234 | 686 |
| plus amino acids | 101 | 94,772 | 199 | 6,168 |

Cells (3T3-4a and primary fetal rat hepatocytes [FRL]) were arrested in amino acid deficient medium according to Holley and Kiernan (1974) and Paul and Walter (1975). After addition of the limiting amino acid $[^3H]$thymidine was added for 2 hours and parallel cultures were incubated for 60 minutes with $[^3H]$inositol. TCA precipitable radioactivity was determined and is expressed as cpm/culture.

by inhibiting ODC with α-hydrazino-ornithine (αHO) and S-adenosyl-methionine decarboxylase with methylglyoxalbis-(guanylhydrazone) (MGBG) (Corti *et al.*, 1974) DNA synthesis initiation is not affected (Table VII). Under the conditions used in this experiment in the presence of both inhibitors neither spermidine nor spermine levels changed. Putrescine levels increased due to some residual synthesis, which cannot be converted into spermidine or spermine. It appears, therefore, that ODC induction in response to serum that leads to increased polyamine levels (Table VII) is not an essential requirement for the initiation of DNA synthesis. Presumably polyamine pools are high enough to allow DNA synthesis initiation even if polyamine synthesis is inhibited (Table VII). When the inhibitor MGBG is added to growing cultures, cell growth ceases after the cells have approximately doubled in number (Fig. 5). Cell growth resumes when either spermidine or spermine is added to the cultures in the presence of the inhibitor. Therefore, although MGBG leads to inhibition of polyamine synthesis it does not appear to have toxic side effects since addition of spermidine or spermine leads to cell growth at a rate similar to that observed in control cultures (Fig. 5). Putrescine does not lead to continuation of growth when added in the presence of MGBG. It is concluded that in quiescent cells polyamine levels, and in particular spermidine and spermine levels are sufficiently high to allow cells to initiate DNA synthesis and to progress through at least one cycle. When the pool sizes of spermidine and spermine are low enough cell growth ceases.

**TABLE 7**

*Initiation of DNA Synthesis in the Presence or Absence of Polyamine Synthesis Inhibitors in Rat Embryo Fibroblasts*

| | Polyamine levels nmoles/$10^6$ cells | | | [$^3$H] thymidine incorporation into TCA precipitable material cpm/$10^5$ cells |
|---|---|---|---|---|
| | putrescine | spermidine | spermine | |
| non-stimulated | 0.32 ± 0.06 | 3.85 ± 0.19 | 2.02 ± 0.24 | 340 ± 5.0 |
| stimulated + 1 μM MGBG + 1 mM αHO | 0.53 ± 0.06 | 3.77 ± 0.10 | 2.01 ± 0.24 | 6025 ± 160 |
| stimulated | 0.35 ± 0.06 | 11.60 ± 0.25 | 3.44 ± 0.31 | 6190 ± 350 |

Secondary rat embryo fibroblasts (REF) were cultured and arrested in $G_1$ by serum deprivation according to Frank *et al.* (1972). Arrested cultures were stimulated by the addition of dialyzed fetal calf serum (10%, v/v) in the presence or absence of polyamine synthesis inhibitors (MGBG [methylglyoxalbisguanylhydrazone], 1 μM; αHO [hydrazino ornithine], 1 mM). The cellular polyamine content was determined 24 hours after serum stimulation by the amino acid analyzer technique described by Kremzner (1973). DNA synthesis was measured by [$^3$H] thymidine incorporation into TCA precipitable material.

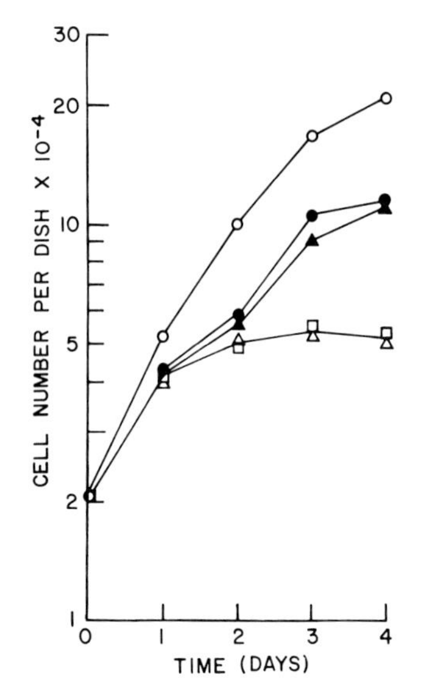


Fig. 5. Inhibition of growth of rat embryo fibroblasts by MGBG and its reversal by polyamines. Secondary rat embryo fibroblasts (REF) were cultured in 10% dialyzed fetal calf serum (v/v) according to Frank *et al.* (1972). After plating the cells, some cultures received 10 $\mu$M MGBG. Twenty-four hours later individual polyamines (1 $\mu$M) were added to the cultures containing MGBG and cells counted with a Coulter Counter. ○—○, untreated; △—△, MGBG; □—□, MGBG + putrescine; ▲—▲, MGBG + spermidine; ●—●, MGBG + spermine.

As mentioned earlier, the culture medium used for studies involving serum factors and hormones is supplemented with hypoxanthine. The growth stimulatory effect of purines is due to an accelerated recruitment of $G_1$-arrested cells to initiate DNA synthesis in response to serum (Brooks, 1975). The increase of ATP levels in $G_1$ observed after serum stimulation in 3T3 or FRL cells (F. Grummt and D. Paul, manuscript in preparation) is accelerated by exogenous hypoxanthine. Purine triphosphate levels control the initiation frequency of pre-rRNA transcription in the nucleoli (Grummt *et al.,* 1976). Since an increased rate of rRNA synthesis is a prerequisite for $G_1 \rightarrow S$ transition (Green, 1974), the increase of purine triphosphate levels in response to serum stimulation of $G_1$-arrested cells is necessary for the initiation of DNA synthesis. It appears that PRPP synthesis limits the availability of purine (and pyrimidine) nucleotides. Therefore, it is of interest that the activity of the rate-limiting enzyme for purine biosynthesis, PRPP synthetase (E.C. 2.7.6.1) is stimulated by cGMP, which is an allosteric effector of the enzyme, whereas cAMP antagonizes the stimulatory effect of cGMP competitively (Green and Martin, 1974). Since in 3T3 cells cGMP levels rise and cAMP levels fall within a few minutes after serum stimulation (Seifert and Rudland, 1974), it is conceivable that *de novo* purine biosynthesis could be stimulated by activation of PRPP synthetase that could lead to increased rates of rRNA synthesis within 10 minutes (see above).

It appears that the receptiveness of cells to growth factors is modulated by purines (Brooks, 1975), i. e. the requirements of cells for growth factors are decreased in the presence of purines. *In vivo*, purines are released into the circulation by purinergic nerves (Su, 1975), e. g. into the portal vein, therefore increasing *local* purine levels in tissues such as the liver. Perhaps this could lead to decreased requirements for cell-specific factors that initiate DNA synthesis in cells, e. g. in the liver after partial hepatectomy. Therefore, the nutritional requirements of cells seem to be as important for the regulation of the cell cycle as serum factors and hormones.

How do the results obtained in tissue culture relate to growth control in vivo? Do such oversimplified tissue culture model systems with man-made 3T3 cells (Todaro and Green, 1963) reflect control mechanisms that are operative *in vivo*? Previous experiments had shown that insulin is one of the growth factors required by FRL cells, whereas glucagon antagonized this stimulatory effect (Leffert, 1974; Paul and Walter, 1975). These results are not easily compatible with findings *in vivo*. For example, Bucher and Swaffield (1975) showed that in the rat insulin is a "permissive" but not DNA synthesis initiating factor in liver regeneration. Furthermore, experiments by Short *et al.* (1972) had shown that glucagon, $T_3$, heparin and amino acids, when infused into the tail vein of an intact rat, lead to DNA synthesis initiation and cell division in the liver. Also, glucagon could be successfully replaced with dibutiryl-cAMP (Short *et al.*, 1975) which is a potent growth inhibitor for all cells in culture. Thus, it appears that results obtained *in vivo* and in culture do not lead to similar conclusions. It is possible that glucagon exerts its activity *in vivo* indirectly. Therefore, it would not be possible to assay the activity in cultured cells. It is known that such complications exist. For example, growth hormone induces ODC in the liver of hypophysectomized rats (Eloranta and Raina, 1975). Since growth hormone does not induce ODC activity in FRL cultures (H. T. Rupniak, unpublished) it must be concluded that the effect of growth hormone *in vivo* is indirect.

We believe that it is necessary to dissect the complex networks of hormonal control systems into their components and to study the responses of cells to individual serum factors and hormones. Such studies can only be done in serum-free culture systems in order to be able to understand how hormones and factors act synergistically in controlling cell growth in culture.

## ACKNOWLEDGMENTS

We thank Margaret Henahan, Javier Villela and Jay Duerr for excellent technical assistance. This work was supported by research grants CA15087, GM20101, NIH Core Grant CA14195 from the U.S. Public Health Service, by a fellowship of the Deutsche Forschungsgemeinschaft, Bad Godesberg (to H.-J. R.) and by a grant of the Gildred Foundation.

## REFERENCES

Brooks, R. F. (1975). *J. Cell. Physiol.* **86**, 369–378.

Bucher, N. L. R. and Swaffield, M. N. (1975). *Proc. Nat. Acad. Sci. U.S.A.* **72**, 1157–1160.

Clarke, G. D., Stoker, M. G. P., Ludlow, A. and Thornton, M. (1970). *Nature* **227**, 798–801.

Corti, A., Dave, C., Williams-Ashman, H. F., Mihich, E. and Scherone, A. (1974). *Biochem. J.* **139,** 351–357.
Eloranta, T. and Raina, A. (1975). *FEBS Letters* **55,** 22–24.
Frank, W., Ristow, H.-J. and Schwalle, S. (1972). *Exp. Cell Res.* **70,** 390–396.
Frank, W. (1973). *Z. f. Naturforsch.* **28C,** 322–326.
Gospodarowicz, D. (1974). *Nature* **249,** 123–125.
Green, C. D. and Martin, D. W. (1974). *Cell* **2,** 241–245.
Green, H. (1974). In "Control of Proliferation in Animal Cells" (B. Clarkson and R. Baserga, eds.), pp. 743–755. Cold Spring Harbor.
Grummt, I., Smith, V. A. and Grummt, F. (1976). *Cell* **7,** 439–446.
Hayashi, I. and Sato, G. (1976). *Nature* **259,** 132–134.
Hoffmann, R., Ristow, H.-J., Packowsky, H. and Frank, W. (1974). *Eur. J. Biochem.* **49,** 317–322.
Holley, R. W. and Kiernan, J. A. (1974). *Proc. Nat. Acad. Sci. U.S.A.* **71,** 2942–2945.
Holley, R. W., Baldwin, J. H., Kiernan, J. A. and Messmer, T. O. (1976). *Proc. Nat. Acad. Sci. U.S.A.,* **73,** 3229–3232.
Kremzner, L. T. (1973). In "Polyamines in Normal and Neoplastic Growth" (D. H. Russel, ed.), pp. 27–33. Raven Press, New York.
Klebe, R. J. (1974). *Nature* **250,** 248–251.
Lapetina, E. G. and Michell, R. W. (1975). *FEBS Letters* **31,** 1–18.
Leffert, H. L. and Paul, D. (1972). *J. Cell Biol.* **52,** 559–568.
Leffert, H. L. (1974). *J. Cell Biol.* **62,** 792–801.
Lipton, A., Klinger, I., Paul, D. and Holley, R. W. (1971). *Proc. Nat. Acad. Sci. U.S.A.* **68,** 2799–2801.
Lipton, A., Paul, D., Henahan, M., Klinger, I. and Holley, R. W. (1972). *Exp. Cell Res.* **74,** 466–470.
Mukherjee, A. S., Washburn, L. L. and Banerjee, M. R. (1973). *Nature* **246,** 159–161.
Paul, D., Lipton, A. and Klinger, I. (1971). *Proc. Nat. Acad. Sci. U.S.A.* **68,** 645–648.
Paul, D., and Walter, S. (1975). *J. Cell Physiol.* **85,** 113–123.
Penman, S., Vesco, C. and Penman, M. (1968). *J. Mol. Biol.* **34,** 49–55.
Rozengurt, E. and Heppel, A. (1975). *Proc. Nat. Acad. Sci. U.S.A.* **72,** 4492–4495.
Seifert, W. and Rudland, P. S. (1974). *Proc. Nat. Acad. Sci. U.S.A.* **71,** 4290–4294.
Short, J., Brown, R. F., Husakawa, A., Gilbertson, J. R., Zemel, R. and Lieberman, I. (1972). *J. Biol. Chem.* **247,** 1757–1761.
Short, J., Tsukada, K., Rubert, W. A. and Lieberman, I. (1975). *J. Biol. Chem.* **250,** 3602–3605.
Su, C. (1975). *Pharmacol. Exptl. Ther.* **195,** 159–166.
Thrash, C. R. and Cunningham, D. D. (1973). *Nature* **242,** 399–401.
Todaro, G. J. and Green, H. (1963). *J. Cell Biol.* **17,** 299–303.

# III. Factors Affecting Nerve Cell Differentiation and Function

# Nerve Growth Factor: Studies on the Localization, Regulation and Mechanism of Its Biosynthesis

Edward A. Berger* and Eric M. Shooter

*Department of Neurobiology*
*Stanford University School of Medicine*
*Stanford, California 94305*

## I. INTRODUCTION

Embryonic development involves a complex interplay of cellular communication, both at the level of direct cell contact as well as through long-range chemical interactions. Perhaps nowhere is this more evident than in the vertebrate nervous system, where synaptogenesis, pattern formation and ultimately neuronal function must depend upon the coordinate regulation of cellular division, differentiation, morphogenesis and motility. In the early 1950's the discovery of a protein factor with specific growth-stimulating effects on nerve cells gave new focus to the study of cellular communication in the developing nervous system. This so-called Nerve Growth Factor, or NGF, was originally detected in certain transplantable sarcomas (Bueker, 1948; Levi-Montalcini and Hamburger, 1951), but is now known to exist in the normal tissues of nearly all vertebrates (Winick and Greenberg, 1965; Levi-Montalcini and Angeletti, 1968a). During the past two decades, a wealth of studies pioneered by Levi-Montalcini and her colleagues has led to the generally accepted view that NGF plays a fundamental role in the development and maintenance of the sympathetic nervous system, and in the development of the sensory ganglia as well (Levi-Montalcini and Angeletti, 1968a and 1968b).

The richest known NGF source is the submaxillary gland of the adult male mouse, and since its original isolation and purification (Cohen, 1960), the NGF from this tissue has been the subject of intensive structural analysis. At neutral pH, the activity is isolated as part of a specific high molecular weight complex called 7S NGF (Figure 1) (Varon et al., 1967; Varon et al., 1968). The complex contains three classes of subunits, referred to as $\alpha$, $\beta$ and $\gamma$, which may be

*Present Address: Department of Biology, B-022, University of California, San Diego, La Jolla, California 92093

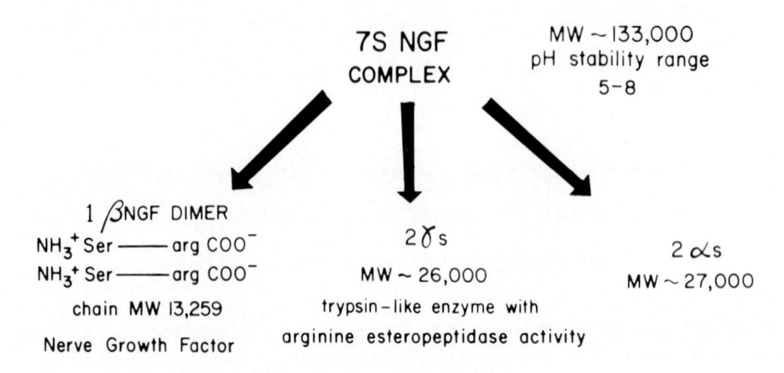


Fig. 1. The subunit structure of the 7S NGF complex.

dissociated, purified and recombined to regenerate the 7S complex (Shooter and Varon, 1970). All the nerve growth-stimulating activity is associated with the β-subunit, which is a dimer of two identical chains (M.W. 13,259) held together noncovalently. The sequence of these chains has been determined (Angeletti and Bradshaw, 1971) and βNGF has recently been crystallized (Wlodawer et al., 1975). The γ-subunit is a potent arginyl-esteropepidase, catalyzing the hydrolysis of synthetic arginine esters and amides (Greene et al., 1969). No enzymatic activity has yet been found for the α-subunit. At least one function of the 7S complex appears to be protection, since βNGF is less susceptible to proteolytic cleavage in submaxillary gland extracts when in the 7S form (Moore et al., 1974; Mobley et al., 1974). However, the association of a growth factor with a specific trypsin-like esteropeptidase speaks to the likelihood of additional enzymatic functions of the subunits which have thus far escaped identification.

Purified βNGF has pronounced morphological and biochemical effects on responsive nerve cells both *in vivo* and *in vitro*. In addition to the well known stimulation of axonal outgrowth from sympathetic or sensory ganglia in culture, NGF has been reported to enhance a variety of metabolic processes such as cell enlargement and proliferation, glucose metabolism, lipid synthesis, RNA and protein synthesis (Levi-Montalcini and Angeletti, 1968a), cyclic AMP production (Nikodijevic, 1975; but see also Frazier et al., 1973), uptake of certain small molecules (Varon, 1975), and assembly of microtubules (Levi-Montalcini et al., 1974). Perhaps most remarkable of all, NGF is apparently endowed with the ability to attract growing sympathetic and sensory axons (Bueker, 1948; Charlwood et al., 1972; Levi-Montalcini, 1975). We are thus dealing with a molecule capable of eliciting a plurality of effects related to neuronal growth, differentiation, and survival, yet an understanding of how these varied functions are integrated into the overall pattern of neuronal development and maintenance remains obscure. Our knowledge of NGF biosynthesis is particularly deficient, and we feel that the role of this factor cannot be fully ascertained without a precise description of where, when and how it is produced. We have therefore undertaken a detailed study of βNGF biosynthesis, with particular regard to questions of the sites of βNGF production, the regulation of synthesis and the mechanisms of processing. We anticipate that such studies will shed some new light on the functions of the 7S complex and on the overall role of NGF in the development and maintenance of the nervous system.

## II. MEASUREMENT OF βNGF BIOSYNTHESIS

The approach used in our studies is outlined in Figure 2. Submaxillary glands are labeled with radioactive amino acids, and βNGF (including newly synthesized material) is isolated by immunoprecipitation with anti-βNGF. The washed immunoprecipitates are analyzed on SDS polyacrylamide gels, and the radioactivity at the βNGF position is taken as a measure of βNGF synthesis.

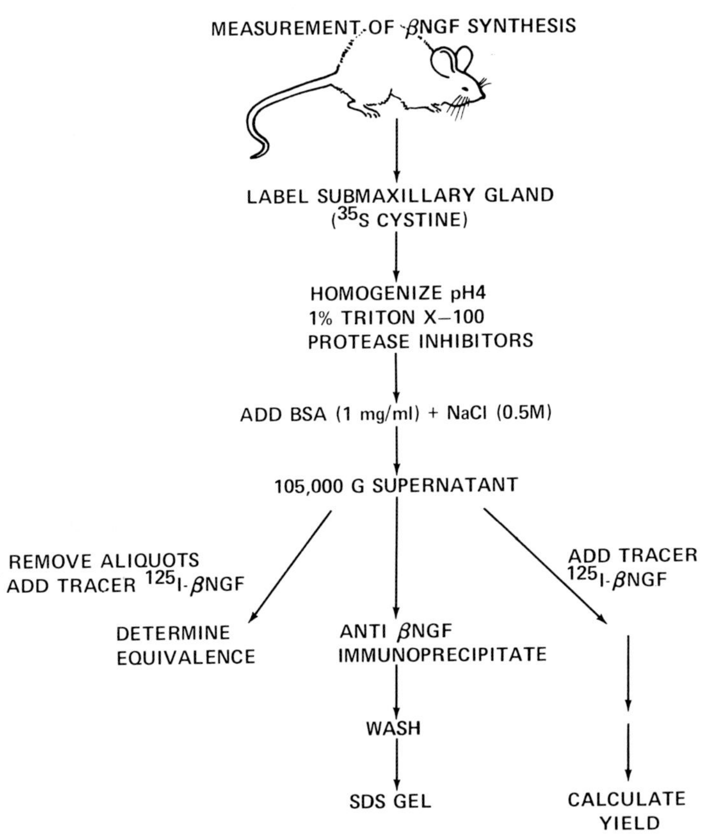


Fig. 2. Measurement of βNGF synthesis.

Typically, gland pieces are labeled in culture, but synthesis may alternatively be studied *in vivo* by direct injection of the isotope into the desired tissue. [$^{35}$S]-L-cystine was chosen because of the relatively high half-cystine content of βNGF (Angeletti and Bradshaw, 1971) and because this amino acid provides distinct advantages for peptide mapping (see below). Homogenates are prepared at pH 4 for several reasons: 1) the 7S complex dissociates at this pH (Varon et al., 1968), thereby avoiding problems of coprecipitation of the α- and γ-subunits, and of incomplete precipitation of βNGF when in the 7S complex; 2) the esteropeptidase activity of the γ-subunit is inactive below pH 5; 3) βNGF is soluble and stable at this pH (Varon et al., 1968); 4) many other proteins are insoluble under these conditions, thereby reducing the background during immunoprecipitation. Bovine serum albumin is added to block nonspecific adsorption of βNGF to the glassware and to the particulate matter, and NaCl is required for quantitative immunoprecipitation at pH 4.

The homogenates are centrifuged for 1 hour at 105,000 g and the equivalence points of the supernatants are determined by immunotitration from aliquots of supernatant to which have been added tracer quantities of [$^{125}$I]-βNGF. The equivalent amounts of anti-βNGF antiserum are then added to fresh aliquots of supernatant, and the immunoprecipitates are extensively washed and electrophoresed on SDS gels (Laemmli, 1970). The gels are cut into 2 mm slices and counted.

By this method, βNGF is recovered with 75-80% yield, determined by performing the identical isolation with separate aliquots of supernatant containing tracer [$^{125}$I]-βNGF. The background contamination is quite low (approximately 0.1% of the non-specific radioactivity appears in the immunoprecipitate).

## III. THE SITES OF βNGF BIOSYNTHESIS

In the adult male mouse, the bulk of the βNGF is found in the submaxillary gland (Cohen, 1960) and by immunofluorescent staining procedures appears confined to the tubular cells of this tissue (Levi-Montalcini and Angeletti, 1968a). Several lines of evidence argue against βNGF uptake by the gland from the serum (Caramia et al., 1962; Burdman and Goldstein, 1965; Levi-Montalcini and Angeletti, 1968b), and the decline in circulating βNGF levels after sialectomy (Cohen, 1960; Hendry and Iversen, 1973) suggests that the submaxillary gland is indeed an endocrine organ which produces and secretes βNGF into the blood. Conclusive proof of this hypothesis requires direct demonstration of βNGF synthesis by the submaxillary gland.

### A. *βNGF Synthesis by Isolated Submaxillary Glands*

Submaxillary gland slices were incubated for 5 hours in medium containing [$^{35}$S]-L-cystine. Immunoprecipitates were obtained from the 105,000 g supernatant and analyzed on SDS gels as described above. As shown in Figure 3,

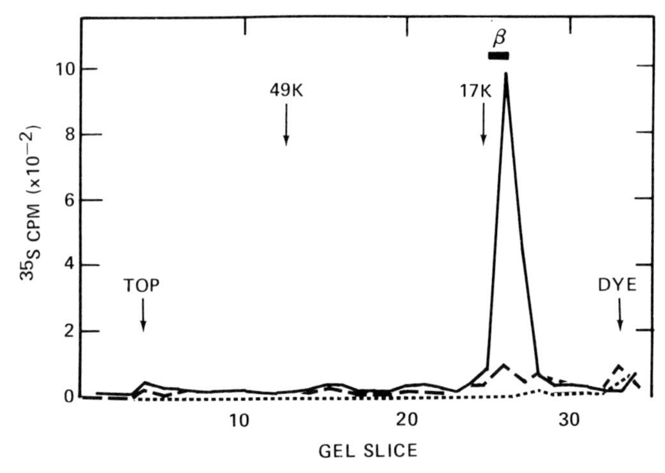


Fig. 3. SDS gel pattern of immunoprecipitates from submaxillary gland homogenates. Castrated adult male mice were injected daily with 1 mg testosterone proprionate on days one through four. On day four, the submaxillary glands were removed, cut into small pieces, and incubated in a cystine-free medium to which had been added [$^{35}$S]-L-cystine (2 mc/ml). After shaking for 5 hours at 37°, the glands were homogenized in buffer containing 0.05 M Na acetate, pH 4.0, 0.5 M NaCl, 1% Triton X-100, Bovine serum albumin (1 mg/ml), sodium azide (0.2 mg/ml), pancreatic trypsin inhibitor (50 μg/ml), and phenylmethylsulfonyl fluoride (5 mM). The 105,000 g supernatants were prepared and titrated with anti-βNGF antiserum as described in the text and shown in Fig. 2. The appropriate volumes of antiserum were added to a fresh aliquot of extract, and the immunoprecipitates were extensively washed and electrophoresed on SDS gels containing 15% acrylamide and 0.4% N,N′-methylenebisacrylamide. 2 mM gel slices were counted for radioactivity. The position of βNGF, indicated by the solid bar, was determined from the $R_f$ of pure βNGF run on a separate gel. Symbols: (————) immunoprecipitate with anti-βNGF; (— — —) immunoprecipitate from the supernatant of the first immunoprecipitation by the addition of carrier βNGF plus equivalent anti-βNGF; (· · ·) immunoprecipitate obtained by the addition of ferritin plus anti-ferritin to a separate aliquot of homogenate.

a washed immunoprecipitate obtained with anti-βNGF contains a single major peak of radioactivity which comigrates with purified βNGF. This peak is absent from two controls: 1) An immunoprecipitate obtained by the addition of anti-βNGF plus carrier βNGF to the supernatant of the first immunoprecipitation, and 2) An immunoprecipitate obtained from a separate aliquot of homogenate by the addition of ferritin plus anti-ferritin. Immunoprecipitates from the medium after incubation contained little radioactivity at the βNGF position. Similar labeling patterns have been obtained *in vivo* by direct injection of labeled amino acids into the submaxillary gland (Berger et al., 1975).

The identity of the major labeled species was established by comparing its tryptic map with that of purified βNGF. Figure 4 indicates that the carboxymethylated cystine-containing peptides of the [$^{35}$S]-unknown exactly comigrate on isoelectric focusing gels with those of purified βNGF, thus confirming the identity of the two molecules. As expected from the amino acid

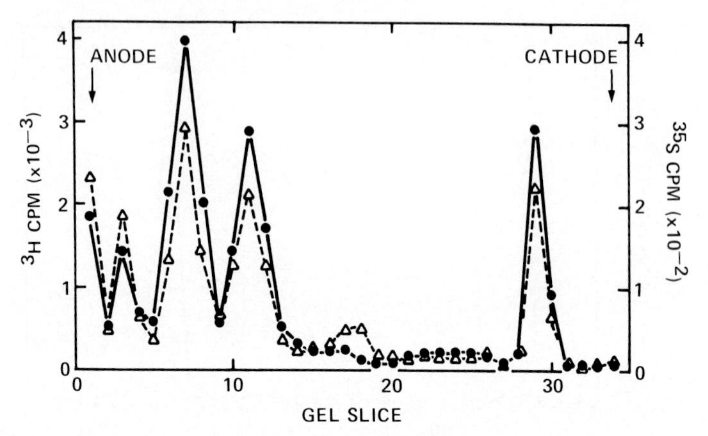


Fig. 4. Isolectric focusing of tryptic peptides of βNGF and the [$^{35}$S]-unknown. Submaxillary glands were labeled and homogenized and the anti-βNGF immunoprecipitate obtained as described in the Legend for Fig. 3. The immunoprecipitate was subjected to preparative SDS gel electrophoresis and the major [$^{35}$S]-peak was isolated, then reduced and carboxymethylated with iodoacetic acid. Purified βNGF was reduced and carboxymethylated with [$^3$H]-iodoacetic acid, then combined with the [$^{35}$S]-labeled material. The mixture was digested with trypsin, and the resulting peptides were subjected to isoelectric focusing in a polyacrylamide gel using a pH gradient of 3 to 6. 2mM slices were counted differentially for [$^3$H] and [$^{35}$S]. Symbols: (△——△) [$^3$H]; (●——●) [$^{35}$S].

sequence of βNGF (Angeletti and Bradshaw, 1971), there are 5 major half-cystine-containing peptides.

We conclude that submaxillary glands synthesize βNGF both *in vitro* and *in vivo*. The relative labeling of βNGF is quite low, representing approximately 0.2% of the TCA-precipitable radioactivity in the 105,000 g supernatant from the submaxillary gland of an adult male mouse.

*B. βNGF Synthesis by other tissues*

We next wished to test whether the βNGF synthesis observed in the submaxillary gland is merely a baseline level typical of all tissues or whether βNGF production is truly a specialized function of the submaxillary gland. We therefore measured synthesis in a variety of cultured tissues by the same methods. The results shown in Table I indicate that of all the tissues examined, βNGF synthesis could only be detected in the submaxillary gland under the conditions employed. The list includes nervous tissue (brain, spinal cord) as well as organs which, like the submaxillary gland, receive heavy sympathetic innervation (vas deferens, adrenal medulla). These results do not preclude the possibility that other tissues produce low but physiologically significant quantities of βNGF. In fact, NGF synthesis has been reported in primary cell and organ cultures of several tissues (Johnson et al., 1972; Young et al., 1975;

**TABLE 1**

*NGF Synthesis by Various Tissues in Culture*

| Tissue | % of TCA precipitable cpm which are in βNGF |
|---|---|
| Submaxillary Gland[a] | 0.23 |
| Liver | <0.002 |
| Kidney | <0.001 |
| Spleen | <0.003 |
| Lung | <0.003 |
| Heart | <0.005 |
| Adrenal* | <0.002 |
| Brain* | <0.007 |
| Spinal Cord* | <0.004 |
| Submaxillary Gland[b] | 0.21 |
| Testes* | <0.002 |
| Vas Deferens* | <0.009 |

The designated tissues were labeled and homogenized, and the immunoprecipitates electrophoresed as described in the legend for Fig. 3. Where the total amount of βNGF in the 105,000 g supernatant was less than 1 μg/ml, carrier βNGF was added prior to the addition of antiserum. Values were obtained by calculating all the radioactivity at the βNGF position. Only in the submaxillary gland immunoprecipitates were there discrete peaks at βNGF.

[a]Castrated adult male mice were injected daily with 1 mg testosterone proprionate. On day 4, the designated tissues were removed for incubation.

[b]Sham operated, untreated animals.

*The media were tested following the incubations by adding carrier βNGF plus antiserum, and electrophoresing the immunoprecipitates. In no case was labeled βNGF detected in the medium.

Harper et al., 1976), as well as in various transformed cells (Longo and Penhoet, 1974; Oger et al., 1974; Murphy et al., 1975). Glial cells have also been suggested as possible sources of NGF (Longo and Penhoet, 1974; Varon, 1975). However, the synthesis in the submaxillary gland far exceeds that measured elsewhere, and we therefore conclude that in male mice the bulk of the submaxillary gland βNGF is produced *in situ* and is the ultimate source for βNGF in the circulation.

## IV. HORMONAL REGULATION OF βNGF BIOSYNTHESIS

In the mouse, the βNGF content of both the submaxillary gland and the serum are regulated by testosterone. Thus, the levels in male animals greatly exceed those in females (Cohen, 1960; Hendry, 1972); the levels in males rise sharply during puberty (Hendry, 1972); castration of adult males causes a pronounced decline in βNGF content (Caramia et al., 1962; Ishii and Shooter, 1975); and, the βNGF levels in females and castrated males are markedly elevated by administration of testosterone (Caramia et al., 1962; Ishii and Shooter, 1975). Kinetic analyses suggest that testosterone acts at the level of βNGF synthesis rather than output from the gland (Ishii and Shooter, 1975), yet a direct effect of steroids on βNGF synthesis has not been demonstrated.

### *A. βNGF Synthesis Under Different Hormonal States*

To test directly the effects of testosterone on βNGF synthesis, we examined synthesis in cultured submaxillary glands from animals under different hormonal states. As shown in Table II, castration of adult male mice causes a marked reduction in relative βNGF biosynthesis, and testosterone treatment of castrated animals boosts the synthetic level to slightly above normal. βNGF synthesis is below detectable limits in submaxillary glands from adult females. Thus, the reported effects of hormonal state on βNGF levels correlate with direct measurements of synthesis, and we conclude that testosterone controls the gland βNGF content at least in part through control of βNGF production. We have yet to examine the effects of steriods on βNGF degradation and release.

## V. THE NATURE OF THE INITIAL BIOSYNTHETIC PRODUCT

A final feature of NGF biosynthesis emerges from a closer analysis of its structure: 1) As discussed above, βNGF may be isolated in complex with two other subunits, one of which (the γ-subunit) is a potent arginyl-esteropeptidase (Greene et al., 1969). 2) The COOH-terminal residue of each βNGF chain is arginine (Angeletti and Bradshaw, 1971; Moore et al., 1974). 3) The COOH-terminal arginine appears critical for maintaining the 7S complex, since removal

**TABLE 2**

*NGF Synthesis in Cultured Mouse Submaxillary Glands from Animals under Different Hormonal States*

| | % of TCA Precipitable CPM which are in βNGF | Relative Synthesis |
|---|---|---|
| Normal Males | 0.22 | 100 |
| Castrated Males | 0.07 | 31 |
| Castrated Males + Testosterone | 0.25 | 114 |
| Normal Females | <0.01 | <4 |

Castrated animals were used at least 21 days after operation. Where indicated, animals were injected daily with 1 mg testosterone proprionate. On day 4, the submaxillary glands were removed. Conditions of labeling, homogenization, immunoprecipitation, and electrophoresis are described in the legend for Fig. 3.

of this residue with carboxypeptidase B has little effect on either biological activity (Moore et al., 1974) or structure (Wlodawer et al., 1975), yet completely prevents recombination into the 7S complex (Moore et al., 1974). Furthermore, the arginyl-esteropeptidase activity of the γ-subunit is inhibited in 7S NGF (Greene et al., 1969) suggesting that in the complex the COOH-terminal arginine of βNGF is situated in the active site of the γ-subunit. 4) A strikingly analogous situation exists for Epidermal Growth Factor (EGF) which likewise can be purified from the mouse submaxillary gland in complex with a specific arginyl-esteropeptidase (EGF Binding Protein) (Taylor et al., 1970). The arginyl-esteropeptidases from the NGF and EGF complexes have very similar physical, chemical and immunological properties but EGF-Binding Protein will not substitute for γ in the formation of a 7S-type complex (Server and Shooter, 1976). This suggests that the two enzymes perform analogous yet specific functions. As with NGF, the COOH-terminal amino acid of EGF is arginine (Taylor et al., 1970) and this residue is required for complex formation (Server et al., 1976) but not for biological activity (Savage et al., 1972). The esteropeptidase activity of EGF-Binding Protein is likewise inhibited in the complex (Server et al., 1976). 5) βNGF has structural and functional similarities to insulin and proinsulin (Bradshaw et al., 1975).

These findings, coupled with the observations that many polypeptide hormone precursors are processed by cleavage at specific arginine residues (Steiner et al., 1974) has led to the prediction (Angeletti and Bradshaw, 1971; Moore et al., 1974) that βNGF is initially synthesized as a higher molecular weight precursor (pro-βNGF) with an additional polypeptide length at its COOH-terminal end, and that this precursor is ultimately cleaved at a particular arginine residue by the action of the γ-subunit (Fig. 5). The active site of the arginyl-esteropeptidase then remains complexed with the new COOH-terminal arginine of βNGF, and in conjunction with the α-subunit forms the 7S complex which is the storage form of NGF in the gland. According to this model then, the γ-subunit of the 7S complex is a highly specific protease which functions in the biosynthetic processing of the final βNGF molecule from a higher molecular weight precursor. An analogous proposal has been put forth for EGF and for EGF Binding Protein (Taylor et al., 1970).

### A. *Kinetics of βNGF Synthesis*

Fig. 6 shows the immunoprecipitates obtained from extracts of submaxillary glands labeled *in vitro* for various lengths of time. With relatively short pulses (10 and 25 minutes) the major labeled peak migrates at an approximate molecular weight of 22,000 (22K), substantially larger than βNGF (13,000). As time proceeds, the total radioactivity in this species reaches a plateau (Fig. 7). By contrast, the radioactivity at the βNGF position initially lags behind but then

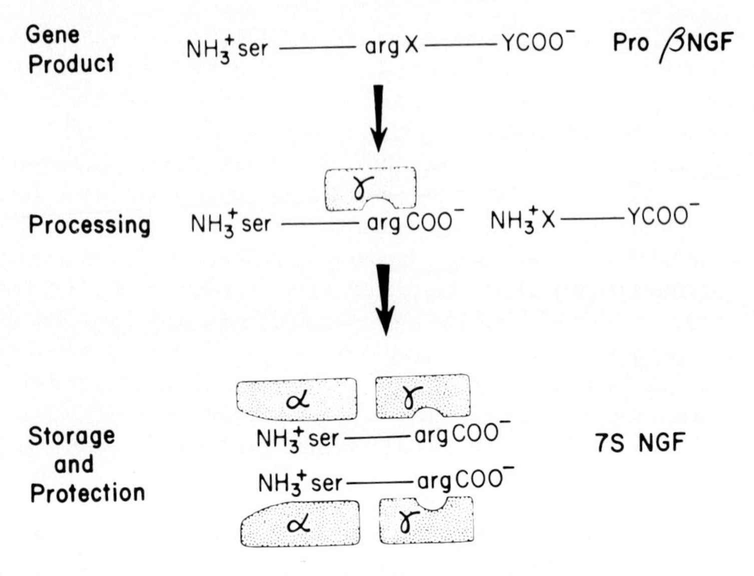


Fig. 5. Precursor Hypothesis for βNGF Biosynthesis.

increases sharply such that by 4 hours it far exceeds the label at the 22K position.

The precursor nature of the 22K peak is further suggested by the pulse-chase experiment shown in Fig. 8. Glands were labeled for 10 minutes and then either homogenized immediately or transferred for various lengths of time to a medium containing a large excess of unlabeled cystine and no additional label. The immunoprecipitates from the homogenates were analyzed as usual. Initially, there is a major labeled species at about 22K, and a smaller peak at the βNGF position. After an initial increase during the early part of the chase, radioactivity in the 22K peak declines and ultimately disappears, concomitant with a sharp rise in radioactivity at the βNGF position. These changes occur under conditions where further incorporation of label into TCA-precipitable material is effectively halted (data not shown), suggesting that we are indeed observing conversion of the 22K species to βNGF during the chase period.

*B. Effects of the γ-subunit on the 22K peak*

Fig. 9 indicates that such a conversion can occur *in vitro* as well as in the intact tissue. A washed immunoprecipitate from glands labeled for 30 minutes was either run directly on the SDS gel (Fig. 9A) or incubated with purified γ-subunit prior to electrophoresis (Fig. 9B). The results indicate that the bulk of the radioactivity at 22K can be converted to the same molecular weight as βNGF by treatment with the γ-enzyme. We have obtained similar conversions with 22K

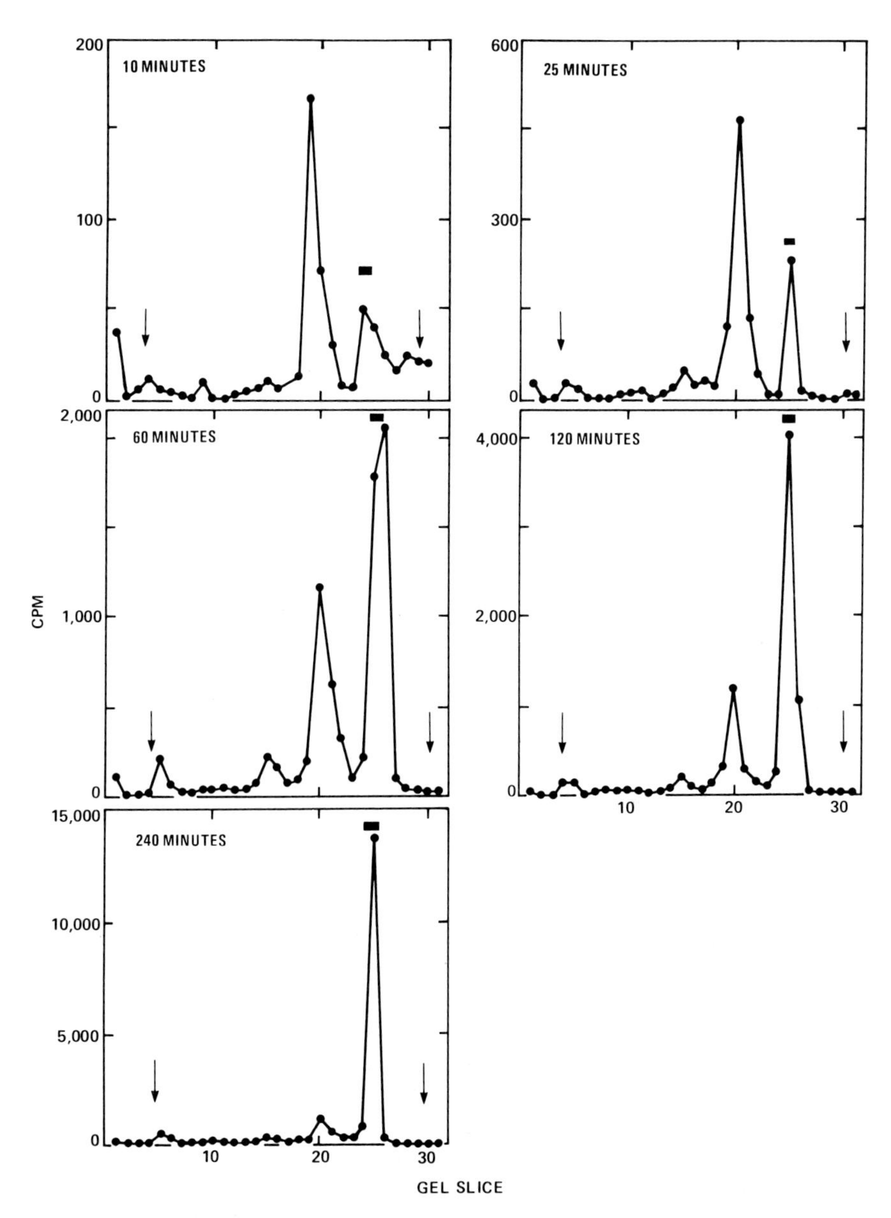


Fig. 6. Kinetics of Incorporation of [$^{35}$S]-L-cystine. Submaxillary glands were labeled for the indicated times. Treatment of the animals, labeling and homogenization of glands, immunoprecipitation and electrophoresis were as described in the Legend for Fig. 3.

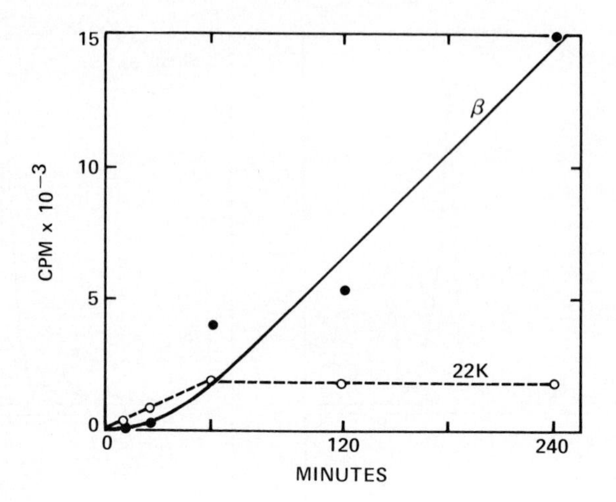


Fig. 7. Kinetics of labeling of the 22K peak and βNGF. Data were obtained from the experiments shown in Fig. 6 by calculating the radioactivity in either the 22K peak or in the βNGF peak at each indicated time.

material purified from preparative SDS gels after removal of the SDS (not shown).

Additional confirmation that the 22K species is indeed a precursor of βNGF derives from preliminary tryptic mapping studies with material purified from SDS gels. Using similar methods to those discussed earlier for characterizing the [$^{35}$S]-βNGF, we have found that the 22K species possesses all the cystine-containing peptides of βNGF and possibly two additional ones (data not shown).

We thus conclude that the 22K species meets the criteria of a true biosynthetic precursor to βNGF, and that the γ-enzyme is capable of cleaving this molecule to yield a product with the same molecular weight as βNGF.

Additional studies will be required to examine the specificity of the conversion reaction, and to determine if the additional polypeptide length is indeed at the COOH-terminal end of βNGF as predicted by the model shown in Fig. 5.

## VI. CONCLUSION

The studies reported here only begin to explore the biosynthesis of NGF, and many intriguing questions remain. They clearly establish that at least in the adult male mouse, the submaxillary gland is the primary site of βNGF production. It remains to be seen if this organ is a major site of βNGF synthesis in female mice and in other species, and if so whether the βNGF is rapidly

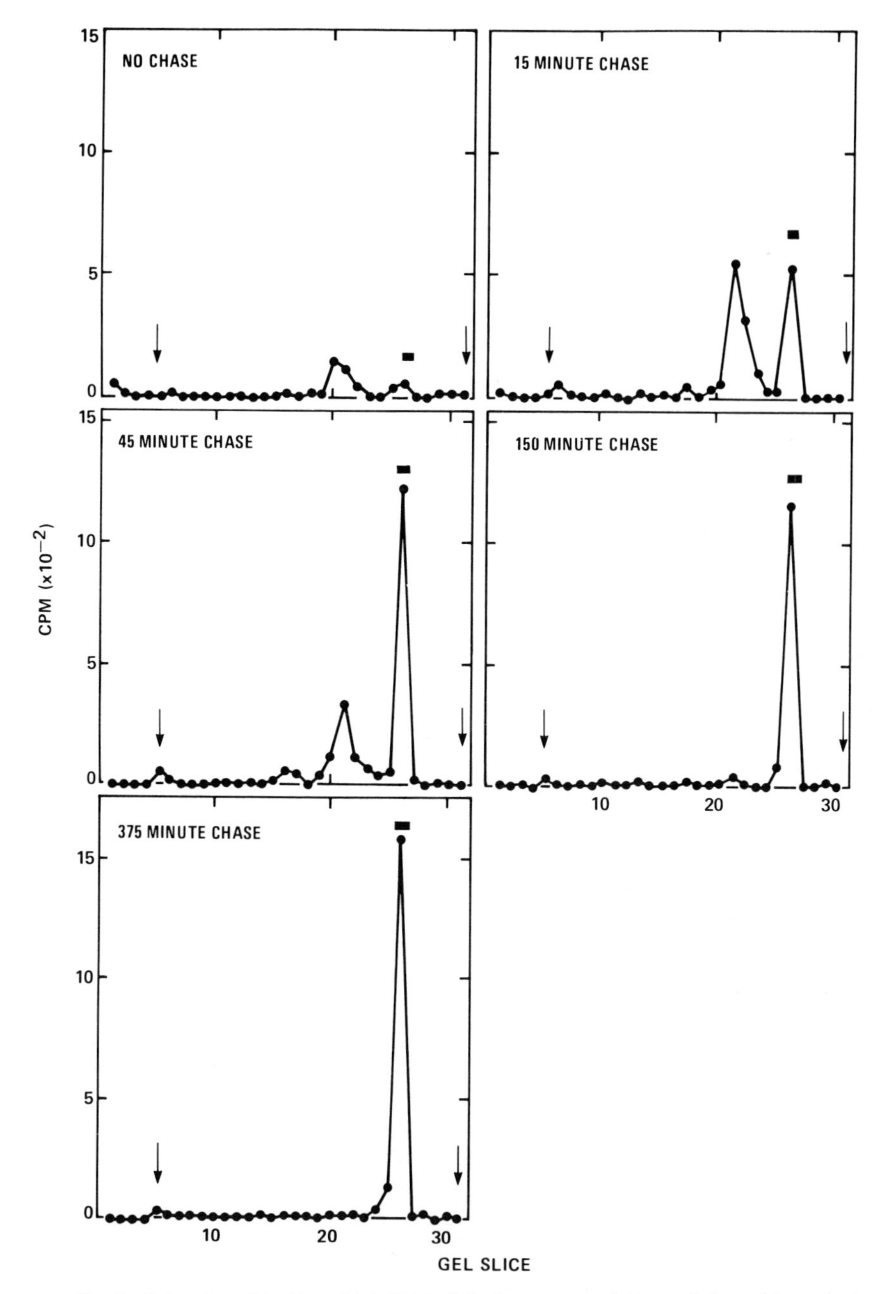


Fig. 8. Pulse-chase kinetics of labeling of the immunoprecipitates. Submaxillary gland pieces were labeled for 10 minutes in medium containing [$^{35}$S]-L-cystine (4 mc/ml) as described in the Legend for Fig. 3 and then either homogenized directly (no chase), or chased for the indicated times by transferring to fresh medium containing 115 $\mu$g/ml unlabeled L-cystine and no additional [$^{35}$S]-L-cystine. Treatment of the animals, homogenization of the glands, immunoprecipitation, and electrophoresis were as described in Fig. 3.

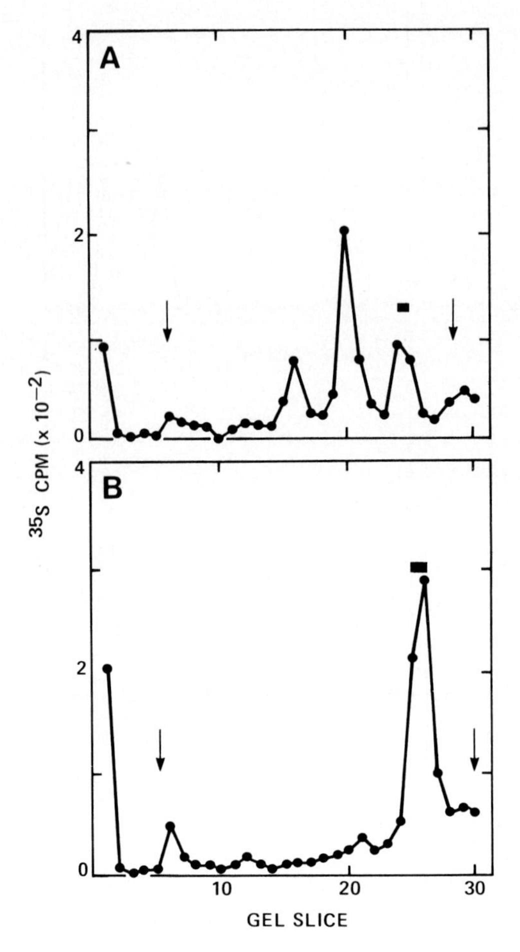


Fig. 9. Effect of the $\gamma$ subunit on the immunoprecipitate. Submaxillary glands were labeled for 30 minutes. Treatment of the animals, labeling and homogenization of the glands, and immunoprecipitation were as described in the Legend for Fig. 3. A. Washed immunoprecipitate electrophoresed directly, as in Fig. 3. B. Washed immunoprecipitate incubated for 4 hr, 37° with 50 $\mu$g of purified $\gamma$-subunit in 0.05 M Tris, pH 8.1 in a total volume of 0.1 ml. Electrophoresis as in Fig. 3.

released or degraded rather than accumulated. Our findings of course do not preclude the possibility that other tissues, such as glial cells and sympathetic end organs, might produce small but physiologically important quantities of $\beta$NGF.

Also unresolved is the question of $\beta$NGF synthesis in sialectomized mice, in which circulating levels initially decline but are eventually restored to their normal values without concomitant regeneration of the submaxillary gland (Hendry and Iverson, 1973). Direct tests of the synthetic capacities of tissues other than the submaxillary gland in such animals will be required to clarify this question.

Perhaps most intriguing of all is the question of $\beta$NGF production in the developing embryo. There are now several indications that NGF can influence the direction of fiber outgrowth from sympathetic and sensory ganglia both *in vivo* and *in vitro*. For example, NGF-producing sarcomas implanted into chick embryos become heavily infiltrated with sensory fibers (Bueker, 1948), and Levi-Montalcini has recently reported (1975) that injections of NGF into the rombencephalon of newborn rats or mice will divert sympathetic fibers into the central nervous system. In culture, fibers from embryonic sensory ganglia grow preferentially towards an NGF-containing capillary (Charlwood et al., 1972). These findings, coupled with the intriguing observations that NGF can be specifically taken up by both sympathetic (Hendry et al., 1974; Stoeckel and Thoenen, 1975) and sensory (Stoeckel et al., 1975; Stoeckel and Thoenen, 1975) nerve terminals and transported retrogradely back to the cell body, have led to the hypothesis (Hendry and Iversen, 1973; Stoeckel and Thoenen, 1975) that NGF synthesis and release by an effector organ during development might ultimately determine that organ's density of innervation. Thus, it becomes imperative to test whether tissues which ultimately become heavily innervated by sympathetic or sensory neurons synthesize and release βNGF *during the period of innervation*. A positive correlation would lend strong support to the concept of NGF as a critical trophic messenger between target organ and growing neuron.

Our discovery of pro-βNGF confirms earlier predictions and suggests a role for at least one of the other subunits of the 7S complex. The ability of the γ-subunit to cleave the 22K species suggests that at least one of its physiological functions is to generate βNGF from a higher molecular weight precursor. These findings assume added significance in view of the recent discoveries of at least three other growth factors in the submaxillary gland, namely two mesenchymal growth factors (Attardi et al., 1967; Weimar and Haraguchi, 1975) and a thymocyte transformation factor (Naughton et al., 1969). Each of these factors appears associated with arginyl-esteropeptidase activity (Weimar and Haraguchi, 1975), though none has been characterized in sufficient detail to know whether they exist as specific high molecular weight complexes with distinct arginyl-esteropeptidase enzymes, as is the case for NGF and EGF. But if it turns out that they do, then it is likely that the processes discussed here for NGF, namely synthesis by the submaxillary gland, regulation by steroid hormones, and processing of biosynthetic precursors by specific arginyl-esteropeptidases, will apply to a whole collection of factors regulating a variety of survival, growth and developmental functions.

## ACKNOWLEDGMENTS

This work was supported by NINCDS Research Grant # NS04270, by a Joseph P. Kennedy Jr. Foundation Fellowship in the Neurosciences, and by a Sloan Foundation Grant-in-Aid.

## REFERENCES

Angeletti, R. H. and Bradshaw, R. A. (1971). *Proc. Natl. Acad. Sci. U.S.A.* **68,** 2417–2420.
Attardi, D. G., Schlesinger, M. J. and Schlesinger, S. (1967). *Science* **156,** 1253–1255.
Berger, E. A., Ishii, D. N., Server, A. C. and Shooter, E. M. (1975). "Proc. 6th Int. Cong. Pharm." (L. Ahtee, ed.) Vol. 2, pp. 239–248. Pergamon Press, Oxford.
Bradshaw, R. A., Frazier, W. A., Pulliam, M. W., Szutowicz, A., Jeng. I., Hogue-Angeletti, R. A., Boyd, L. F. and Silverman, R. E. (1975). "Proc. 6th Int. Cong. Pharm." (L. Ahtee, ed.), Vol. 2, pp. 231–238. Pergamon Press, Oxford.
Bueker, E. D. (1948). *Anat. Record* **102,** 369–390.
Burdman, J. A. and Goldstein, M. N. (1965). *J. Exp. Zool.* **160,** 183–188.
Caramia, F., Angeletti, P. U. and Levi-Montalcini, R. (1962). *Endocrinology* **70,** 915–922.
Charlwood, K. A., Lamont, D. M. and Banks, B. E. C. (1972). *In* "Nerve Growth Factor and Its Antiserum" (E. Zaimis, ed.) pp. 102–107. Athlone Press, London.
Cohen, S. (1960). *Proc. Natl. Acad. Sci U.S.A.* **46,** 302–311.
Frazier, W. A., Ohlendorf, C. A., Boyd, L. F., Aloe, L., Johnson, E. M., Ferrendelli, J. A. and Bradshaw, R. A. (1973). *Proc. Natl. Acad. Sci. U.S.A.* **70,** 2448–2452.
Greene, L. A., Shooter, E. M. and Varon, S. (1969). *Biochemistry* **8,** 3735–3741.
Harper, G. P., Pearce, F. L. and Vernon, C. A. (1976). *Nature* **261,** 251–253.
Hendry, I. A. (1972). *Biochem. J.* **128,** 1265–1272.
Hendry, I. A. and Iversen, L. L. (1973). *Nature (London)* **243,** 500–504.
Hendry, I. A., Stoeckel, K., Thoenen, H. and Iversen, L. L. (1974). *Brain Res.* **68,** 103–121.
Ishii, D. N. and Shooter, E. M. (1975). *J. Neurochem.* **25,** 843–851.
Johnson, D. G., Silberstein, S. D., Hanbauer, I. and Kopin, I. J. (1972). *J. Neurochem.* **19,** 2025–2029.
Laemmli, U. K. (1970). *Nature* **227,** 680–685.
Levi-Montalcini, R. (1975). "Proc. 6th Int. Cong. Pharm." (L. Ahtee, ed.), Vol. 2, pp. 221–230. Pergamon Press, Oxford.
Levi-Montalcini, R. and Angeletti, P. U. (1968a). *Physiol. Rev.* **48,** 534–569.
Levi-Montalcini, R. and Angeletti, P. U. (1968b). *In* "Ciba Foundation Symposium: Growth of the Nervous System" (G. E. W. Wolstenholme and M. O'Connor, eds.), pp. 126–142. J. & A. Churchill Ltd., London.
Levi-Montalcini, R. and Hamburger, V. (1951). *J. Exp. Zool.* **116,** 321–362.
Levi-Montalcini, R., Revoltella, R. and Calissano, P. (1974). *Rec. Prog. Horm. Res.* **30,** 635–669.
Longo, A. M. and Penhoet, E. E. (1974). *Proc. Natl. Acad. Sci. U.S.A.* **71,** 2347–2349.
Mobley, W. C., Moore, J. B. Schenker, T. and Shooter, E. M. (1974). *Mod. Prob. Ped.* **13,** 1–12.
Moore, J. B., Mobley, W. C. and Shooter, E. M. (1974). *Biochemistry* **13,** 833–839.
Murphy, R. A., Pantazis, N. J., Arnason, B. G. W. and Young, M. (1975). *Proc. Natl. Acad. Sci. U.S.A.* **72,** 1895–1898.
Naughton, M. A., Koch, J., Hoffman, H., Bender, V., Hogopian, H. and Hamilton, E. (1969). *Exp. Cell Res.* **57,** 95–103.
Nikodijevic, B., Nikodijevic, O., Yu, M. W., Pollard, H. and Guroff, G. (1975). *Proc. Natl. Acad. Sci. U.S.A.* **72,** 4769–4771.
Oger, J., Arnason, B. G. W., Pantazis, N., Lehrich, J. and Young, M. (1974). *Proc. Natl. Acad. Sci. U.S.A.* **71,** 1554–1558.
Savage, C. R., Inagami, T. and Cohen, S. (1972). *J. Biol. Chem.* **247,** 7612–7621.
Server, A. C. and Shooter, E. M. (1976). *J. Biol. Chem.* **251,** 165–173.
Server, A. C., Sutter, A. and Shooter, E. M. (1976). *J. Biol. Chem.* **251,** 1188–1196.

Shooter, E. M. and Varon, S. (1970). *In* "Protein Metabolism of the Nervous System" (A. Lajtha, ed.), pp. 419–438. Plenum Press, New York.

Steiner, D. F., Kemmler, W., Tager, H. S. and Peterson, J. D. (1974). *Fed. Proc.* **33,** 2105–2115.

Stoeckel, K. and Thoenen, H. (1975). "Proc. 6th Int. Cong. Pharm." (L. Ahtee, ed.), Vol. 2, pp. 285–296. Pergamon Press, Oxford.

Stoeckel, K., Schwab, M. and Thoenen, H. (1975). *Brain Res.* **89,** 1–14.

Taylor, J. M., Cohen, S. and Mitchell, W. M. (1970). *Proc. Natl. Acad. Sci. U.S.A.* **67,** 164–171.

Varon, S. (1975). "Proc. 6th Int. Cong. Pharm." (L. Ahtee, ed.), Vol. 2, pp. 275–284. Pergamon Press, Oxford.

Varon, S., Nomura, J. and Shooter, E. M. (1967). *Biochemistry* **6,** 2202–2209.

Varon, S., Nomura, J. and Shooter, E. M. (1968). *Biochemistry* **7,** 1296–1303.

Weimar, V. L. and Haraguchi, K. H. (1975). *Physiol. Chem. and Physics* **7,** 7–21.

Winick, M. and Greenberg, R. E. (1965). *Pediatrics* **35,** 221–228.

Wlodawer, A., Hodgson, K. O. and Shooter, E. M. (1975). *Proc. Natl. Acad. Sci. U.S.A.* **72,** 777–779.

Young, M., Oger, J., Blanchard, M. H., Asdourian, H., Amos H. and Arnason, B. G. (1975). *Science* **187,** 361–362.

# Nerve Growth Factor as a Mediator of Information Between Effector Organs and Innervating Neurons

H. Thoenen, M. Schwab and U. Otten

*Department of Pharmacology*
*Biocenter of the University*
*Basel, Switzerland*

## I. INTRODUCTION

The ontogenetic development of integrated neuronal systems is regulated by a well coordinated sequence of epigenetic events involving cell-cell interaction, production and release of low and high molecular messengers from neuronal and non-neuronal cells and their interaction with corresponding specific receptive sites. The same mechanisms which produce reversible plastic modulations in fully differentiated systems can be responsible for irreversible imprinting effects during critical periods of ontogenetic development and represent an integral part in the formation of the gross neuronal outline. Thus, the detailed analysis of mechanisms of neuronal plasticity in differentiated systems may provide valuable information on the events taking place during critical phases of ontogenetic development.

In the following we will focus on a very specific aspect of neuronal function, namely the influence of effector organs on innervating neurons. The peripheral sympathetic nervous system serves as a representative model. There also, the same mechanisms can have a modulatory function in the fully differentiated system and an imprinting one during ontogenetic development.

It has been known for a long time that effector organs have a remarkable influence on the innervating neurons. In his classical experiments Hamburger (1938) has shown that in chicken embryos the development of the motor and sensory system of the spinal cord corresponds to the volume of the effector organs to be supplied. For instance, if a wing bud is removed at an early embryonic stage, the development of the corresponding dorsal root ganglia and the motor area in the spinal cord is markedly impaired. Furthermore, the transplantation of a wing bud to the contralateral side enhances the development of the corresponding motor and sensory system according to the volume of transplanted tissue.

Although the influence of effector organs on innervating neurons is well documented under many experimental conditions, the underlying mechanisms remained a matter of speculation until recent investigations in the peripheral sympathetic nervous system opened up aspects which may give a better understanding of at least one mechanism of interaction between effector organs and innervating meurons. The importance of effector organs on innervating adrenergic neurons was demonstrated very impressively by transplantation experiments. Transplantation of organs which either have a dense or sparse adrenergic innervation to sites which are able to reinnervate the transplant with adrenergic fibres (e.g. the anterior eye-chamber or the posterior diencephalon) showed that the density and pattern of adrenergic innervation is determined by the transplanted tissue and not by the site of transplantation (for references see Burnstock, 1969; Olson and Malmfors, 1970; Chamley *et al.*, 1973; Björklund *et al.*, 1974; Moore et al., 1974). Moreover, Björklund et al. (1974) have also shown that the local administration of nerve growth factor (NGF) or previous incubation of the transplants in a NGF-containing medium, produced an increase in the density of innervation. In contrast, incubation of transplants in a medium containing monospecific antibodies to NGF had the opposite effect i.e. a marked impairment of the ingrowth of adrenergic fibres. These observations, together with the evidence that NGF is not only synthesized by the salivary gland of adult male mice but also by other tissues (Levi-Montalcini and Angeletti, 1968; Burnham *et al.*, 1972; Hendry, 1972; Johnson *et al.*, 1971; Hendry and Iversen, 1973; Oger *et al.*, 1974; Varon *et al.*, 1974; Young *et al.*, 1975), prompted the hypothesis that NGF might act as a messenger between effector organs and innervating adrenergic neurons. In order to evaluate the validity of this hypothesis the following questions had to be answered.

a) Is NGF taken up by adrenergic nerve terminals and does it reach the perikaryon by retrograde axonal transport?
b) Is this retrograde transport specific both with respect to the molecules transported and with respect to the neurons which transport them?
c) Is the moiety of NGF reaching the cell body by retrograde axonal transport responsible for the specific biological effect of NGF on the cell body i.e. the selective induction of enzymes characteristic for adrenergic neurons?
d) Is there evidence for a relationship between the density of adrenergic innervation of an organ and the formation and/or storage of NGF?

## II. EVIDENCE FOR RETROGRADE AXONAL TRANSPORT OF NGF IN ADRENERGIC NEURONS

After unilateral injection of labelled NGF into the anterior eye chamber there is a preferential accumulation of radioactivity in the superior cervical ganglion of the injected side (Fig. 1). The time-lag between the injection and the

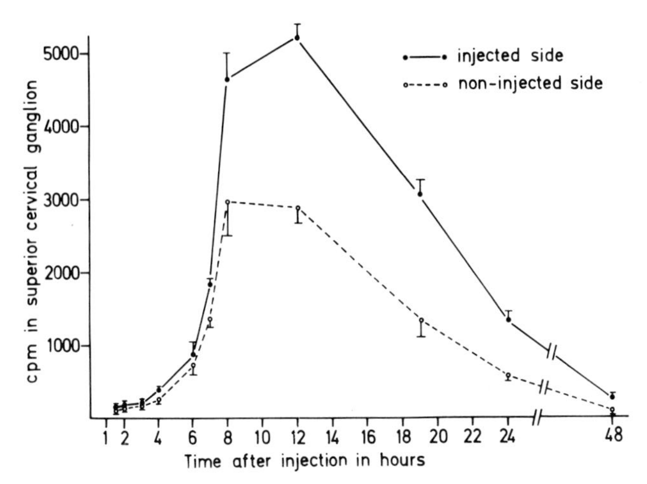


Fig. 1 Time-course of accumulation of radioactivity in superior cervical ganglia after unilateral injection of $^{125}$I-NGF into the anterior eye-chamber (from Stöckel *et al.*, 1976).

occurrence of the first statistically significant difference of the accumulation of labelled NGF between injected and non-injected side together with the distance between the nerve terminals and the corresponding cell bodies allows the calculation of the approximate rate of retrograde axonal transport (Hendry *et al.*, 1974). Interestingly, the rate of transport is the same for both rats and mice and both young and adult animals (Stöckel and Thoenen, 1975). The average transport rate is 2.5 – 3 mm/hour. The major part of the radioactivity accumulating in the rat superior cervical ganglion after retrograde axonal transport was shown to be intact NGF i.e. more than 90% of the extracted radioactivity was bound to solid phase monospecific antibodies to NGF. Moreover, the radioactivity present in the supernatant of homogenates of sympathetic ganglia ran to the position of $\beta$-NGF in polyacrylamide gel electrophoresis.

That the preferential accumulation of $^{125}$I-NGF in the superior cervical ganglion of the injected side really results from retrograde axonal transport is supported by the observation that transection of the postganglionic adrenergic fibres or the injection of colchicine prior to $^{125}$I-NGF administration abolished the preferential accumulation on the injected side. The latter observation also demonstrates that the retrograde axonal transport is sensitive to colchicine, as is the rapid orthograde axonal transport. Moreover, electronmicroscopic autoradiography has shown that after intraocular injection of $^{125}$I-NGF the label is exclusively located over the postganglionic adrenergic fibres (Iversen *et al.*, 1975; Schwab and Thoenen, 1976) connecting the site of injection with the superior cervical ganglia (Fig. 2). The concept of retrograde axonal transport is also supported by the light microscopic autoradiographic observation that after

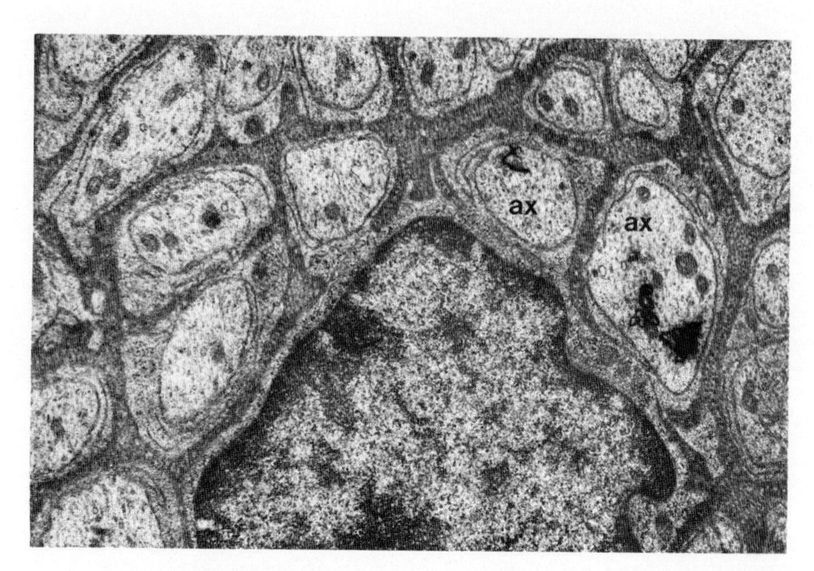


Fig. 2. Electron microscopic autoradiogram of postganglionic axons in the rat superior cervical ganglion 14 hours after injection of $^{125}$I-NGF into the anterior eye chamber and the submandibular gland. The label is localized within the axons (ax). – Magnification 14400 ×

unilateral injection of $^{125}$I-NGF into the anterior eye-chamber or the submaxillary gland the preferential accumulation of label in the superior cervical ganglion is restricted to a limited number of adrenergic neurons (Fig. 3). The selective labelling of single neurons by retrograde axonal transport of NGF provides a possibility of determining which neurons within the sympathetic ganglia supply a specific organ or even particular areas within this organ (Iversen *et al.*, 1975).

Recent experiments have also provided evidence that NGF accumulating in sympathetic ganglia after intravenous injection of $^{125}$I-NGF results predominantly from retrograde transport (Stöckel *et al.*, 1976). This can be deduced from the fact that the time-course of the level of radioactivity present in all organs studied after intravenous injection of $^{125}$I-NGF reflects the time-course in the blood plasma with a single exception, the sympathetic ganglia. Fig. 4 shows that the initial decay rate of $^{125}$I-labelled NGF in the blood plasma after intravenous injection is rather fast and corresponds to the half-life of 30 min, as determined in previous experiments (Thoenen *et al.*, 1974). The time-course of the radioactivity in the heart is very similar to that in the blood. In contrast, in the superior cervical ganglion, after a small gradual increase within the first hour following intravenous injection the level remains virtually unchanged up to 4 hours. Thereafter a dramatic increase, up to 10-fold, occurs within the next 4 hours representing the moiety of NGF reaching the adrenergic cell bodies by

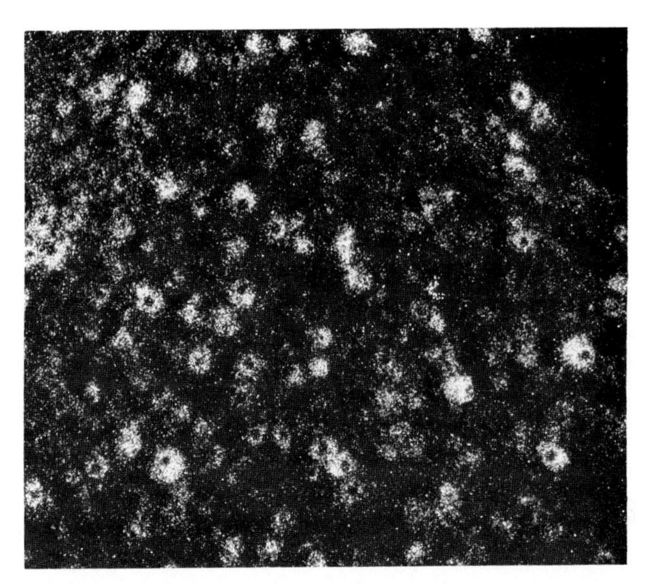

Fig. 3. Light microscopic autoradiogram (dark field illumination) of a semithin section through the rat superior cervical ganglion 14 hours after injection of $^{125}$I-NGF into the anterior eye chamber and the submaxillary gland. A limited population of neurons appears heavily labelled whereas other nerve cells remain unlabelled. Within the neuron, the label is strictly confined to the cytoplasm; the nucleus is free of radioactivity. – Diffuse labelling corresponds to radioactivity localized in axons and dendrites. – Magnification 200×

retrograde axonal transport. This interpretation was further supported by the observation that the time-lag between intravenous injection and the marked secondary increase of $^{125}$I-NGF accumulation could be reduced if the postganglionic adrenergic fibres were transected a week before the injection of $^{125}$I-NGF and the transected nerve fibres were allowed to recover and to sprout into the immediate surrounding. The freshly regenerating nerve fibres were able to take up the labelled NGF and since the distance between the site of uptake and the cell body was markedly reduced there was also a reduction in the time-lag between the injection and the marked secondary increase.

In very recent electronmicroscopic studies the subcellular distribution of retrogradely transported NGF was analysed in more detail by histochemical and autoradiographic procedures. The ultrahistochemical localization was based on the fact that in peripheral adrenergic neurons of the adult rat horseradish peroxidase alone is not taken up by the nerve terminals to an appreciable amount and transported retrogradely to the perikaryon if it is injected in the same relatively low concentrations as NGF (Stöckel *et al.*, 1974a). Therefore, the enzymatic activity of horseradish peroxidase can be used to localize at the electron microscopic level a coupling product of NGF with horseradish

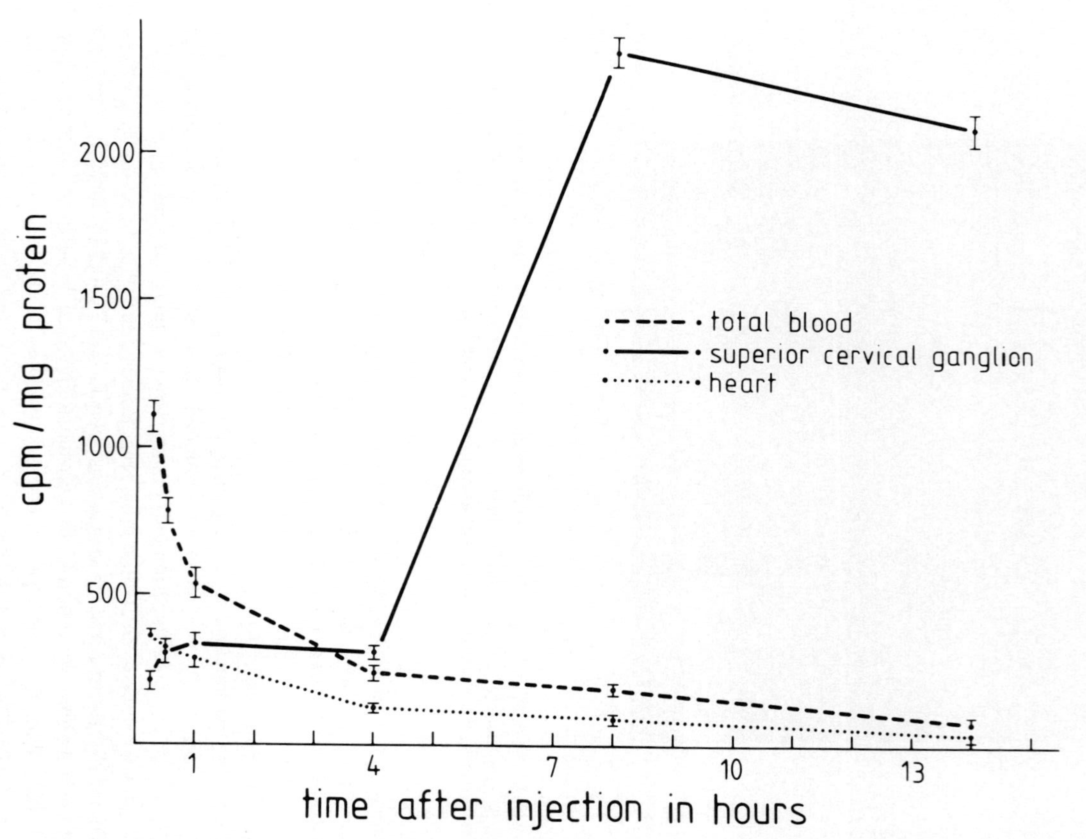


Fig. 4. Time-course of the accumulation of radioactivity in blood, heart and superior cervical ganglia after intravenous injection of $^{125}$I-NGF (From Stöckel *et al.*, 1976).

peroxidase which is transported in a similar way as NGF. In the axons the reaction product is virtually exclusively located in vesicles and in cisternae of smooth endoplasmic reticulum. In the perikaryon the coupling product was also localized in the endoplasmic reticulum, in vesicles and in addition in dense and multivesicular bodies i.e. lysosomal structures (Fig. 5A). The subcellular localization of the coupling product was in perfect agreement with the statistical analysis of the electronmicroscopic autoradiograms (Fig. 5B) (Schwab and Thoenen, 1976).

With respect to the mechanism of the selective induction of tyrosine hydroxylase (TH) and dopamine β-hydroxylase (DBH) by NGF (Thoenen *et al.*, 1971) which can be blocked with actinomycin (Stöckel *et al.*, 1974b) it is note worthy that neither the ultrahistochemical nor the electronmicroscopic autoradiographic studies provided evidence for the transfer of NGF to the cell nucleus (Fig. 3,5). Thus, the regulation of the induction of the two enzymes does not seem to result from a direct regulatory action of NGF on the chromatin.

## III. SPECIFICITY OF THE RETROGRADE AXONAL TRANSPORT OF NGF IN ADRENERGIC NEURONS

In order to investigate whether the uptake of NGF by adrenergic nerve terminals and its subsequent retrograde axonal transport is determined by simple physico-chemical properties of the NGF molecule or whether this retrograde axonal transport depends on more specific properties we compared the retrograde transport of NGF with that of a large number of molecules which varied over a large range of molecular weights and isoelectric points (Stöckel et al., 1974a). The molecular weights varied between 6000 and 500,000 and the isoelectric points between 4.5 and 10. None of these macromolecules showed a selective accumulation in the superior cervical ganglion on the side of their intraocular injection (Fig. 6). Of particular interest was the absence of a detectable retrograde axonal transport of cytochrome C, a molecule which has very similar properties to NGF, both with respect to molecular weight and isoelectric point. The specificity of the retrograde transport of NGF was even further evidenced by the fact that the oxidation of the tryptophan moieties of the NGF molecule reduced the retrograde transport in proportion to the extent of their oxidation. The impairment of the retrograde transport was accompanied by a corresponding reduction of the fiber outgrowth in chicken dorsal root ganglia (Stöckel *et al.*, 1974a).

It should be stated very clearly that these experiments do not exclude the possibility that very small amounts of all these macromolecules are transported retrogradely. However, the quantities possibly transported are too small to be detected by methods which easily allow the demonstration of the retrograde transport of NGF.

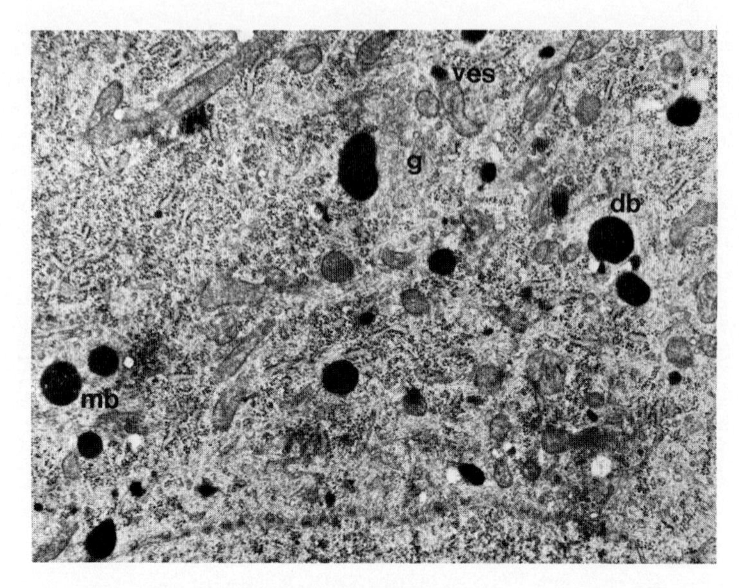


Fig. 5A. Sector of the cytoplasm of an adrenergic neuron in the rat superior cervical ganglion 14 hours after injection of a NGF-horseradish peroxidase coupling product into the anterior eye-chamber and the submaxillary gland. The histochemical reaction product (deep black precipitates) is localized in secondary lysosomes, dense bodies (db) and multivesicular bodies (mb), in vesicles (ves) and segments of the smooth endoplasmic reticulum. The Golgi cisternae (g) are free of peroxidase activity. – Magnification 16200 ×

## IV. RESTRICTION OF RETROGRADE AXONAL TRANSPORT OF NGF TO ADRENERGIC AND SENSORY NEURONS

After it had been established that the retrograde axonal transport of NGF in adrenergic neurons is highly selective and that even small changes in the chemical constitution of the NGF molecule markedly impaired the retrograde axonal transport the question arose as to whether this selectivity is determined by properties of the adrenergic neurons which are common to all neurons or confined to those which respond to NGF.

Under experimental conditions which allowed an unequivocal demonstration of retrograde axonal transport of tetanus toxin in spinal motoneurons no evidence for a corresponding transport of NGF could be obtained (Stöckel *et al.*, 1975b). However, after subcutaneous injection of NGF a highly preferential accumulation of radioactivity in the dorsal root ganglia of the corresponding spinal segments could be demonstrated (Stöckel *et al.*, 1975a). As in the peripheral adrenergic neurons this preferential accumulation could be abolished by transecting the nerve fibres of the corresponding cutaneous segment or by administration of colchicine prior to the injection of the $^{125}$I-labelled NGF. Here also autoradiographic studies revealed that the label was not equally

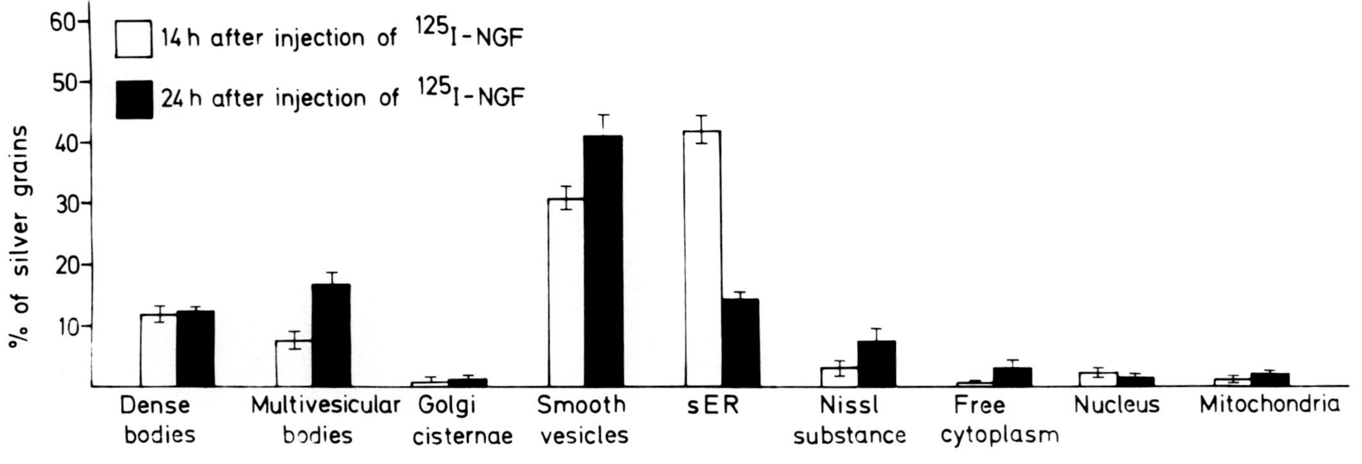


Fig. 5B. Percentage distribution of silver grains over adrenergic neurons in electron microscopic autoradiograms of the rat superior cervical ganglia 14 hours and 24 hours after injection of $^{125}$I-NGF into the anterior eye-chamber and the submaxillary gland. Values for various organelles represent the mean ± S.E.M. of 3–4 animals for each group. The results correspond closely to the histochemical localization of the NGF-horseradish peroxidase as illustrated in Fig. 5A.

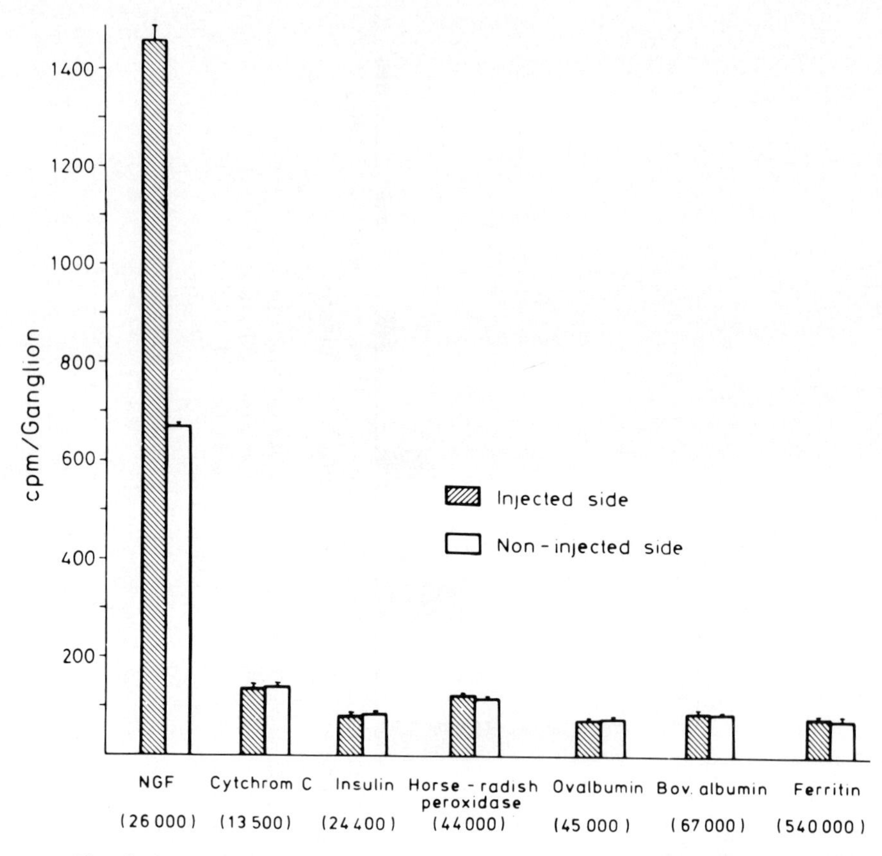


Fig. 6. Accumulation of radioactivity in superior cervical ganglia 14 hours after unilateral injection of $^{125}$I-NGF and various other $^{125}$I-labelled proteins into the anterior eye-chamber (From Stöckel *et al.*, 1974).

distributed over all the neurons of the dorsal root ganglia. For instance after injection of $^{125}$I-NGF into the forepaw of the rat not more than 10% of the cell bodies in the dorsal root ganglia $C_6$ and $C_7$ were labelled (Stöckel *et al.*, 1975a). Interestingly, the selective labelling was confined to the large neuron type. With the information available so far it cannot be decided whether this preferential labelling is determined by specific properties of the neuronal cell membrane of the large neurons or whether the selectivity is determined by the topographical localization of the nerve terminals. Moreover, the $^{125}$I-NGF reaching the dorsal root ganglia by retrograde transport was not further transferred to the spinal cord via the dorsal roots. Thus, there seems to be no direct connection between the retrograde transport system from the periphery and the orthograde one to the spinal cord.

In contrast to the peripheral sympathetic nervous system which responds to NGF throughout the whole life cycle the response of the dorsal root ganglia is restricted to a short period of the ontogenetic development (Levi-Montalcini, 1966; Herrup *et al.,* 1974). In spite of this limited duration of the response the sensory neurons seem to preserve the membrane properties necessary to bind NGF as a prerequisite for its subsequent retrograde transport. The preservation of this capacity for retrograde axonal transport in rodents is of special interest with respect to the observation of Herrup and collaborators (Herrup and Shooter, 1975; Herrup *et al.,* 1974) that the specific high affinity binding of NGF to mechanically dissociated cells of chicken dorsal root ganglia was correlated with the time of appearance and disappearance of the response to NGF by fibre outgrowth. Their observations very strongly suggest a causal relationship between the specific binding of NGF to the cell surface and the biological response. It cannot be decided whether the preservation of the retrograde transport in rodents throughout the whole life cycle and the temporarily limited high affinity binding in chicken results from a species difference or from a difference between cell body and nerve terminals.

## V. FUNCTIONAL SIGNIFICANCE OF THE MOIETY OF NGF REACHING THE PERIKARYON BY RETROGRADE AXONAL TRANSPORT

In order to further evaluate the concept that NGF acts as a messenger between effector organs and innervating neurons it had to be established that the moiety of NGF reaching the perikaryon is responsible for the specific effect on the neuron. At least in the peripheral adrenergic neuron there is good evidence that NGF reaching the cell body by retrograde axonal transport is responsible to a large extent for a well defined biological effect of NGF, namely the selective induction of TH and DBH, enzymes which are characteristic for adrenergic neurons and catalyze the rate limiting steps in the synthesis of the adrenergic transmitter norepinephrine (Paravicini *et al.,* 1975). If high doses of NGF are unilaterally injected into the anterior eye chamber and the submaxillary gland the subsequent induction of TH observed 48 hours later in the superior cervical ganglia is about 2-fold higher on the injected as compared to the contralateral side (Fig. 7). Admittedly, this difference is not very impressive but it has to be borne in mind that a considerable part of the injected NGF escapes into the circulation and reaches the contralateral sympathetic ganglia both directly and via retrograde axonal transport. Moreover, it has to be borne in mind that the experimental procedure (anesthesia and surgery) of unilateral injection into the anterior eye chamber and the salivary gland evokes a neuronally-mediated TH induction on both injected and non-injected side. This can be deduced from the fact that unilateral injection of cytochrome c – a molecule with a very similar molecular weight and isoelectric point to NGF – produces a TH induction in the superior cervical ganglion of about 20% on both injected and non-injected sides. If this non-specific neuronally-mediated TH induction is deducted from the NGF-mediated induction and if one assumes that all the nerve terminals on one side were exposed to the same concentration of NGF as those of the iris a ratio

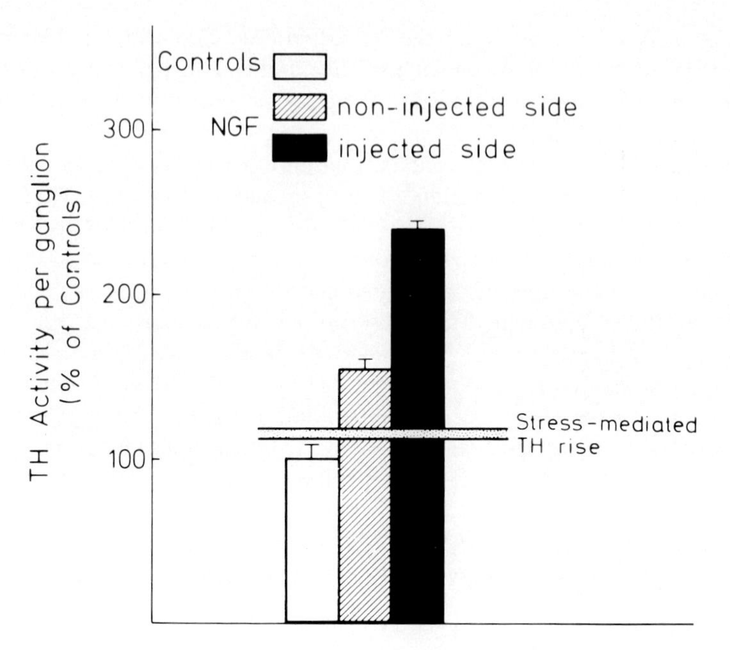


Fig. 7. Effect of unilateral injection of β-NGF into the anterior eye-chamber and submaxillary gland on TH activity in the superior cervical ganglion determined 48 hours after injection. The "stress-mediated" increase in TH activity was evaluated by unilateral injection of cytochrome c, a molecule which has a very similar molecular weight and isoelectrical point as β-NGF (According to Paravicini *et al.*, 1975).

of about 20:1 between injected and non-injected side were to be expected (Stöckel and Thoenen, 1975).

Although the experimental evidence presented above strongly suggests that NGF itself acts as a messenger between the adrenergic nerve terminals and the cell body it is possible that the binding of NGF to the cell membrane of the nerve terminal or a close non-neuronal receptor could initiate the formation of a (second) messenger which, similar to NGF, reaches the perikaryon by retrograde axonal transport and which is responsible for the specific enzyme induction rather than the retrogradely transported NGF.

## VI. EVIDENCE FOR THE FUNCTIONAL IMPORTANCE OF RETROGRADELY TRANSPORTED NGF DURING ONTOGENETIC DEVELOPMENT

During the first two postnatal weeks the peripheral sympathetic nervous system of the rat and mouse is still in a process of rapid ontogenetic development (for references see Black, 1974; Thoenen, 1972). This is manifested both at the morphological and biochemical level i.e. within this time period the ganglionic neuroblasts are transformed into fully differentiated neurons and there is also a rapid increase in the level of enzymes which are characteristic for adrenergic neurons, namely TH and DBH. The comparison of the effect of NGF during this critical period of ontogenetic development as compared to its effect in adult animals offers a representative example for the fact that the same mechanisms which have a modulatory function in fully differentiated neuronal systems can have an imprinting function during ontogenetic development. In adult rodents the transection of the postganglionic fibres of the superior cervical ganglion results in relatively small morphological changes in the perikaryon and in a temporary impairment of the ganglionic transmission which becomes apparent by electrophysiological changes (Matthews and Nelson, 1975; Purves, 1975) and by blockade of trans-synaptic enzyme induction (Brimijoin and Molinoff, 1971). However, all these effects are reversible and also the reinnervation of the effector organs by adrenergic fibres reaches the pretransection stage. Interestingly, Purves and Nja (1976) have shown that the electrophysiological consequences of axotomy can largely be prevented by administration of NGF.

In contrast to the adult animals the transection of the postganglionic adrenergic fibres in newborn animals results not only in a reduction of TH levels to 10–15% of controls but also to a corresponding reduction of the number of adrenergic neurons (Hendry, 1975a). The fact that systemic administration of NGF for 2 to 3 weeks after transection completely prevents the deleterious effect of transection (Hendry, 1975b) strongly suggests that the destruction of the adrenergic neurons in the early postnatal stage results from an interruption of the normal supply of NGF from the periphery. This assumption is in agreement with the observation that the administration of monospecific antibodies to NGF to adult animals evokes only minor transient ultramorphological and biochemical effects (Angeletti *et al.,* 1971; Bjerre *et al.,* 1975) whereas in newborn animals the administration of NGF antibodies results in an extensive destruction of the peripheral sympathetic nervous system (for references see Levi-Montalcini 1966; Levi-Montalcini and Angeletti, 1968).

The interpretation that the deleterious effect of axotomy in newborn animals results from the interruption of NGF supply from the periphery is supported by the observation that the same effect as with surgical interruption can be obtained by chemical destruction of the adrenergic nerve terminals with 6-hydroxydopamine (Levi-Montalcini *et al.,* 1975). This amine has the characteristic property of destroying adrenergic nerve terminals in adult and newborn animals (for references see Thoenen and Tranzer, 1973). Like surgical axotomy the effect of 6-hydroxydopamine is reversible in adult animals and leads to an irreversible destruction of the whole neurons if animals are treated in the early postnatal phase. As with surgical axotomy the destruction of the cell bodies can be prevented by administration of NGF (Levi-Montalcini *et al.,* 1975). The destructive effect of 6-hydroxydopamine on the nerve terminals remains unaltered demonstrating that NGF does not have a non-specific general protective effect against 6-hydroxydopamine, but its systemic administration replaces the normal supply from the periphery which was interrupted by the destruction of the nerve terminals.

## VII. PRODUCTION OF NGF IN ADRENERGICALLY INNERVATED EFFECTOR ORGANS

By far the weakest link in the chain of arguments favoring the role of NGF as a mediator of information between effector organs and innervating neurons is the evidence for differences in the synthesis and/or storage of NGF between densely and sparsely innervated organs. There is no doubt that NGF is not exclusively synthesized in the submaxillary gland of adult mice (Levi-Montalcini and Angeletti, 1968; Hendry and Iversen, 1973; Oger *et al.,* 1974; Varon, 1975a; Young *et al.,* 1975). However, the unequivocal demonstration that organs densely innervated with adrenergic neurons produce larger amounts of NGF than organs with a sparse or absent innervation is confronted by a series of methodological difficulties. The reliable determination of NGF in effector organs requires not only a very sensitive immunoassay but the interpretation of the results obtained by the commonly used radioimmunoassays (competition for the binding sites of monospecific antibodies) is complicated by the fact that serum of both rats and mice as well as homogenates of adrenergically innervated organs contain macromolecules which bind labelled NGF with a sufficiently high affinity to compete with the binding sites of the solid phase antibodies (Suda and Thoenen, unpublished observations). These macromolecules may not only be of importance for pitfalls in NGF immunoassays but they may also play a role as regulators for the storage and release of NGF. It has also to be borne in mind that the determination of the amount of NGF present in an effector organ gives only very limited information on the rate of synthesis. This is particularly true in view of the very effective uptake and (rapid) retrograde axonal transport by adrenergic and sensory nerve fibres. Moreover, it remains to be established to what extent NGF present in effector organs originates from de novo synthesis or from cleavage from a precursor protein.

## VIII. SITE AND MECHANISM OF ACTION OF NGF

The response of adrenergic and sensory neurons to NGF involves not only enhanced general growth but also an enhanced differentiation. The enhanced differentiation is manifested by an acceleration of the formation of neuronal processes in sensory and adrenergic neurons and a selective induction of TH and

DBH in adrenergic neurons (Levi-Montalcini and Angeletti, 1968; Thoenen *et al.*, 1971). The recent observation that the injection of NGF into the cerebrospinal fluid prompts peripheral sympathetic fibres to change their direction of growth and to sprout into the central nervous system – such a diversion of sympathetic fibres is never seen under normal conditions – provides most direct evidence for a chemotactic effect (Levi-Montalcini, 1975).

In view of this multiplicity of effects of NGF there arises the question as to whether all these effects are mediated by the same primary site of action with variable secondary pathways depending on the target cell and its stage of development. At the present stage of knowledge we do not even know whether the effects of NGF are mediated by a second messenger formed in consequence to the binding of NGF to cell surface receptors or whether this binding represents nothing more than the necessary prerequisite for the transfer of NGF into the interior of the cell in order to reach the proper site of action. The fact that there is a close correlation between the time-course of appearance and disappearance of specific binding sites on dissociated cells of chicken dorsal root ganglia and the response of these ganglia to NGF with fibre outgrowth would suggest an action via surface receptors (Herrup, *et al.*, 1974; Herrup and Shooter, 1975). This assumption is further supported by the observation that NGF covalently bound to sepharose beads is able to evoke fibre outgrowth in chicken dorsal root ganglia (Frazier *et al.*, 1973). However, these experiments are open to the same criticism as those concerning sepharose-bound insulin for which it has been shown that the amount of insulin released from the sepharose beads is sufficient to account for the whole biological effect (Davidson *et al.*, 1973). The assumption of a primary action of NGF via surface receptors is also supported by the observation that the early stimulation of RNA synthesis in chicken dorsal root ganglia by NGF results from a very rapid (less than 10 min lag) stimulation of a most probably membrane-mediated precursor accumulation (Varon, 1975a). However, the selective induction of TH and DBH in adrenergic neurons seems to result from NGF which enters the cell not only in experiments in which NGF reaches the perikaryon by retrograde axonal transport but also during the initiation of selective enzyme induction in organ cultures. In the latter case also, the selective induction of TH and DBH is associated with an transfer to NGF into the interior of the cells. Unfortunately, neither colchicine nor cytochalasin B could prevent the uptake of NGF into the perikaryon from the cell surface. Thus, it cannot be decided whether the internalization of NGF is an absolute prerequisite for the initiation of selective enzyme induction.

The selective induction of TH and DBH in adrenergic neurons can be blocked by actinomycin suggesting a regulation at the transcription level (Stöckel *et al.*, 1974a). Interestingly, neither after retrograde axonal transport nor after uptake into the perikaryon in organ cultures is NGF accumulated in the cell nucleus. Thus, it has to be assumed that any regulation at the transcription level is not effected by NGF itself but by a secondary mechanism initiated by NGF somewhere else. The regulation of TH and DBH induction by NGF via an actinomycin-sensitive mechanism is also of interest from the point of view that fibre outgrowth is not dependent on RNA synthesis but requires protein synthesis (Partlow and Larrabee, 1971; Varon, 1975a,b). Thus, selective enzyme induction and fibre outgrowth do not seem to be mediated via the same site of action. This is also to some extent supported by the fact that the concentrations of NGF necessary to induce TH and DBH in organ cultures of sympathetic ganglia are about 100-times higher than those necessary to initiate fibre outgrowth. In very recent experiments it has been shown that the concentration of NGF necessary to induce TH can be markedly reduced if the sympathetic ganglia are preincubated with glucocorticoids (Otten and Thoenen, 1976a). It is not yet established whether the concentrations of NGF necessary to initiate fibre outgrowth are reduced by glucocorticoids to a similar extent as those necessary for selective enzyme induction. However, in preliminary experiments with chicken dorsal root ganglia no potentiating effect of glucocorticoids on fibre outgrowth was observed.

The potentiating effect of glucocorticoids on selective TH and DBH induction by NGF is also of interest with respect to the neuronally-mediated enzyme induction. Both NGF and an increased activity of preganglionic cholinergic fibres produce a strikingly similar induction pattern of enzymes in terminal adrenergic neurons i.e. a selective induction of TH and DBH whereas other enzymes involved in general cell function such as lactate dehydrogenase or other enzymes involved in synthesis or metabolic degradation of the adrenergic transmitter such as MAO and dopa decarboxylase remain unchanged (for references see Thoenen, 1975). This similarity between the neuronally and NGF-mediated changes in enzyme pattern suggests that both the trans-synaptic enzyme induction which is mediated via nicotinic receptors on the cell surface and the inductive effect of NGF merge to a common final mechanism. This assumption is strongly supported by the observation that not only the NGF effect but also the neuronally-mediated enzyme induction is markedly potentiated by glucocorticoids (Otten and Thoenen, 1975, 1976a,b).

## IX. CONCLUDING REMARKS

After transplantation experiments had shown that the density and pattern of adrenergic innervation is determined by the transplanted effector organ and that preincubation of the transplant with NGF enhanced, whilst preincubation with NGF-antiserum impaired the innervation of the transplant we investigated the working hypothesis that NGF acts as a messenger between effector organs and innervating adrenergic neurons. The validity of this hypothesis was supported by the observation that NGF is taken up with high specificity by adrenergic nerve terminals and is transported retrogradely within the axon to the perikaryon. Moreover, it has been shown that the moiety of NGF reaching the perikaryon via retrograde axonal transport is responsible for the selective induction of TH and DBH in the adrenergic cell body, a characteristic biological effect of NGF. It remains to be established which cells in the effector organs produce NGF (by *de novo* synthesis or by cleavage from a precursor molecule), whether a correlation between the density of innervation and the production of NGF exists, and which are the factors responsible for the regulation of the NGF production.

## ACKNOWLEDGMENTS

This work was supported by the Swiss National Foundation for Scientific Research (Grant Nr. 3.432.74).

## REFERENCES

Angeletti, P. U., Levi-Montalcini, R. and Caramia, F. (1971). *Brain Res.* **27,** 343–355.

Bjerre, B., Wiklund, L. and Edwards, D. C. (1975). *Brain Res.* **92,** 257–278.

Björklund, A., Bjerre, B. and Stenevi, U. (1974). *In* "Dynamics of Degeneration and Growth in Neurons" (K. Fuxe, L. Olson and Y. Zotterman, eds.) pp. 389–409, Pergamon Press.

Black, I. B. (1974). *In* "Dynamics of Degeneration and Growth in Neurons" (K. Fuxe, L. Olson and Y. Zotterman, eds.) pp. 455–467, Pergamon Press.

Brimijoin, S. and Molinoff, P. B. (1971). *J. Pharmacol. Exp. Ther.* **178,** 417–424.

Burnham, P., Raiborn, C. and Varon, S. (1972). *Proc. Nat. Acad. Sci. U.S.A.* **69,** 3556–3560.

Burnstock, G. (1969). *Pharmacol. Rev.* **21,** 247–324.

Chamley, J. H., Goller, I. and Burnstock, G. (1973). *Develop. Biol.* **31,** 362–379.

Davidson, M. B., Van Herle, A. J. and Gerschenson, L. E. (1973). *Endocrinology* **92,** 1442–1446.

Frazier, W. A., Boyd, L. F. and Bradshaw, R. A. (1973). *Proc. Nat. Acad. Sci. U.S.A.* **70,** 2931–2935.

Hamburger, V. (1938). *J. Exp. Biol.* **68,** 449–494.

Hendry, I. A. (1972). *Biochem. J.* **128,** 1265–1272.

Hendry, I. A. (1975a). *Brain Res.* **86,** 483–487.

Hendry, I. A. (1975b). *Brain Res.* **94,** 87–97.

Hendry, I. A. and Iversen, L. L. (1973). *Nature* **243,** 500–504.

Hendry, I. A., Stöckel, K., Thoenen, H. and Iversen, L. L. (1974). *Brain Res.* **68,** 103–121.

Herrup, K., Stickgold, R. and Shooter, E. M. (1974). *Ann. N.Y. Acad. Sci.* **228,** 381–392.

Herrup, K., and Shooter, E. M. (1975). *J. Cell. Biol.* **67,** 118–125.

Iversen, L. L., Stöckel, K. and Thoenen, H. (1975). *Brain Res.* **88,** 37–43.

Johnson, D. G., Gordon, P. and Kopin, I. J. (1971). *J. Neurochem.* **18,** 2355–2362.

Levi-Montalcini, R. (1966). *Harvey Lectures Ser.* **60,** 217–259.

Levi-Montalcini, R. (1975). *In* "Proceedings of the Sixth International Congress of Pharmacology, Helsinki, Finland" (J. Tuomisto and M. K. Paasonen, eds.) Vol. II, pp. 221–230.

Levi-Montalcini, R. and Angeletti, U. (1968). *Physiol. Rev.* **48,** 534–569.
Levi-Montalcini, R., Aloe, L., Magnaini, E., Oesch, F. and Thoenen, H. (1975). *Proc. Nat. Acad. Sci. U.S.A.* **72,** 595–599.
Matthews, M. R. and Nelson, V. H. (1975). *J. Physiol. (Lond.)* **245,** 91–135.
Moore, Y. R., Björklund, A. and Stenevi, U. (1974). *In* "The Neurosciences" Third Study Program. pp. 961–977, 1974, MIT Press.
Oger, J., Arnason, B. G. W., Pantazis, N., Lehrich, J. and Young, M. (1974). *Proc. Nat. Acad. Sci. U.S.A.* **71,** 1554–1558.
Olson, L. and Malmfors, T. (1970). *Acta physiol. Scand. Suppl.* **348,** 1–112.
Otten, U. and Thoenen, H. (1976a). *Brain Res.,* **111,** 438–441.
Otten, U. and Thoenen, H. (1976b). *J. Mol. Pharmacol.,* **12,** 353–361.
Paravicini, U., Stöckel, K. and Thoenen, H. (1975). *Brain Res.* **84,** 279–291.
Partlow, L. M. and Larrabee, M. G. (1971). *J. Neurochem.* **18,** 2101–2118.
Purves, D. (1975). *J. Physiol. (Lond.)* **252,** 429–463.
Purves, D. and Nja, A. (1976). *Nature* **260,** 535–536.
Schwab, M. and Thoenen, H. (1977). *Brain Res.,* **122,** 459–474.
Stöckel, K., Paravicini, U. and Thoenen, H. (1974a). *Brain Res.* **76,** 413–421.
Stöckel, K., Solomon, F., Paravicini, U. and Thoenen, H. (1974b). *Nature* **250,** 150–151.
Stöckel, K. and Thoenen, H. (1975). *In* "Proceedings of the Sixth International Congress of Pharmacology, Helsinki, Finland" (J. Tuomisto and M. K. Paasonen, eds.) Vol. II, pp. 285–296.
Stöckel, K., Schwab, M. and Thoenen, H. (1975a). *Brain Res.* **84,** 1–14.
Stöckel, K., Schwab, M. and Thoenen, H. (1975b). *Brain Res.* **99,** 1–16.
Stöckel, K., Guroff, G., Schwab, M. and Thoenen, H. (1976). *Brain Res.,* **109,** 271–284.
Thoenen, H. (1972). *Pharmacol. Rev.* **24,** 255–267.
Thoenen, H. (1975). *In* "Handbook of Psychopharmacology", pp. 443–475, Plenum Press.
Thoenen, H., Angeletti, P. U., Levi-Montalcini, R. and Kettler, R. (1971). *Proc. Nat. Acad. Sci. U.S.A.* **68,** 1598–1602.
Thoenen, H. and Tranzer, J. P. (1973). *Annu. Rev. Pharmacol.* **13,** 169–180.
Thoenen, H., Hendry, I. A., Stöckel, K., Paravicini, U. and Oesch, F. (1974). *In* "Dynamics of Degeneration and Growth in Neurons" (K. Fuxe, L. Olson and Y. Zotterman, eds.), pp. 315–328, Pergamon Press.
Varon, S. (1975a). *In* "Proceedings of the Sixth International Congress of Pharmacology, Helsinki, Finland" (J. Tuomisto and M. K. Paasonen, eds.), Vol. II, pp. 275–284.
Varon, S. (1975b). *Exp. Neurol.* **48,** 75–92.
Varon, S., Raiborn, Ch. and Burnham, P. (1974). *J. Neurobiol.* **5,** 355–371.
Young, M., Oger, J., Blanchard, M. H., Asdourian, H., Amos, H. and Arnason, B. G. W. (1975). *Science* **187,** 361–362.

# Peptides as Central Nervous System Neurotransmitters

Solomon H. Snyder

*Departments of Pharmacology and Experimental Therapeutics and Psychiatry and Behavioral Sciences Johns Hopkins University School of Medicine Baltimore, Maryland 21205*

## I. INTRODUCTION

For years the biogenic amines, acetylcholine, norepinephrine, dopamine and serotonin, were thought by many to be the only likely neurotransmitters in the mammalian central nervous system. Recent strong evidence from a variety of sources has indicated that the biogenic amines quantitatively are relatively minor neurotransmitters, each accounting for only a few percent of brain synapses. Instead, certain amino acids constitute the major transmitters. In the spinal cord and lower brainstem glycine is the predominant inhibitory neurotransmitter, directly hyperpolarizing membranes by increasing chloride permeability. In higher centers gamma-aminobutyric acid (GABA) is the primary inhibitory transmitter, acting by enhanced chloride conductance. Depending on the brain region under study, glycine and GABA each account for 25–40% of all synapses. The major excitatory neurotransmitters are probably glutamic acid and aspartic acid, though evidence for them is not as strong as for the inhibitory amino acids. A variety of biochemical and histochemical data now suggest that, besides the biogenic amines and amino acids, a wide range of peptides may be neurotransmitters or neuromodulators. Since most of the peptides with apparent synaptic activity have been discovered "accidentally", it is conceivable that numerous presently unknown peptides may have major central roles. This raises the question as to the total number of candidate neurotransmitters in the brain, whether it be 10, 50 or many more. Here we will discuss two of these, substance P and enkephalin, the brain's opioid peptide, in detail. More limited descriptions will be provided of neurotensin and angiotensin.

## SUBSTANCE P

Substance P provides an interesting example of the circuitous route whereby brain peptides have been discovered. While examining various tissues for the presence of acetylcholine, Von Euler and Gaddum (1931) found a substance in intestinal and brain tissue which contracted smooth muscle and did not correspond in its properties to any known substances. The "P" simply refers to "powder." A variety of groups attempted to isolate the material but were unsuccessful for about forty years. While seeking hypothalamic releasing factors, Leeman and Hammerschlag (1967) identified a substance which provoked massive salivation in rats after intravenous injection. Quite independently Lembeck and Starke (1968) observed that substance P activity, which they had been in the process of isolating for many years, also caused salivation; they suggested that the material studied by Leeman and Hammerschlag might in fact represent substance P. Using the saliva inducing activity as a bioassay, Chang and Leeman (1970) isolated substance P and identified it as an undecapeptide with the amino acid sequence H-ARG-PRO-LYS-PRO-GLN-GLN-PHE-PHE-GLY-LEU-MET-OH.

With the availability of synthetic substance P, antibodies have been formed so a radioimmunoassay can be utilized to measure tissue levels (Powell *et al.*, 1973). Substance P is localized primarily to the gastrointestinal tract and the central nervous system, confirming the pioneering observations Von Euler and Gaddum made. This localization may have a developmental basis in the occurrence of neuroectodermal tissue in the gastrointestinal tract. It is thought that several peptides may have such a common embryological origin, also with localizations in intestine and brain (Said and Rosenberg, 1976). Peptides already known to be localized selectively to intestine and brain include substance P, vasoactive intestinal peptide (VIP), enkephalin, somatostatin and pancreozymin.

Early studies using bioassay had suggested a localization of substance P in the dorsal gray matter of the spinal cord and in sensory ganglia. Accordingly, it was hypothesized that substance P might be a transmitter of primary sensory afferent nerves. Using antibodies to synthetic substance P, Hokfelt *et al.*, (1975) demonstrated substance P immunofluorescence in about 15% of sensory afferent nerves with terminals in the skin and in the dorsal gray matter of the spinal cord and cell bodies in the sensory ganglia. After lesions of the dorsal root, substance P content is depleted from the dorsal gray matter (Otsuka *et al.*, 1975) as would be expected for the transmitter of sensory nerves. When applied by iontophoresis substance P is about 250 times more potent than glutamic acid in exciting spinal cord cells, which is striking considering that glutamic acid had been the major candidate as the sensory neurotransmitter.

Thus it seems fairly well established that substance P is one of the sensory neurotransmitters. Substance P displays an uneven distribution throughout the central nervous system, with very high levels in the hypothalamus and substantia nigra but low levels in areas such as the cerebellum and cerebral cortex. Because of its heterogeneous distribution it is likely that substance P containing neurons serve a diversity of functions.

## II. ENKEPHALIN: THE BRAIN'S MORPHINE-LIKE PEPTIDE

Opiate-like peptides in the central nervous system, like substance P, were discovered in a fairly indirect fashion. The major impetus derived from the demonstration in recent years of specific opiate receptor sites in vertebrate brain which mediate pharmacological responses to opiates (Snyder, 1975). In subcellular fractionation studies the opiate receptor appears localized to synaptic membranes as would be expected for a neurotransmitter receptor. Autoradiographic studies show an extremely discrete localization of opiate receptor sites to particular regions throughout the central nervous system (Pert *et al.*, 1976).

The heterogeneous regional localization of the opiate receptor, with highest concentrations in areas related to pain perception and emotional behavior, fits with known actions of opiates. The regional and subcellular studies also suggested that the opiate receptor might interact with some naturally-occurring substance. Its specificity seemed too great for an accidental membrane protein which interacts only with exogenous drugs. To identify such a hypothetical morphine-like factor, Hughes (1975) took advantage of the ability of morphine and other opiates to inhibit electrically induced contractions of smooth muscle such as the guinea pig ileum or mouse vas deferens. He found a substance in brain extracts which, like morphine, inhibited electrically induced contractions of the mouse vas deferens and guinea pig ileum. A different peptide with similar opiate-like effects on smooth muscle was extracted from the pituitary gland (Cox *et al.*, 1975; Lazarus *et al.*, 1975).

In our own laboratory (Pasternak *et al.*, 1975; Snyder 1975) and that of Terenius (1975), opiate receptor binding was used as an assay to identify and then purify the brain's morphine-like factor, which will hereafter be referred to as "enkephalin"*.

Enkephalin activity of brain extracts represents their ability to inhibit the binding of radioactive opiates to the opiate receptor. Since many substances, especially ions, can interfere with opiate receptor binding, careful attention was devoted in purification studies to ensure that only a physiologically relevant substance was being studied. Our primary means of ensuring specificity was to show that the regional distribution of enkephalin activity throughout the brain paralleled that of the opiate receptor with negligible levels in the cerebellum, very high densities in the corpus striatum and hypothalamus and intermediate values in other regions.

Using effects on smooth muscle as a bioassay, Hughes *et al.*, (1975) isolated enkephalin from pig brain and showed it to be a mixture of two pentapeptides,

*Opioid activities of tissue extracts have been referred to with different terminology by various laboratories. The term "endorphin" is used generally to refer to peptides with opioid activities. α-Endorphin and β-endorphin are specific opioid peptides isolated from the pituitary gland. Enkephalin is the opioid peptide which has been isolated only from brain.

Tyr-Gly-Gly-Phe-Met-OH (methionine enkephalin; m-enk) and Tyr-Gly-Gly-Phe-Leu (leucine enkephalin; l-enk). Independently by monitoring influences on opiate receptor binding, we isolated and identified the structures of the same two peptides in bovine brain (Simantov and Snyder, 1976a,b). However in pig brain Hughes *et al.*, (1975) found 4 times more m-enk than l-enk, whereas in bovine brain we observed 4 times more l-enk than m-enk (Simantov and Snyder, 1976a,b).

With synthetic enkephalin it has been possible to prepare antibodies with selectivity for m-enk and l-enk respectively (Table. I). We also succeeded in developing specific and sensitive radioimmunoassays for the two enkephalins (Simantov and Snyder, 1976c). The radioimmunoassays provide discrete measurements of m-enk and l-enk respectively. Using the ability of brain extracts to compete for opiate receptor binding, we can also measure total levels of enkephalin activity but cannot distinguish between the two forms of enkephalin. Relative amounts of enkephalin in different brain regions of the rat measured by the two techniques are fairly similar (Table II), though absolute values by radioimmunoassay tend to be higher than those measured by radioreceptor

**TABLE 1**

*Selectivity of Radioimmunoassays for Methionine Enkephalin and Leucine Enkephalin*

| Compound Added | Conc. (nM) | $^{125}$I-m-enk Bound to anti-m-enk (% of control) | $^{125}$I-l-enk Bound to anti-l-enk (% of control) |
|---|---|---|---|
| None | – | 100 | 100 |
| m-enk | 5 | 82 | 97 |
| | 10 | 70 | 93 |
| | 25 | 57 | 88 |
| | 100 | 25 | 80 |
| l-enk | 5 | 98 | 85 |
| | 10 | 96 | 80 |
| | 25 | 94 | 56 |
| | 100 | 84 | 30 |
| Tyr-Gly-Gly | 100 | 84 | 88 |
| | 500 | 70 | 82 |
| Glucagon | 500 | 92 | 99 |
| | 2000 | 78 | 90 |
| Angiotensin II | 2000 | 92 | 98 |
| Neurotensin | 2000 | 95 | 93 |
| Substance P | 2000 | 98 | 103 |
| Insulin | 2000 | 94 | 101 |

Morphine, naloxone, acetylcholine, norepinephrine, GABA, cyclic AMP, cyclic GMP, and bacitracin showed 0.5% inhibition in 5-1,000 nM concentrations.

Binding of $^{125}$I-m-enk to anti-m-enk or binding of $^{125}$I-l-enk to anti-l-enk was performed in 0.5 ml 0.05 M sodium veronal buffer pH 8.6.

**TABLE 2**

*Regional Distribution of m-enk and l-enk in Rat Brain Determined by Radioimmunoassay and Competition For Opiate Receptor Binding*

| Brain Region | Radioimmunoassay | | | Radioreceptor Assay | |
|---|---|---|---|---|---|
| | m-enk (pmole/gr tissue) | l-enk (pmole/gr tissue) | Ratio m-enk/l-enk | Units/ gm tissue | *Calculated pmole enk/gm tissue |
| Corpus Striatum | 45 ± 7 | 5 ± 0.7 | 9 | 20 ± 3.2 | 33 |
| Hypothalamus | 39 ± 6 | 3 ± 0.5 | 13 | 16 ± 2.0 | 27 |
| Midbrain | 25 ± 4 | 2.2 ± 0.5 | 11 | 10 ± 1.1 | 16.5 |
| Brainstem and Medulla | 18 ± 3 | 1.2 ± 0.5 | 15 | 7.5 ± 1.1 | 12 |
| Hippocampus | 15 ± 3 | <0.5 | >30 | 6.0 ± 0.7 | 10 |
| Cerebral Cortex | 9 ± 3 | 0.7 ± 0.4 | 13 | 2.6 ± 0.5 | 4.4 |
| Cerebellum | <3 | <0.5 | – | <0.5 | – |
| Whole Brain | 10 ± 2 | 0.8 ± 0.4 | 12.5 | 3.6 ± 0.5 | 6 |

*Enkephalin concentrations were calculated according to the ratio of m-enk and l-enk found by the radioimmunoassay assuming $IC_{50}$ of 8 nM for m-enk and 25 nM for l-enk. These figures are uncorrected for 10% recovery.

assay. Interestingly, in most rat brain regions m-enk levels are 10–15 times higher than those for l-enk, while in the hippocampus l-enk is essentially undetectable.

Earlier studies in several species using a radioreceptor assay have shown that the levels of enkephalin in different regions vary in close parallel to those of the opiate receptor (Snyder, 1975). If enkephalin is a neurotransmitter, then the opiate receptor is its physiological synaptic receptor. Using antibodies to synthetic enkephalins, Hokfelt and collaborators have recently visualized enkephalin neurons by immunofluorescence (Elde *et al.*, 1976). Enkephalin is contained in systems of arborizing axons and terminals with numerous varicosites closely resembling the terminal networks of norepinephrine, dopamine and serotonin-containing neurons. Cell bodies for enkephalin containing neurons have not yet been identified. Interestingly, the microscopic mapping of enkephalin terminal systems throughout the brain closely resembles the autoradiographic mapping of opiate receptors. In the spinal cord enkephalin fluorescence is most concentrated in the dorsal gray matter in laminae I, II, V and VII and in the area around the central canal. In the medulla oblongata, fluorescent fibers are most dense in the lateral reticular nucleus, the nucleus ambiguus, the nucleus tractus solitarius, the nucleus of origin of the facial and the hypoglossal nerves. In the pons, the central gray matter contains a dense collection of nerves as does the nucleus parabrachialis dorsalis. In the midbrain the periaqueductal gray matter and zona compacta of the substantia nigra, both enriched in opiate receptors, contain enkephalin terminals. Most hypothalamic nuclei also display enkephalin fluorescence, especially in the ventromedial and dorsomedial nuclei, the periventricular area and the medial preoptic nucleus. The nucleus paratenialis of the thalamus contains a considerable number of enkephalin fibers, while other parts of the thalamus are not as richly endowed. The globus pallidus of the corpus striatum displays many enkephalin fibers, which accords with biochemical evidence that it possesses the highest levels of enkephalin in the brain, as determined by radioreceptor assay (Simantov *et al.*, 1976). The caudate nucleus contain a patchy distribution of enkephalin fibers, reminiscent of the patchy distribution of opiate receptors, with highest densities in the rostral and ventral portions of the nucleus as well as the lateral part of the interstitial nucleus of the stria terminals. The amygdaloid nuclei, well-known for its high concentration of opiate receptors, display substantial enkephalin fibers, especially in the central amygdaloid nucleus. Very little enkephalin flourescence is observed in the cerebral cortex or the pituitary gland.

Since the distribution of enkephalin terminals and opiate receptors coincides so closely, one can draw inferences about the disposition of enkephalin systems in the brain by knowing the detailed localization of opiate receptors. In the spinal cord opiate receptors are highly localized to laminae I and II of the dorsal gray matter, the substantia gelatinosa, which is the first way station in the integration of sensory information (Fig. 1).

Lesions of the dorsal root of the spinal cord in monkeys (LaMotte *et al.*, 1976) result in a depletion of opiate receptors in the dorsal gray matter as

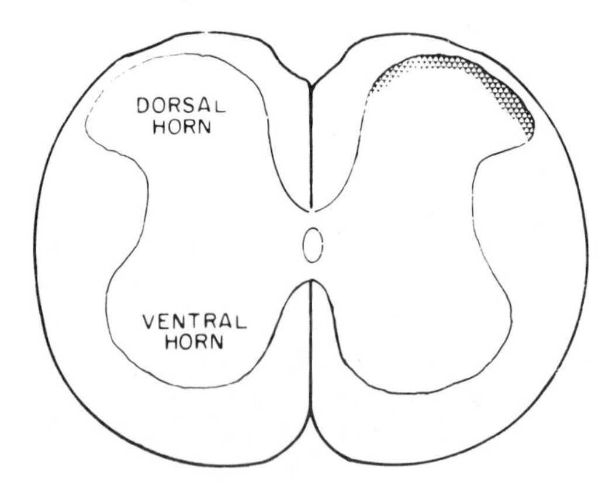


Fig. 1. Opiate receptor autoradiographic grains in the spinal cord of the rat. Animals were injected with 125 $\mu$Ci $^3$H-diprenorphine (13 Ci/mmole) and killed one hour later when essentially all tritium in the brain can be accounted for as unmetabolized diprenorphine. Autoradiography was performed by thaw-mount. The dense area indicates localization of grains to the substantia gelatinosa. Adapted from C. B. Pert, M. J. Kuhar, and S. H. Snyder, (1976).

detected by biochemical or autoradiographic techniques. The time course of this depletion is consistent with the degeneration of the sensory nerve terminals and not with postsynaptic trophic alterations. Thus it is likely that opiate receptors are contained on the nerve terminals of the sensory afferent fibers. Such a presynaptic localization is consistent with the existence of axoaxonic synapses between enkephalin terminals and afferent nerve terminals as occurs for synapses associated with presynaptic inhibition in the spinal cord. A similar presynaptic localization of opiate receptors occurs in the nucleus tractus solitarius and the nucleus ambiguus of the rat's brainstem, where opiate receptors, assayed by autoradiography, disappear after destruction of the vagus nerve in the neck (Atweh and Kuhar, 1977). Similarly, opiate receptors in the terminal nuclei of the inferior accessory optic pathway, which is completely crossed in the rat, disappear after removal of the contralateral eye.

Opioid peptides are not restricted to the central nervous system. As already mentioned, enkephalin occurs in the gastrointestinal tract. Recently opioid peptides have been detected in the pituitary gland. The sequence of two of these, α-endorphin and β-endorphin have been determined (Lazarus *et al.*, 1976; Cox *et al.*, 1976). Interestingly, methionine-enkephalin, α-endorphin and β-endorphin all derive from β-lipotropin, a 91 amino acid containing peptide isolated from pituitary (Li *et al.*, 1965 (Table 3). β-Endorphin and α-endorphin are similar in potency to the enkephalins in competing for opiate receptor binding. Because β-endorphin is less likely than enkephalin to be destroyed by proteolysis, it apparently reaches the brain after parenteral administration, and is 5 times as potent as morphine in producing analgesia when injected intravenously (C.·H. Li, personal communication). Whether β-lipotropin, α-endorphin or βendorphin are natural precursors of the enkephalins is not yet clear, since these large peptides have not been demonstrated unequivocally in the brain. Some insight into the function of the pituitary opioid peptides has been obtained by binding studies demonstrating that the pituitary possesses opiate receptors (Simantov and Snyder, 1976c). The characteristics of the pituitary opiate receptors are fairly similar to those of brain receptors except that enkephalins have less affinity for pituitary than brain receptors. Conceivably the opioid peptides in the pituitary act directly on pituitary opiate receptors to produce effects known to be elicited by opiates, such as release of antidiuretic hormone. Such an action would accord with the observation that opiate receptor binding is much more enriched in the posterior than in the anterior pituitary.

## OTHER PEPTIDES

There appear to be a large number of peptides with possible central nervous functions. In the interest of brevity, we have provided detailed information only on two of these, substance P and the enkephalins. There is also substantial

**TABLE 3**

*Amino Acid Sequence of β-Lipotropin and Its Biologically Active Constituent Peptides*

| | |
|---|---|
| | 1 ... 20 |
| β-Lipotropin: | $NH_2$-GLU-LEU-ALA-GLY-ALA-PRO-PRO-GLU-PRO-ALA-ARG-ASP-PRO-GLU-ALA-PRO-ALA-GLU-GLY-ALA-ALA-ALA-ARG-ALA |
| | 37 |
| β-Lipotropin: (continued) | GLU-LEU-GLU-TYR-GLY-LEU-VAL-ALA-GLU-ALA-GLN-ALA-ALA-GLU-LYS-LYS-ASP-GLU-GLY-PRO-TYR-LYS |
| β-MSH | ASP-GLU-GLY-PRO-TYR-ARG- |
| | 47 ... 61 |
| β-Lipotropin: (continued) | MET-GLU-HIS-PHE-ARG-TRY-GLY-SER-PRO-PRO-LYS-ASP-LYS-ARG-TYR-GLY-GLY-PHE-MET-THR-SER-GLU-LYS-SER- |
| β-MSH: (continued) | MET-GLU-HIS-PHE-ARG-TRY-GLY-SER-PRO-PRO-LYS-ASP |
| $ACTH_{4-10}$: | MET-GLU-HIS-PHE–ARG-TRY-GLY |
| α-Endorphin: | TYR-GLY-GLY-PHE-MET-THR-SER-GLU-LYS-SER |
| β-Endorphin: | TYR-GLY-GLY-PHE-MET-THR-SER-GLU-LYS-SER |
| Methionine Enkephalin: | TYR-GLY-GLY-PHE-MET |
| | 76 ... 91 |
| β-Lipotropin: (continued) | GLN-THR-PRO-LEU-VAL-THR-LEU-PHE-LYS-ASN-ALA-ILE-VAL-LYS-ASN-ALA-HIS-LYS-LYS-GLY-GLN-OH |
| α-Endorphin: (continued) | GLN-THR-PRO-LEU-VAL-THR |
| β-Endorphin: (continued) | GLN-THR-PRO-LEU-VAL-THR-LEU-PHE-LYS-ASN-ALA-ILE-VAL-LYS-ASN-ALA-HIS-LYS-LYS-GLY-GLN-OH |

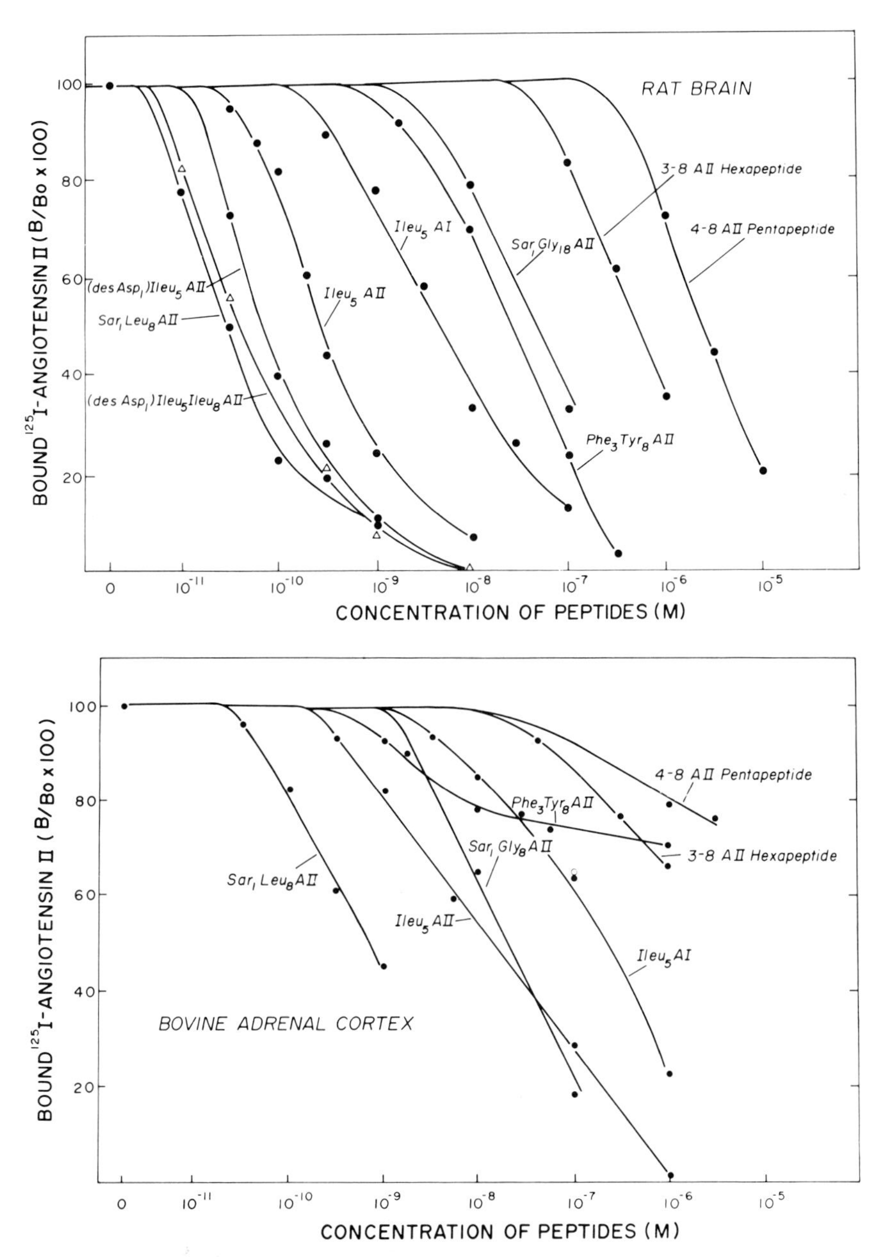


Fig. 2. Displacement of $^{125}$I-angiotensin II bound to rat brain (A) and bovine adrenal cortex (B) membranes by angiotensin peptides $^{125}$I-Angiotensin II (0.05 nM) was incubated with membranes and bound radioactivity in the presence of increasing concentrations of various angiotensin peptides was assayed by filtration. $B_O$ = binding in absence of unlabeled peptide, B = binding in presence of unlabeled peptides. (From Bennett and Snyder, 1976)

evidence for a role of angiotensin II and neurotensin in the brain. Direct administration of small doses of angiotensin II into the brain can influence cardiovascular reflexes and drinking behavior. Renin and angiotensin converting enzyme exist in the brain. However, angiotensin II itself has not been directly demonstrated in brain tissue. Recently we detected angiotensin II receptor binding in brain tissue similar to levels to receptor binding in the adrenal cortex, the peripheral tissue most enriched in angiotensin receptors (Bennett and Snyder, 1976). Indeed, in a screen of various tissues, only brain and adrenal cortex demonstrated detectable levels of angiotensin II receptor binding. Substrate specificity of receptors in the brain and adrenal cortex are quite similar (Fig. 2). Taken together these data suggest a natural role for angiotensin in the brain.

Neurotensin is a tridecapeptide whose amino acid sequence is <glu-leu-tyr-glu-asn-lys-pro-arg-arg-pro-tyr-ile-leu-OH (Carraway and Leeman, 1973, 1975) Neurotensin was isolated from hypothalamic extracts as an apparent biproduct of the isolation of substance P. It was detected as a substance which in small doses could elicit hypotension, increased vascular permeability, pain sensation, increased hematrocrit, cyanosis, morphine-inhibitable stimulation of ACTH secretion, increased LH secretion, increased FSH secretion, hyperglycemia and a variety of smooth muscle effects including contraction of the rat uterus and guinea pig ileum. Thus far neurotensin has been detected only in the brain. Radioimmunoassay shows striking regional variations with levels in the hypothalamus about 30 times higher than in the cerebral cortex and 200 times higher than in the cerebellum. We have detected high affinity specific neurotensin receptor binding in the brain but not in any peripheral tissues (Uhl *et al.*, 1976).

The immunohistochemical localization of somatostatin and TRH to specific neurons in the brain suggests that these compounds may also be neurotransmitter candidates (Hokfelt et al., 1975a,b). For TRH, specific receptor binding has been demonstrated in brain and pituitary but in no peripheral organs (Burt and Snyder, 1975).

## REFERENCES

Atweh, S. and Kuhar, M. J. (1977). *Brain Res.*, **124,** 53–67.

Bennett, J. P., Jr. and Snyder, S. H. (1976). *J. Biol. Chem.,* **251,** 7423–7430.

Burt, D. R. and Snyder, S. H., (1975). *Brain Res.* **93,** 309–328.

Carraway, R. and Leeman, S. E. (1973). *J. Biol. Chem.* **248,** 6854–6861.

Carraway, R. and Leeman, S. E. (1975). *J. Biol. Chem.* **250,** 1907–1911.

Chang, M. M. and Leeman, S. E. (1970). *J. Biol. Chem.* **245,** 478–479.

Cox, B. M., Opheim, K. E., Teschemacher, H. and Goldstein, A. (1975). *Life Sci.* **16,** 1777–1782.

Cox, B. W., Goldstein, A. and Li, C. H. (1976). *Proc. Nat. Acad. Sci. U.S.A.* **73,** 1821–1823.

Elde, R., Hokfelt, T., Johansson, O. and Terenius, L. (1976). *Neuroscience,* **1,** 349–355.

Hokfelt, T., Efendic, S., Johansson, O., Luft, R. and Arimura, A. (1975a). *Brain Res.* **80,** 165–169.

Hokfelt, T., Fuxe, K., Johansson, O., Jeffcoate, S. and White, N. (1976b). *Neuroscience Letters* **1,** 133–139.

Hokfelt, T., Kellerth, J.-O., Nelson, G. and Pernow, B. (1975). *Brain Res.* **100,** 235–252.

Hughes, J. (1975). *Brain Res.* **88,** 1–14.

Hughes, J., Smith, T., Kosterlitz, H. W., Fothergill, L. A., Morgan, B. and Morris, H. R. (1975). *Nature* **258,** 577–579.

Lamotte, C., Pert, C. B. and Snyder, S. H. (1976). *Brain Res.* **112,** 407–412.

Lazarus, L. H., Ling, N. and Guillemin, R. (1976). *Proc. Nat. Acad. Sci., U.S.A.* **73,** 2156–2159.

Leeman, S. E. and Hammerschlag, R. (1967). *Endocrinology* **81,** 803–810.

Lembeck, F. and Starke, K. (1968). *Arch. Exptl. Path. and Pharmacol.* **259,** 375–385.

Li, C. H., Barnafi, L., Chretien, M. and Chung, D. (1965). *Nature* **208,** 1093–1094.

Otsuka, M., Konishi, S. Takahashi, T. (1975). *Fed. Proc.* **34,** 1922–1928.

Pasternak, G. W., Goodman, R. and Snyder, S. H. (1975). *Life Sciences* **16,** 1765–1769.

Pert, C. B., Kuhar, M. J. and Snyder, S. H. (1976). *Proc. Nat. Acad. Sci., U.S.A.* **73,** 3729–3733.

Powell, D., Leeman, S., Tregear, G. W., Niall, H. D. and Potts, J. T. (1973). *Nature New Biol.* **241,** 252–254.

Said, S. I. and Rosenberg, R. N. (1976). *Science* **192,** 907–908.

Simantov, R., Kuhar, M. J., Pasternak, G. W. and Snyder, S. H. (1976). *Brain Res.* **106,** 189–197.

Simantov, R. and Snyder, S. H. (1976a). *Life Sci.* **18,** 781–788.

Simantov, R. and Snyder, S. H. (1976b). *Proc. Nat. Acad. Sci., U.S.A.,* **73,** 2515–2519.

Simantov, R. and Snyder, S. H. (1976c). *In* "Opiates and Opioid Peptides" (Ed. by H. W. Kosterlitz) North Holland, Amsterdam, pp. 41–48.

Snyder, S. H. (1975). *Nature* **257,** 185–189.

Terenius, L. and Wahlstrom, A. (1975). *Acta Physiol. Scand.* **94,** 74–81.

Uhl, G. R., Bennett, J. P., Jr. and Snyder, S. H. (1977). *Brain Res.,* in press.

Von Euler, U. S. and Gaddum, J. H. (1931). *J. Physiology* **72,** 74–87.

# IV. Cell Interactions in Blood Cell Development

# Factors Affecting the Differentiation of Blood Cells

J. E. Till, G. B. Price, S. Lan and E. A. McCulloch

*Ontario Cancer Institute, and*
*Departments of Medical Biophysics and Medicine*
*University of Toronto*
*Toronto, Ontario*
*Canada*

## I. INTRODUCTION

Little is known of the mechanisms by which progeny cells are able to express traits not seen in their progenitors, although the events underlying such expression are central to the process of differentiation. At an intracellular level, it is widely assumed that differentiation is associated with activation or inactivation of specific genetic information and that regulation of the process is possible at many levels, including transcription and possibly translation (Davidson, 1968; Watson, 1976). An alternative hypothesis, largely based on studies of the development of antibody diversity, is that clonal evolution, associated with genetic change and selection, occurs in embryogenesis and may continue in post-natal life (Edelman, 1974; Watson, 1976). These intracellular processes, though fascinating, are difficult to manipulate experimentally and are not readily available at targets for therapeutic intervention. Another approach involves a concentration on intercellular mechanisms responsible for the regulation of proliferation and differentiation in specific subpopulations of cells in organ systems; these intercellular mechanisms are likely to be more approachable for the investigator and more accessible to the physician.

Intercellular interactions usually take the form of a transfer of molecular factors from one cell to another. Cell culture systems provide convenient means for the identification and analysis of the factors involved, the cells that release them and the cells that receive them. This approach has been applied to several biologically active molecules, including hormones with known activities *in vivo* and "culture factors" with biological activities that are established in culture but not necessarily *in vivo*.

Cultures of blood forming cells have been used to investigate both categories of factors. The glycoprotein hormone erythropoietin stimulates erythropoietin responsive cells to proliferate in culture and form colonies (Stephenson *et al.*,

1971; Axelrad *et al.,* 1974); alternatively, the action of the hormone on biochemical events such as hemoglobin synthesis can be detected (Goldwasser, 1975). In contrast, granulopoiesis in culture is dependent upon bioactive molecules for whom evidence of activity *in vivo* remains scanty and controversial (Robinson, 1974; van Bekkum and Dicke, 1972). For other hemopoietic cells, the cell culture technology required for detailed studies of cell factor interrelationships is still under development (Metcalf *et al.,* 1975a, b; Fibach *et al.,* 1976; Sredni *et al.,* 1976; Nakeff and Daniels-McQueen, 1976). Particularly, pluripotent stem cells, the ancestors of both erythropoietic and granulopoietic progenitors, have not as yet been demonstrated to form colonies in culture.

The purpose of this paper is to examine the present status of factor-cell relationships for hemopoietic cells in culture and to explore potential avenues for extension of the work. Particular emphasis will be placed on possible approaches to the development of a culture assay for pluripotent stem cells. Finally, some evidence will be considered that human leukemic cells may retain some of the intercellular regulatory mechanisms characteristic of normal hemopoiesis.

## II. CELL BIOLOGY OF EARLY HEMOPOIESIS

It is widely accepted that the origin of adult hemopoietic development is in a class of stem cells, with a capacity for extensive proliferation including self-renewal, and able to differentiate at least along erythropoietic, granulocytic and megakaryocytic pathways. These pathways are assumed to be regulated independently of each other and of their pluripotent ancestors because each is headed by an early committed progenitor; these are separated from their pluripotent ancestors by an unknown number of differentiation events. As a consequence of these steps, the committed progenitors are limited in their capacity for either proliferation or differentiation, but have acquired increased sensitivity to regulatory mechanisms specific for the appropriate pathway. Such mechanisms often appear to be based on passage of bioactive molecules from one cell to another. Cell culture evidence for such interactions is available from studies of granulopoiesis; it is beginning to emerge for erythropoiesis, and it is urgently required at the level of pluripotent stem cells (for reviews of earlier data, see McCulloch *et al.,* 1974; McCulloch and Till, 1971).

### *A. Granulopoiesis in Culture*

Granulopoietic colony formation in culture (Bradley and Metcalf, 1966; Pluznik and Sachs, 1965) is observed when hemopoietic cells are immobilized in viscid or semi-solid medium and are cultivated in the presence either of certain stimulators (colony stimulating activity or CSA) or cells that produce such factors. Granulopoietic progenitors (CFU-C) represent a cell class with marked heterogeneity in size and proliferative potential (McCulloch and Till, 1971); they may be considered as a spectrum of differentiated progeny derived from pluripotent stem cells and committed to granulopoiesis. In human marrow, these cells co-exist with factor-producing cells. The populations may be separated by a simple adherence procedure and for routine bioassays, nonadherent cells are used (Messner *et al.,* 1973). When such nonadherent cells are cultured in the presence of an optimal concentration of colony stimulating activity, their frequency of colony formation provides an estimate of the number of granulopoietic progenitors present; for human marrow, this value is usually in the range of 40-100 colonies/$10^5$ nucleated cells. This estimate represents a lower limit of the true frequency, since some cells with the potential to form granulopoietic colonies may fail to do so under the conditions used, and will not be detected. At less than optimal concentrations of colony stimulating activity, an approximately linear dose response curve is observed relating colony number and added exogenous CSA or number of CSA producing cells. This dose response relationship is utilized for quantitative assays of either factors or of factor-producing cells.

Using the assay, four active molecular species have been identified in media conditioned by human peripheral blood leukocytes (Price *et al.,* 1973, 1975a). Three of these are proteins with molecular weights of approximately 15000, 35000 and 90000. The fourth factor appears to be a hydrophobic peptide with a molecular weight of less than 1300; this factor has only marginal granulopoietic stimulating capacity, but, as will be discussed later, may have a role in myelopoiesis at a stage earlier than granulopoietic progenitor cells.

### *B. Erythropoiesis and Erythropoietin*

Although the existence of committed progenitors of erythropoiesis had been postulated on the basis of physiological and genetic experiments, such cells were identified only with the development of cell culture methods that permitted the growth of erythropoietic colonies in cultures with erythropoietin. Two main classes of erythropoietic colony forming cells in culture have been described (Stephenson *et al.,* 1971; Axelrad *et al.,* 1974); the first (CFU-E) responds to relatively low concentrations of erythropoietin by proliferating to form small colonies of erythroid cells. The second class (BFU-E) responds to higher concentrations of erythropoietin by the formation of "bursts" composed of several clusters of cells located in close proximity (Fig. 1). The data presently available are consistent with the view that BFU-E are more closely related to pluripotent stem cells than are CFU-E; the former are considered to represent a stage in the hierarchy of erythropoietic differentiation analogous to CFU-C in granulopoietic differentiation while the latter (CFU-E) represent a later stage of erythropoietic differentiation, perhaps immediately preceding the emergence of

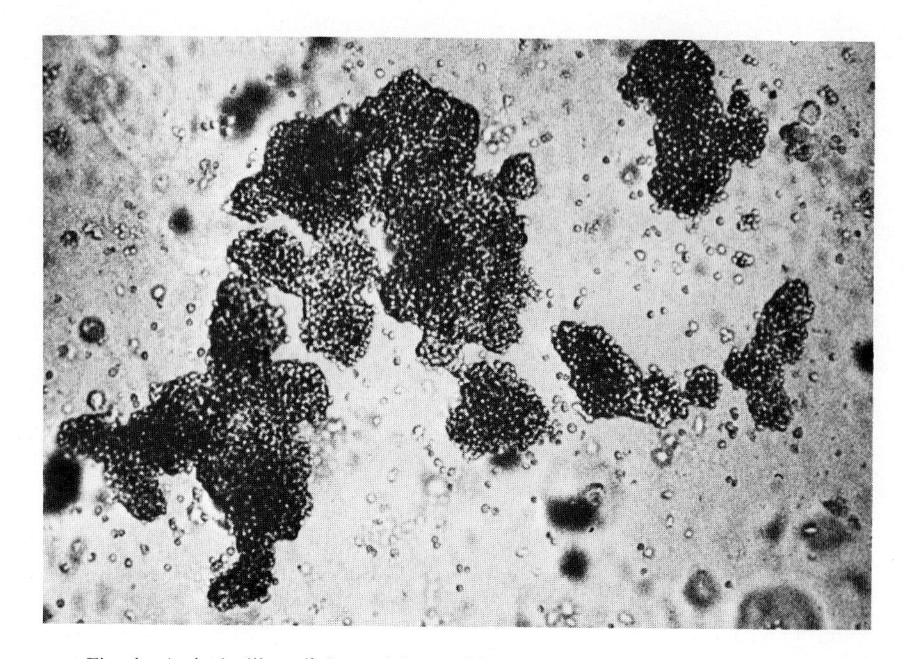

Fig. 1. A single "burst" formed by proliferation and differentiation of an erythropoietic progenitor cell (BFU-E, Axelrad *et al.*, 1974) derived from human marrow, after 14 days in cell culture in the presence of 2.5 units/ml of erythropoietin and 10% leukocyte conditioned medium (× 56).

morphologically recognizable erythroblasts (Gregory *et al.*, 1973; Gregory, 1976).

The stimulator of erythropoietic colonies in culture, erythropoietin, has long been recognized as the primary physiological regulator of erythropoiesis *in vivo* (Krantz and Jacobson, 1970). The existence of an erythropoietin-responsive progenitor cell was predicted on the basis of these studies, and the cell culture work summarized above has confirmed this prediction. It should be noted that the formation of erythropoietic "bursts" in culture requires concentrations of erythropoietin much greater than those considered physiological. This indicates a need for caution in attributing *in vivo* significance to data obtained under such culture conditions. For example, the requirement for a high concentration of erythropoietin may be simply a reflection of suboptimal culture conditions, and may provide no meaningful information about mechanisms *in vivo*. Alternatively, the requirement for high concentrations of erythropoietin in culture may signal the existence *in vivo* of mechanisms which serve to provide a cellular milieu analogous to that of cultures at high erythropoietin concentration.

### C. *Pluripotent Stem Cells and Approaches to their Assay in Culture*

At present, pluripotent stem cells can be detected only in rodents by the spleen colony assay, a technique based on colony formation by pluripotent cells in the spleens of irradiated or genetically anemic animals (Till and McCulloch, 1961; McCulloch *et al.,* 1964). This *in vivo* technique does not permit extensive manipulation of the cellular environment of the kind essential for detailed analysis of cell interactions. Nonetheless, evidence for short-range cell interactions in the regulation of pluripotent stem cells have been obtained from studies of mice of genotype $Sl/Sl^{d}$ (McCulloch *et al.,* 1965), from studies of mice exposed to part-body irradiation (Gidali and Lajtha, 1972), and from analyses of the variation in the differentiation patterns of the progeny of stem cells proliferating in different locations within the blood forming system and in different species (Trentin *et al.,* 1974). Factors influencing the regulation of stem cell proliferation have been identified by methods that depend upon observing changes in the proliferative state of such cells following exposure to active agents. Byron (1975) has used changes in the inactivation of proliferating stem cells by incorporated tritiated thymidine to identify three mechanisms involved in the modulation of murine stem cell proliferation. These are androgenic steriods, cholinergic receptor stimulators, and $\beta_1$-adrenergic receptor stimulators. Evidence for the first two has also been obtained from studies *in vivo*. Analogous work on granulopoietic progenitors in culture (Moore *et al.,* 1974) has not revealed the presence of $\beta$-adrenergic receptors on this class of cell, as expected if this type of stimulation plays a role in the regulation of pluripotent stem cells but not of their more differentiated granulopoietic progeny.

It is evident from the above that the development of knowledge about stem cell regulation is hindered by the lack of a suitable culture assay. A number of approaches to the development of such an assay are available. Dexter and Lajtha (1974) have demonstrated the maintenance and proliferation of stem cells in long-term cultures. This approach, particularly if adapted to human hemopoietic populations, might provide a description of the relevant culture conditions for a colony assay. Another approach is based on an attempt to detect pluripotent stem cells by their ability to give rise to granulopoietic progenitor cells in liquid cultures. Under some conditions, a net increase in CFU-C/ml of culture has been observed (Sutherland *et al.,* 1971; Iscove *et al.,* 1972). Such increases could be the result of self-renewal of CFU-C pre-existent in the cultures, or recruitment of new CFU-C, formed as a result of differentiation from more primitive progenitors, presumably pluripotent stem cells. Cell separation experiments have provided some evidence that, for both human and murine cell populations, the cell class responsible for the net increase in CFU-C in liquid cultures is not identical with pre-existing CFU-C (Sutherland *et al.,* 1971; Iscove *et al.,* 1972). Rather, increase of CFU-C in liquid cultures appears to derive either from a subclass of the original CFU-C population or another class of progenitors. Although the data cannot be interpreted confidently in terms of properties of pluripotent stem cells, they indicate a possible approach to the study of such populations.

An example of how this approach may be applicable to the study of the effect of cell culture factors on stem cells is provided by observations on the effects of leukocyte conditioned media on the increase of CFU-C in liquid cultures (Niho *et al.*, 1975). Crude leukocyte conditioned media were able to promote a 3-fold increase in numbers of CFU-C after 7 days in liquid culture, but no such increase was observed in the presence of dialyzed leukocyte conditioned media, even though dialysis had little or no effect on the capacity of leukocyte conditioned medium to stimulate granulopoietic colony formation (Fig. 2). Preliminary experiments have provided evidence that dialysed conditioned medium that had been reconstituted by the addition of low molecular weight colony stimulating activity regained its ability to promote an increase in CFU-C in liquid culture (G. B. Price, unpublished). These data are compatible with a role for low molecular weight factors at a stage in granulopoiesis earlier than CFU-C. The hydrophobicity and low molecular weight of this material (Price *et al.*, 1973) suggest that it may have properties analogous to the cyclic peptides that are known to be membrane active such as valinomycin (Ovchinnikov, 1974). This analogy raises the possibility that low molecular weight CSA might be able to act as an ionophore in biological membranes. Whatever their mechanism of action may be, the low molecular weight factors should serve as useful probes in the development of methods for the identification of primitive hemopoietic cells.

Pluripotent stem cells might also be recognized by their ability to give rise to erythropoietic progenitors such as BFU-E. When erythroid "bursts" were sampled at intervals and tested for their content of recognizable granulopoietic cells, there was no evidence to indicate a pluripotent origin for any of these bursts (S. Lan, unpublished). A modification of this approach is based on the finding that leukocyte conditioned media can influence the development of erythroid progenitors (Aye, 1976). It involves the addition of erythropoietin at increasing time intervals to cells cultured in the presence of leukocyte conditioned media. As shown in Fig. 3, erythropoietic progenitors appeared to survive somewhat longer in the presence of leukocyte conditioned media than in dialysed conditioned media or in growth medium alone. It should be noted that all the cultures were scored for the presence of colonies and bursts at the same time after initiation of the cultures (14 days), independent of whether erythropoietin was added on day 1 or day 7. The bursts seen in cultures given erythropoietin on day 7 did not differ noticeably in size from those that received erythropoietin at an earlier time. These results suggest that at least some early steps in the formation of erythroid bursts may be able to occur in such cultures even in the absence of added erythropoietin. It remains to be seen whether the successful culture of pluripotent stem cells will require the addition of new stimulatory factors or the removal of inhibitors that have not yet been identified, or whether a further refinement of culture methods presently

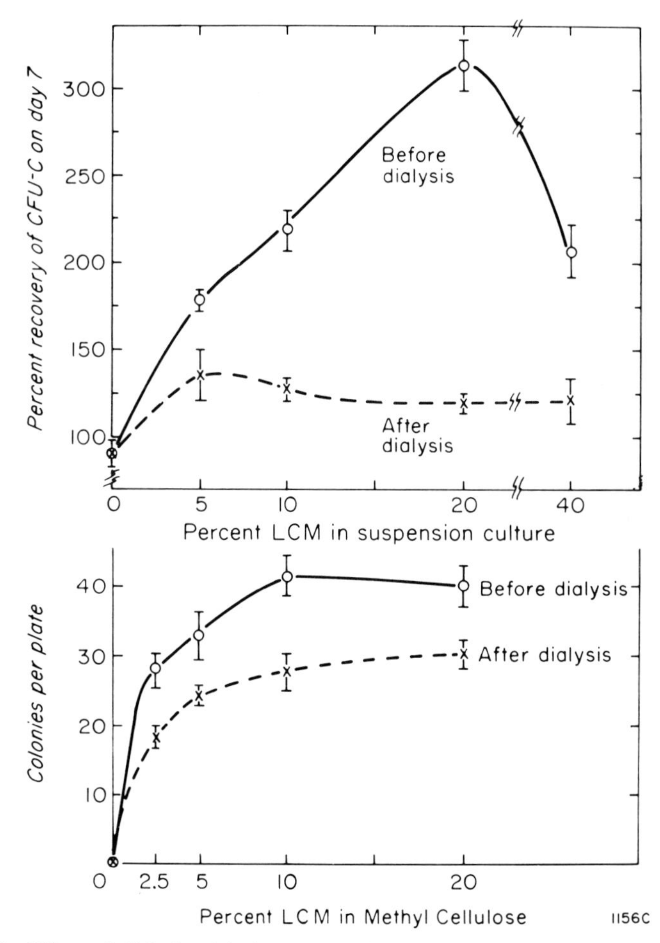


Fig. 2. Effect of dialysis of leukocyte conditioned medium (LCM) on its ability to promote an increase in numbers of granulopoietic progenitor cells (CFU-C) in populations of non-adherent human marrow cells after 7 days in suspension culture (upper panel). The dialysed and undialysed leukocyte conditioned media were also titrated for colony stimulating activity (lower panel). Bioassays for CFU-C were carried out in cultures containing 0.8% methyl cellulose; for details of the experimental conditions used, see Niho *et al.* (1975). Reproduction of this figure is by permission of Grune and Stratton, Inc.

available will suffice. Conditions must be sought which will permit a net increase in the numbers of erythroid progenitor cells in culture, as can be achieved for granulopoietic progenitors (Sutherland *et al.*, 1971; Iscove *et al.*, 1972).

The approaches to the identification of pluripotent stem cells described above are all somewhat indirect. A more direct approach is to attempt to identify specific markers on such cells. Recently van den Engh and Golub (1974)

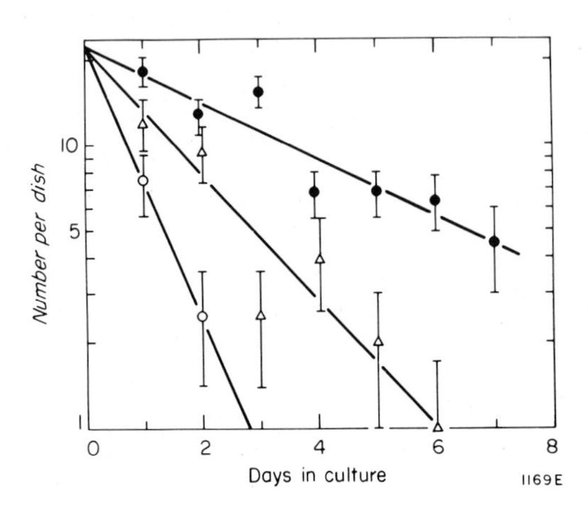


Fig. 3. Effect of prior culture in the absence of erythropoietin on the number of erythropoietic colonies/bursts. Normal, non-adherent human marrow cells were plated in duplicate at $5 \times 10^4$ cells per 35 mm culture dish. The cultures were exposed for various periods of time either to 10% leukocyte conditioned medium (•), to 10% dialysed leukocyte conditioned medium (△), or to growth medium alone (○), prior to the addition of 2.5 units/ml of erythropoietin. All cultures were scored for colonies/bursts on day 14, i.e., the cultures given erythropoietin on day 7 were scored 7 days later. The slopes of the 3 curves differ significantly from each other on the basis of a t-test ($P < 0.001$).

have described an anti-brain antiserum which inactivates murine pluripotent stem cells but shows no apparent effect on granulopoietic progenitors. If this antiserum is identifying stem cell-associated markers, then fluorescence-activated cell sorting (Bonner *et al.*, 1972) provides a way in which these markers could be exploited in the development of a stem cell assay. It should be possible to accomplish this aim even if the markers are not uniquely specific for stem cells.

## III. THE MEMBRANE BIOLOGY OF EARLY HEMOPOIESIS

It was suggested by Rubin (1966) a decade ago that, for cells growing in culture, one might expect molecules associated with the cell surface membrane to be released readily into the culture medium. This suggestion stimulated the investigation of surface membranes as a source of culture factors, especially since a location in association with the cell membrane would be suitable for molecules mediating intercellular information transfer. To test the hypothesis, high molecular weight species of CSA were sought in preparations of surface membranes obtained from human peripheral blood leukocytes. The membranes were solubilized with sodium dodecyl sulfate and tested for colony stimulating activity. Such activity was found in association with surface membrane fractions but not with the nuclear or soluble fractions (Price *et al.*, 1975b). This evidence for an association of high molecular weight CSA with the surface membrane of CSA producing cells provides support for the view that these factors may be membrane active. A direct test of this view would require purified preparations of labelled CSA of high specific activity which could be used to search for specific interactions between labelled molecules of demonstrated biological activity and constituents of the surface membranes of granulopoietic progenitor cells. This approach has been undertaken from material active on mouse cells (Stanley *et al.,* 1975) but has not yet been attempted for high molecular weight material active on human cells.

Another approach to the membrane biology of molecules with colony stimulating activity is to investigate their possible relationship to other membrane associated molecules. Products of the major histocompatibility gene complex are known to be membrane associated, although their functions in the membrane are not understood. Anti-HLA antisera were used to search for a relationship between HLA antigens and high molecular weight CSA. Preliminary evidence (Till *et al.,* 1976b) was obtained that these anti-HLA sera could inhibit the stimulation of granulocytic colony formation by solubilized membrane fractions from cells of a known HLA type. Although these preliminary results suggested the presence of an anti-CSA activity in these anti-HLA sera, the possibility that a non-specific inhibitory activity was also present in the sera could not be ruled out. More convincing evidence for a relationship between high molecular weight CSA and HLA antigen was obtained from analogous experiments carried out using an anti-human $\beta_2$-microglobulin (anti-$\beta_2$m) antiserum. Human $\beta_2$-microglobulin ($\beta_2$m) is excreted in urine and material from this source has been isloated and extensively purified (Moller, 1974). It is also a component of HLA antigens (Poulik *et al.,* 1973; Grey *et al.,* 1973; Nakamuro *et al.,* 1973; Peterson *et al.,* 1974; Moller, 1974) so anti-$\beta_2$m antisera can be used to detect this component of HLA antigens. A monospecific anti-$\beta_2$m antiserum was able to inhibit the stimulation of granulocytic colony formation by human leukocyte conditioned medium (Price *et al.,* 1976). Control sera including the pre-immunization serum showed no such inhibition. Addition of leukocyte conditioned medium, and of purified human $\beta_2$m (generously provided by Dr. M. D. Poulik) resulted in a restoration of colony-stimulating activity, presumably because of competition for anti-$\beta_2$ antibody. The amount of $\beta_2$m required to restore colony-stimulating activity agreed with the amount expected on the basis of the known titer of the anti-$\beta_2$m serum. These results support the view that it was the anti-$\beta_2$ activity, rather than activity against a non-cross-reacting contaminant, that was responsible for the inhibitory effect of the antiserum.

These results provide evidence for $\beta_2$m-like determinants in molecules with granulopoietic colony stimulating activity. They indicate the existence of a structural similarity between CSA and HLA antigens, thus supporting the view that the species of CSA are membrane-associated. They also provide evidence for a relationship between molecules with colony-stimulating activity and the $\beta_2$m-like determinants that have been detected on other molecular species believed to be involved in cell interactions (Table I).

The experiments utilizing anti-$\beta_2$m antisera described above do not establish identity between $\beta_2$m and the cross-reactive determinants on the high molecular weight species of CSA. Strong evidence for identity would require a demonstration of colony-stimulating activity in purified preparations of human urinary

**TABLE I**

*Molecules Containing $\beta_2$ Microglobulin-like Determinants*

1. HLA antigens (Transplantation Reviews **21**, 1974).
2. Thymus leukemia antigens (Vitetta *et al.*, 1975a).
3. Products of the *T/t* locus (Vitetta *et al.*, 1975b).
4. Allogeneic effect factor (Armerding *et al.*, 1975).

$\beta_2$m. So far such tests have not yielded clear evidence of activity, but it is possible that the purified human urinary $\beta_2$m had been partially denatured or degraded and had lost biological activity for this reason. It is likely that a comparison of amino acid sequences will be required to clarify the nature of the apparent relationship between human $\beta_2$m and molecules with granulopoietic colony-stimulating activity.

In addition to their great theoretical interest, bioactive factors derived from cells in culture may provide useful probes for detecting heterogeneity within cell populations. An example of this application is provided by studies using low molecular weight CSA. This species derived from non-leukemic leukocytes proved to be different from that derived from a subpopulation of leukemic sources (Price et al., 1974). Because of its hydrophobicity, the low molecular weight CSA is readily purified to homogeneity as indicated by thin layer chromatography. This purified material can then be iodinated with radioiodine. When hemopoietic cell suspensions were incubated with such radioactive preparations of colony stimulating activity, a proportion lost colony forming capacity, presumably as a result of damage inflicted by radiation emitted from radioactivity associated with low molecular weight-CSA. Unlabelled material from a subpopulation of leukemic sources did not compete with radioactive material from non-leukemic sources, even though material from different members of the subpopulation of leukemic sources competed with each other as did material from different non-leukemic sources. This is evidence for biologically meaningful heterogeneity in preparations of low molecular weight CSA from different sources; this heterogeneity should be borne in mind if the low molecular weight factors prove to be useful probes for the identification of primitive hemopoietic cells (see above).

## IV. IMPLICATIONS FOR HUMAN MYELOGENOUS LEUKEMIA

There is convincing cytological evidence that leukemic stem cells in both acute and chronic myelogenous leukemia are pluripotent with capacity for both granulopoietic and erythropoietic differentiation (Blackstock and Garson, 1974; Rastrick, 1969). Accordingly, both forms of leukemia may be considered as the result of transformation occurring at the level of normal pluripotent stem cells. It is feasible to ask what effect the leukemic transformation has upon intercellular information transfer mechanisms of the type discussed above. Data have been obtained from two sources: first, the techniques of granulopoietic and erythropoietic colony formation in culture have been applied to populations derived from patients with leukemia, the second, proliferation in cultures of leukemic blast cells has been examined for evidence of persisting intracellular regulatory mechanisms.

### *A. Granulopoietic Colony Formation in Cultures of Cells from Patients with Leukemia*

Marrow or peripheral blood cells from patients with leukemia form colonies in culture with evidence of either complete or abortive granulopoietic differentiation (for recent data see van Bekkum and Dicke, 1972 and Robinson, 1974). Some of these colonies have been shown on the basis of chromosomal markers to belong to a leukemic clone (Duttera *et al.*, 1972; Moore and Metcalf, 1973; Aye *et al.*, 1974a); nonetheless, their development in culture is as dependent upon the presence of colony stimulating activity as that of granulopoietic progenitors of non-leukemic origin. The pattern of colony formation in leukemia is heterogeneous; cells from patients with chronic myelogenous leukemia usually exhibit increased colony formation while in acute leukemia a continuous spectrum is observed from very low to very high levels of colony formation (van Bekkum and Dicke, 1972; Robinson, 1974). In contrast to this varied pattern in the progenitor population, a common theme was observed when molecular species with colony stimulating activity were examined. Media taken from cultures of cells from patients with either acute leukemia or chronic leukemia in relapse were found to contain only one of the three high molecular weight-CSA species found in equivalent media conditioned by non-leukemic leukocytes (Price *et al.*, 1975a). In contrast, when membranes of leukemic leukocytes were examined, they were found to contain all three high molecular weight-CSA species (Price, 1974; Till *et al.*, 1976a), even when media conditioned by the same cell population contained only a single molecular weight species. This change in the ease with which membrane-associated components are liberated into culture media does not appear to be secondary to abnormal cell distributions in leukemia. Examination of patients with idiopathic acquired sideroblastic anemia showed a similar pattern of a single species of high molecular weight CSA detectable in conditioned media prior to development of leukemia (Senn *et al.*, 1976), although the distribution of cell types in the population of peripheral leukocytes was not strikingly abnormal. These data, therefore, are consistent with the view that the leukemic phenotype is associated with an abnormality in the release of high molecular weight species of CSA into culture media. If these culture observations are symbolic of events *in vivo*, then intercellular regulatory mechanisms persist in leukemia, but the leukemic transformation may affect the availability of factors involved in regulatory interactions with progenitor cells.

### B. *Cellular Interactions in Cultures of Leukemic Blast Cells*

As an alternative to applying established colony techniques to cells from patients with leukemia, proliferation was examined in cultures of blast cells from the peripheral blood of such patients. In these experiments cell proliferation was detected by measuring incorporation of tritiated thymidine into acid insoluble material. Only a minority of the cell population was able to incorporate tritiated thymidine under the culture conditions used, and the incorporation showed a radiation sensitivity characteristic of cell proliferation (Aye et al., 1974b). Factors present in leukocyte conditioned medium were able to enhance incorporation, and the potency of the leukocyte conditioned medium was much greater if the conditioned medium had been prepared in the presence of phytohemagglutinin (PHA). On the basis of the evidence presently available, it appears that PHA itself is not able to stimulate thymidine incorporation in a manner analogous to the stimulation of lymphocytes by PHA. Instead, the effect of PHA appears to be indirect, in that it stimulates the production of factors able to enhance the proliferation of a minority class of cells present in the leukemic cell population (Aye et al., 1974b, 1975; McCulloch and Till, 1977). Preliminary characterization of a factor involved in this interaction between PHA-stimulated cells and autologous thymidine-incorporating cells indicates that it differs from the four species of granulopoietic colony stimulating activities discussed above (Aye, 1974; G. B. Price, unpublished). The relationship of the cellular interactions observed in leukemic cell populations to those observed in populations from normal individuals has not yet been determined. If the cellular interactions observed in cultures of leukemic leukocytes represent persistence in the leukemic population of normal regulatory mechanisms, then manipulations of these residual mechanisms could be useful in the management of leukemia.

If the proliferating cells detected by the assay for thymidine incorporation are able to multiply for several cell generations, it should be possible to develop a colony assay for these cells. Evidence for colony formation in culture by cells derived from leukemic marrow has been reported by Dicke *et al.* (1976). Evidence was obtained that the cells in such colonies are of leukemic origin. Also, the culture method used by these authors involved stimulation of the cell populations by PHA, in a manner analogous to that described above for leukemic cell populations studied using the thymidine incorporation assay. Preliminary studies in our laboratory have provided evidence for colony formation in cultures of leukemic peripheral leukocytes stimulated either by leukocyte conditioned medium prepared in the presence of PHA, or by partially-purified preparations of factors able to promote incorporation of tritiated thymidine (R. Buick, G. B. Price and E. A. McCulloch, unpublished). The latter partially-purified preparations were active in promoting colony formation by leukemic peripheral leukocytes in the absence of detectable granulopoietic colony formation. PHA alone did not stimulate colony formation by leukemic peripheral leukocytes under these culture conditions. Further studies of the factors present in these preparations should contribute to an understanding of regulatory mechanisms affecting normal blood cell development in culture, and any aspects of such regulatory mechanisms that are retained in leukemic cell populations.

## ACKNOWLEDGMENTS

The work described in this paper was supported by the Ontario Cancer Treatment and Research Foundation, the Medical Research Council and the National Cancer Institute of Canada. S. Lan is a Fellow of the Medical Research Council of Canada.

## REFERENCES

Armerding, D., Kubo, R. T., Grey, H. M. and Katz, D. H. (1975). *Proc. Nat. Acad. Sci. U.S.A.* **72,** 4577–4581.

Axelrad, A. A., McLeod, D. L., Shreeve, M. M. and Heath, D. S. (1974). *In* "Hemopoiesis in Culture, Second International Workshop" (W. A. Robinson, ed.) pp. 226–234. U.S. Government Printing Office, Washington, D.C.

Aye, M. T. (1974). Ph.D. Thesis, University of Toronto, pp. 126–134.

Aye, M. T. (1976). Eighteenth Annual Meeting of the American Society of Hematology, Dallas, Texas, Dec. 6–9, 1975, p. 68 (abstract).

Aye, M. T. (1974a). *Exp. Hemat.* **2,** 362–371.

Aye, M. T. (1974b). *Blood* **44,** 205–219.

Aye, M. T. (1975). *Blood* **45,** 485–493.

Blackstock, A. M. and Garson, O. M. (1974). *Lancet* **2,** 1178–1179.

Bonner, W. A., Hulett, H. R., Sweet, R. G. and Herzenberg, L. A. (1972). *Rev. Sci. Inst.* **43,** 404–409.

Bradley, T. R. and Metcalf, D. (1966). *Aust. J. Exp. Biol. Med. Sci.* **44,** 287–299.

Byron, J. W. (1975). *Exp. Hemat.* **3,** 44–53.

Davidson, E. H. (1968). "Gene Activity in Early Development". Academic Press, New York.

Dexter, T. M. and Lajtha, L. G. (1974). *Brit. J. Haemat.* **28,** 525–530.

Dicke, K. A., Spitzer, G. and Ahearn, M. J. (1976). *Nature* **259,** 129–130.

Duttera, M. J., Bull, J. M. C., Whang-Peng, J. and Carbone, P. P. (1972). *Lancet* **1,** 715–717.

Edelman, G. M. (1974). "Cellular Selection and Regulation in the Immune Response". Raven Press, New York.

Fibach, E., Gerassi, E. and Sachs, L. (1976). *Nature* **259,** 127–129.

Gidali, J. and Lajtha, L. G. (1972). *Cell Tissue Kinet.* **5,** 147–157.

Goldwasser, E. (1975). *Federation Proc.* **34,** 2285–2292.

Gregory, C. J. (1976). *J. Cell. Physiol.,* **89,** 289–302.

Gregory, C. J., McCulloch, E. A. and Till, J. E. (1973). *J. Cell. Physiol.* **81,** 411–420.

Grey, H. M., Kubo, R. T., Colon, S. M., Poulik, M. D., Cresswell, P., Springer, T., Turner, M. J. and Strominger, J. L. (1973). *J. Exp. Med.* **138,** 1608–1612.

Iscove, N. N., Messner, H., Till, J. E. and McCulloch, E. A. (1972). *Ser. Haemat.* **5,** 37–49.

Krantz, S. B. and Jacobson, L. O. (1970). "Erythropoietin and the Regulation of Erythropoiesis". University of Chicago Press, Chicago.

McCulloch, E. A., Siminovitch, L. and Till, J. E. (1964). *Science* **144,** 844–846.

McCulloch, E. A., Siminovitch, L., Till, J. E., Russell, E. S. and Bernstein, S. E. (1965). *Blood* **26,** 399–410.
McCulloch, E. A. and Till, J. E. (1971). *Amer. J. Path.* **65,** 601–619.
McCulloch, E. A., Mak, T. W., Price, G. B. and Till, J. E. (1974). *Biochim. Biophys. Acta.* **355,** 260–299.
McCulloch, E. A. and Till, J. E. (1977). *Blood,* **49,** 269–280.
Messner, H. A., Till, J. E. and McCulloch, E. A. (1973). *Blood* **42,** 701–710.
Metcalf, D., Warner, N. L., Nossal, G. J. V., Miller, J. F. A. P., Shortman, K., and Rabellino, E. (1975a). *Nature* **255,** 630–632.
Metcalf, D., MacDonald, H. R., Odartchenko, N. and Sordat, B. (1975b). *Proc. Nat. Acad. Sci. U.S.A.,* **72,** 1744–1748.
Moller, G. (1974). *Transplantation Revs.* **21.** Munksgaard, Copenhagen.
Moore, M. A. S. and Metcalf, D. (1973). *Int. J. Cancer* **11,** 143–152.
Moore, M. A. S., Spitzer, G., Williams, N. and Metcalf, D. (1974). *In* "Hemopoiesis in Culture, Second International Workshop" (W. A. Robinson, ed.), pp. 303–311. U.S. Government Printing Office, Washington, D.C.
Nakamuro, N., Tanigaki, N. and Pressman, D. (1973). *Proc. Nat. Acad. Sci. U.S.A.* **70,** 2863–2865.
Nakeff, A. and Daniels-McQueen, S. (1976). *Proc. Soc. Exp. Biol. Med.* **151,** 587–590.
Niho, Y., Till, J. E. and McCulloch, E. A. (1975). *Blood* **45,** 811–821.
Ovchinnikov, Y. A. (1974). *FEBS Letters* **44,** 1–21.
Peterson, P. A., Rask, L. and Lindblom, J. B. (1974). *Proc. Nat. Acad. Sci. U.S.A.* **71,** 35–39.
Pluznik, D. and Sachs, L. (1965). *J. Cell. Comp. Physiol.* **66,** 319–324.
Poulik, M. D., Bernoco, M., Bernoco, D. and Ceppellini, R. (1973). *Science* **182,** 1352–1355.
Price, G. B. (1974). *In* "The Cell Surface: Immunological and Chemical Approaches" (B. D. Kahan and R. A. Reisfeld, eds.), pp. 237–239, Plenum Press, New York.
Price, G. B., McCulloch, E. A. and Till, J. E. (1973). *Blood* **42,** 341–348.
Price, G. B., Senn, J. S., McCulloch, E. A. and Till, J. E. (1974). *J. Cell Physiol.* **84,** 383–395.
Price, G. B., Senn, J. S., McCulloch, E. A. and Till, J. E. (1975a). *Biochem. J.* **148,** 209–217.
Price, G. B., McCulloch, E. A. and Till, J. E. (1975b). *Exp. Hemat.* **3,** 227–233.
Price, G. B., McCulloch, E. A. and Till, J. E. (1976). J. Immunol., **117,** 416–418.
Rastrick, J. M. (1969). *Brit. J. Haemat.* **16,** 185–191.
Robinson, W. A. (1974). "Hemopoiesis in Culture, Second International Workshop". U.S. Government Printing Office, Washington, D.C.
Rubin, H. (1966). *In* "Major Problems in Developmental Biology" (M. Locke, ed.), pp. 315–337. Academic Press, New York.
Senn, J. S., Pinkerton, P. H., Price, G. B., Mak, T. W. and McCulloch, E. A. (1976). *Brit. J. Cancer* **33,** 299–306.
Sredni, B., Kalechman, Y., Michlin, H. and Rozenszajn, L. A. (1976). *Nature* **259,** 130–132.
Stanley, E. R., Hansen, G., Woodcock, J. and Metcalf, D. (1975). *Federation Proc.* **34,** 2272–2278.
Stephenson, J. R., Axelrad, A. A., McLeod, D. L. and Shreeve, M. M. (1971). *Proc. Nat. Acad. Sci. U.S.A.* **68,** 1542–1546.
Sutherland, D. J. A., Till, J. E. and McCulloch, E. A. (1971). *Cell Tissue Kinet.* **4,** 479–490.
Till, J. E. and McCulloch, E. A. (1961). *Radiat. Res.* **14,** 213–222.
Till, J. E., Mak, T. W., Price, G. B., Senn, J. S. and McCulloch, E. A. (1976a). *In* "Modern Trends in Human Leukemia II" (R. Neth, ed.), pp. 33–45. J. F. Lehmanns Verlag, Munich.

Till, J. E., Price, G. B., Senn, J. S. and McCulloch, E. A. (1976b). *In* "Growth Kinetics and Biochemical Regulation of Normal and Malignant Cells" Proceedings of the 29th Annual Symposium on Fundamental Cancer Research, Houston, Texas, in press.

Trentin, J. J., Rauchwerger, J. M. and Gallagher, M. T. (1974). *In* "Control of Proliferation in Animal Cells" (B. Clarkson and R. Baserga, eds.), pp. 927–932. Cold Spring Harbor Laboratory, Cold Spring Harbor, N.Y.

van Bekkum, D. W. and Dicke, K. A. (1972). "In Vitro Culture of Hemopoietic Cells" Radiobiological Institute TNO, Rijswijk.

van den Engh, G. J. and Golub, E. S. (1974). *J. Exp. Med.* **139,** 1621–1627.

Vittetta, E. S., Uhr, J. W. and Boyse, E. A. (1975a). *J. Immunol.* **114,** 252–254.

Vittetta, E. S., Artzt, K., Bennett, D., Boyse, E. A. and Jacob, F. (1975b). *Proc. Nat. Acad. Sci. U.S.A.* **72,** 3215–3219.

Watson, J. D. (1976). "Molecular Biology of the Gene", 3rd Edition, W. A. Benjamin, Menlo Park, Calif.

# Factors Controlling the Proliferation of Haemopoietic Stem Cells In Vitro

T. M. Dexter, T. D. Allen and L. G. Lajtha

*Paterson Laboratories,*
*Christie Hospital and Holt Radium Institute,*
*Manchester M20 9BX.*
*U.K.*

## I. INTRODUCTION

A variety of clonal assay systems are available which permit the growth of granulocytic, erythroid, lymphoid and megakaryocytic precursor cells *in vitro* (Bradley and Metcalf, 1966; Stephenson *et al,* 1971; Metcalf *et al,* 1975a; Fibach *et al,* 1976; Sredni *et al,* 1976; Metcalf *et al,* 1975b). From these systems useful information is accumulating on the nature of the colony forming cells and the characteristics of the various stimulating factors. However, evidence indicates that the various cell types making up the haemopoietic system share a common ancestor (Wu *et al,* 1968; Edwards *et al,* 1970) – the pluripotent stem cells (CFU-S) characterised by Till and McCulloch (1961). Obviously the factors controlling proliferation and differentiation of these pluripotent stem cells are of fundamental importance in the control of haemopoiesis. We have recently described a method whereby proliferation of stem cells can be maintained *in vitro* for several months (Dexter and Lajtha, 1976; Dexter and Testa, 1976) and in this communication characterise the system further.

## II. MATERIALS AND METHODS

### *A. Liquid Culture System*

To establish the cultures $10^7$ pooled femoral bone marrow cells from 6–8 week old virgin $BDF_1$ female mice are inoculated into 4 oz flat bottomed, screw capped glass bottles (United Glass, London) in 10 ml of Fischer's medium supplemented with 20–25% horse serum (Flow Laboratories); 500 units/ml penicillin and 50 $\mu$g/ml streptomycin sulphate. The cultures are gassed with a mixture of air + 5% $CO_2$ and incubated at 37°C. At weekly intervals the cultures are fed by removal of half the growth medium (5ml) and addition of an equal volume of fresh growth medium. Over a three week period the numbers of non-adherent cells decline (as do also CFU-S and CFU-C) and a monolayer of adherent cells (fibroblastoid, epitheloid and phagocytic mononuclear cells) becomes established. After three weeks the cultures are fed, as before, but with 5ml of growth medium containing a further $10^7$ freshly isolated syngeneic femoral bone marrow cells from comparably aged mice. In such cultures it has been observed that there is proliferation of suspension (non-adherent) cells and production of stem cells for several months (Dexter and Lajtha, 1976). The detailed methodology involved is described by Dexter and Testa (1976).

### *B. CFU-S and CFU-C Assays*

Generally 5 culture bottles are used in each experiment. The pooled suspension cells present in the growth medium removed at weekly intervals are counted directly on a haemacytometer. The growth medium is then centrifuged (800g for 10 min) and the cell pellet is resuspended in a small volume of Fischer's medium. The cells can be smeared directly onto slides for staining and differential morphological analysis or analysed for the presence of CFU-S and CFU-C. CFU-S were assayed according to the technique of Till and McCulloch, 1961. CFU-C content was determined using the single layer agar technique (Bradley and Sumner, 1968). Routinely, embryo or heart conditioned medium was used as a source of CSF activity (Dexter and Testa, 1976).

### *C. Ultrastructural Methods*

Preparations of the entire cultures were made in the following manner for both transmission electron microscopy and scanning electron microscopy: – 10ml of 6% glutaraldehyde (pH 7.4 in M/15 Sorensens phosphate buffer) was gently pipetted into the culture bottles causing as little disturbance to the culture as possible. The fixative was mixed with the 10ml of growth medium, resulting in a final concentration of 3% glutaradldehyde. Fixation was allowed to proceed for 10 min and then the mixture of growth medium and fixative gently decanted and replaced with fresh 3% glutaraldehyde. After a further 30 min fixation the monolayer was washed briefly (5 min) in three changes of buffer alone, and then post fixed in 1% osmium tetroxide in the same buffer. After several buffer washes the monolayer was rinsed with distilled $H_2O$. At this point treatment for those bottles destined for scanning electron microscopy and transmission electron microscopy diverged. For scanning electron microscopy the bottles were broken (in as controlled a manner as possible) into small fragments (approximately 1cm × 1cm) and dehydrated through ascending concentrations of acetone, followed by critical point drying from liquid $CO_2$. The cells were then coated with a 200 Å coating of gold by sputter coating. The stubs were examined in a Cambridge S4-10 Scanning electron microscope at 20 kv accelerating voltage.

For transmission electron microscopy, the culture bottles were maintained intact and dehydrated through ascending concentrations of alcohol, followed by infiltration into Epon resin using propylene oxide. The Epon was polymerised as a thin layer *in situ* on the bottom of the culture bottles and then separated from the glass by breaking the bottles followed by alternate immersion in hot water and liquid nitrogen. Individual areas of cells were selected under the light microscope and sectioned at right angles to the growing surface. The sections were stained with uranyl acetate and lead citrate and examined in an AEI 801A transmission electron microscope.

## III. RESULTS

### *A. Characteristics of Liquid Cultures*

*1. Culture at 37° and Weekly Feeding.* The data from a "typical" experiment are shown in Table I. Cell proliferation is obviously occurring since the numbers of suspension cells approximately double after each feeding. For example at week 5 before feeding, there were $7.2 \times 10^5$ suspended cells/culture. The cultures were then "fed", i.e. half the non-adherent cells were removed, and by week 6 the cell number had recovered to $9.0 \times 10^5$. As the numbers of monolayer or adherent cells remains fairly constant throughout the experiment, the increase in the numbers of suspended cells represents a true proliferation, rather than simply detachment of the monolayer population.

**TABLE 1**

*Maintenance of $CFU_s$ and $CFU_c$ in BM + BM Cultures*

| Weeks cultured | Cell count (Suspension × $10^5$ | CFU-x/culture (suspension) | CFU-c/culture (suspension) | Morphology | | |
|---|---|---|---|---|---|---|
| | | | | B | Gr | Mono |
| 1 | 20.5 | 580 ± 50 | 8,500 ± 420 | 8 | 68 | 24 |
| 2 | 11.0 | 700 ± 35 | 9,500 ± 850 | 19 | 78 | 3 |
| 3 | 11.0 | 193 ± 18 | ND | 10 | 83 | 7 |
| 4 | 9.8 | 370 ± 45 | 5,500 ± 500 | 7 | 83 | 9 |
| 5 | 7.2 | 320 ± 19 | 1,870 ± 300 | 27 | 50 | 23 |
| 6 | 9.0 | 372 ± 60 | ND | 19 | 26 | 53 |
| 7 | 6.9 | 185 ± 12 | ND | 32 | 12 | 56 |
| 8 | 6.2 | 245 ± 18 | ND | 12 | 3 | 85 |
| 9 | 6.0 | ND | 1,400 ± 200 | 0 | 18 | 82 |
| 10 | 5.5 | 180 ± 26 | ND | 3 | 14 | 83 |
| 11 | 5.0 | ND | 202 ± 35 | 2 | 3 | 92 |
| 12 | 5.0 | 10 | 86 ± 12 | 1 | 3 | 95 |
| 13 | 8.5 | 0 | 56 ± 10 | 1 | 2 | 97 |
| 14 | 11.3 | 0 | 0 | 0 | 3 | 96 |

ND = No data
B = Blasts
Gr = Granulocytes (all stages)
Mono = Phagocytic mononuclear cells.

Similarly the pluripotent stem cells (CFU-S) are proliferating at least for 12 weeks, also showing an approximate doubling weekly. Since CFU-S and CFU-C are not detectable amongst the adherent cell population (Dexter *et al,* 1973) it is possible to monitor changes occurring in these stem cell populations by assaying their content in the suspension cell population. These CFU-S are apparently normal, forming erythroid, granulocytic and megakaryocytic colonies in the spleens of potentially lethally irradiated mice and further, if sufficient numbers of stem cells are injected (approximately 40) the cultured cells can protect irradiated mice from haemopoietic death. Granulocyte precursors cells (CFU-C) are also being maintained for several months. There are presumably being produced by differentiation from the pluripotent stem cell – although a limited capacity for self-regeneration is not ruled out. The colonies formed are similar to those produced from freshly isolated bone marrow cells, showing maturation towards granulocytes and macrophages. The formation of these colonies is dependent upon the addition of CSF (in this case, heart conditioned medium). In the absence of CSF, colonies do not form.

In all cultures, extensive granulopoiesis is seen initially, with all maturation stages represented (Figure 1). With time in culture, however, granulocytes decline, macrophages increase (Figure 2) and in many cultures there is also an accumulation of blast cells (Figure 3). The nature of these blast cells is unknown but studies have shown that they possess neither $\theta$ antigen (characteristic of T lymphocytes, Raff 1969) nor surface lg determinants (characteristic of B-lymphocytes, Raff *et al,* 1970). Furthermore they show no myeloperoxidase or Sudan Black activity (characteristic of maturing granulocytes). With further time in culture the numbers of such blast cells decrease and phagocytic

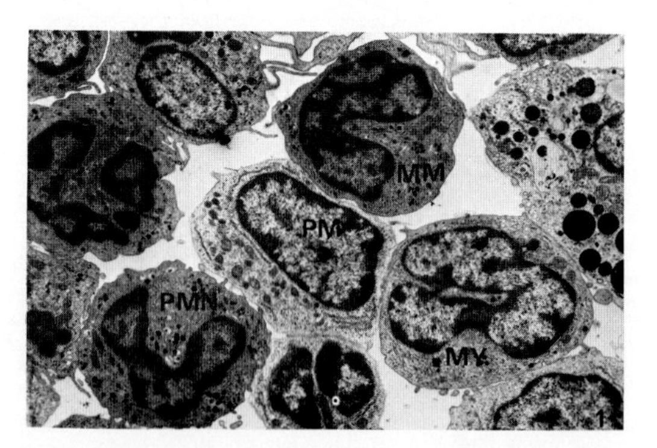


Fig. 1. Section through a pellet of granulocytes collected by gentle agitation of the medium causing them to become detached. All stages from promyelocyte (PM), myclocyte (MY), metamyclocyte (MM), and mature granulocyte (PMN) are present, showing the characteristic increase in nuclear segmentation and cytoplasmic granule population with maturation. Mag. × 3675

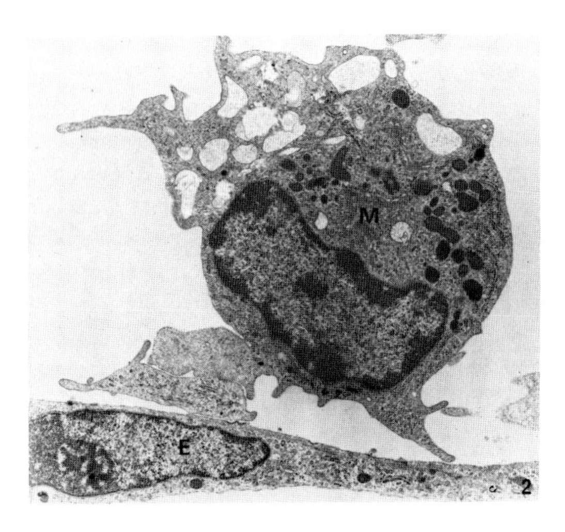


Fig. 2. Section cut at right angles to the growing surface showing a phagocytic mononuclear cell (M) attached to an underlying "epithelial" cell (E). The granules and phagocytic cytoplasmic extensions clearly distinguish the phagocytic cell. There is also a difference in nuclear morphology, the "epithelial" cell possessing a smoother nuclear profile and less condensed marginal chromatin than the phagocytic mononuclear cell. Mag. × 8100

mononuclear cells become the predominant cell type. Proliferation of such mononuclear cells has been observed in some cultures for periods in excess of 30 weeks.

*2. Variations in Extent of Stem Cell Maintenance.* The maintenance of proliferation of stem cells in the liquid culture system is dependent to a large

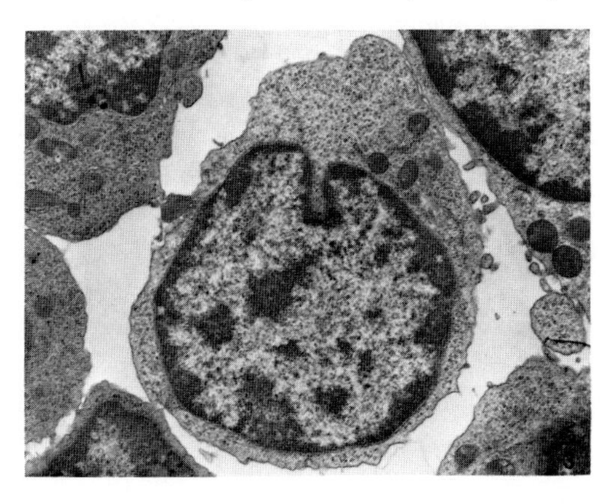

Fig. 3. Section through an undifferentiated cell illustrating the high nucleocytoplasmic ratio and smooth nuclear profile, interpreted to be characteristic of an immature blast (stem) cell. Mag. × 6750

extent on the type and source of serum used. We have found that foetal calf serum (Gibco or Flow Laboratories) does not support the growth of stem cells – neither does horse serum purchased from Gibco. However, approximately 50% of batches of horse serum supplied by Flow Laboratories will support the proliferation of stem cells (Dexter and Testa, 1976). However, using the same batch of sera, we find variations in the length of time in culture of stem cell growth and granulopoiesis. Table II shows that 25% of experiments show stem cell maintenance for only 2–4 weeks. In these cultures, there is a gradual decline in stem cell numbers and a conversion to phagocytic mononuclear cells. More than 50% of experiments show stem cell maintenance for 6–12 weeks, and in these cultures stem cell proliferation, extensive granulopoiesis and a blast cell accumulation (15–80%) occurs. Infrequently, there may be proliferation of stem cells for up to 16 weeks.

*B. The Morphology of the Monolayer and its Role in Stem Cell Maintenance*

Stem cell proliferation is not seen in siliconized bottles (where potentially adherent cells cannot attach) nor in bottles where there is poor development of the adherent population and the variations in stem cell maintenance may in part be dependent upon the quality of monolayer produced. The monolayer consists of 3 separate cell populations. Firstly, attached mononuclear cells which display a characteristic granule population, phagocytic inclusions (Figs. 2 and 5) and an overall tendency in the scanning electron microscope (SEM) to spread along a single axis. Secondly, there is a population of attached cells which exhibit either an epithelial or fibroblastic morphology in the scanning electron microscope and an absence of the characteristic mononuclear granules or phagocytic inclusions. The nuclear morphology also differs significantly from the mononuclear cells as there is less condensed marginal chromatin and a generally smoother nuclear profile in the "epithelial" cells (Figs. 2 and 5). The "epithelial" cells also tend to form a confluent monolayer, which the mononuclear cells do not. The third cell type seems to be responsible for stem cell maintenance which occurs concomitantly with its appearance in the monolayer. This population is

**TABLE 2**

*Experimental Variation in Extents of Stem Cell Maintenance in vitro*

| Number of experiments | CFU-s maintenance (weeks) | CFU-c maintenance (weeks) |
|---|---|---|
| 6 | 3–4 | 4–5 |
| 10 | 6–7 | 7–8 |
| 8 | 9–12 | 10–13 |
| 3 | >13 | >14 |

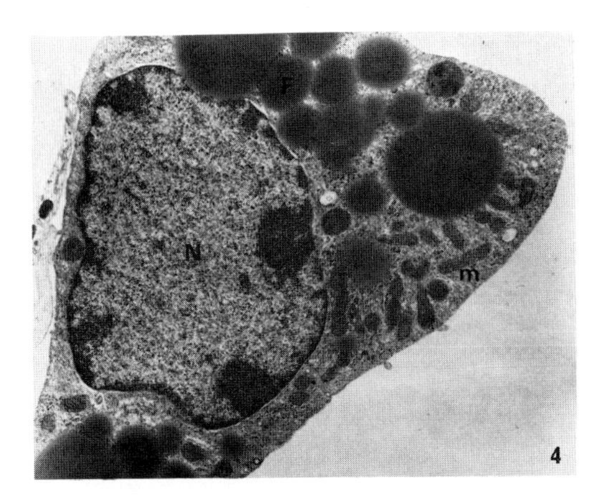


Fig. 4. Section through a 'giant fat' cell, showing in this profile only a small accumulation of fat droplets (F), but illustrating the normal appearance of the nuclear profile (N), cytoplasm, and mitochondria (M). Some coalescence of the fat droplets to form larger ones may be indicated by their aggregation both above and below the nucleus. Mag. × 5670

characterised by its enormous accumulation of storage vacuoles which cause an increase in overall cell size to a range of 50–150 $\mu$m. The material contained within these vacuoles stains intensely with Sudan Black B and is also extremely osmiophilic, indicating a lipid nature. These cells develop over a three week period and form clusters in the "epithelial" monolayer. This clustering in the monolayer, together with the fact that the "epithelial" cells which are regularly observed with a few small fat vacuoles may indicate that the fat containing cells themselves may derive from the "epithelial" cells (Fig. 7). There is also a marked tendency for maturing granulocytes to cluster around the fat cell aggregations, which often form radiating foci. This association between the fat cells and developing granulocytes may well suggest a feeder cell function for stem cell maintenance, or a nurse cell function during granulocyte maturation. The feeder

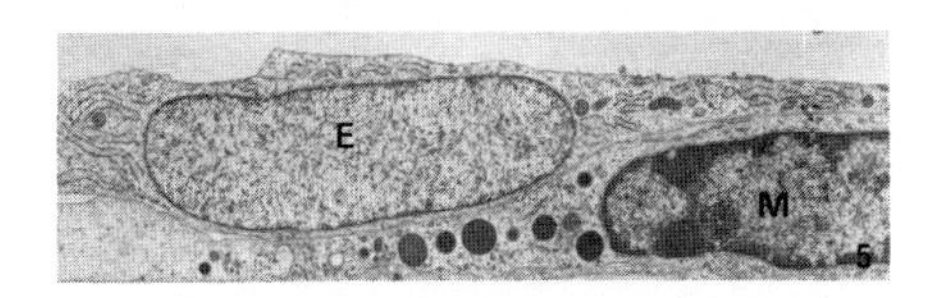


Fig. 5. Section cut at right angles to the growing surface showing an attached mononuclear cell (M) with phagocytic inclusions and condensed marginal chromatin in the nucleus. The overlying "epithelial" cell has little or no condensed marginal chromatin, and an absence of phagocytic inclusions, and the cytoplasm which shows several cisternae of smooth endoplasmic reticulum. Mag. × 4200

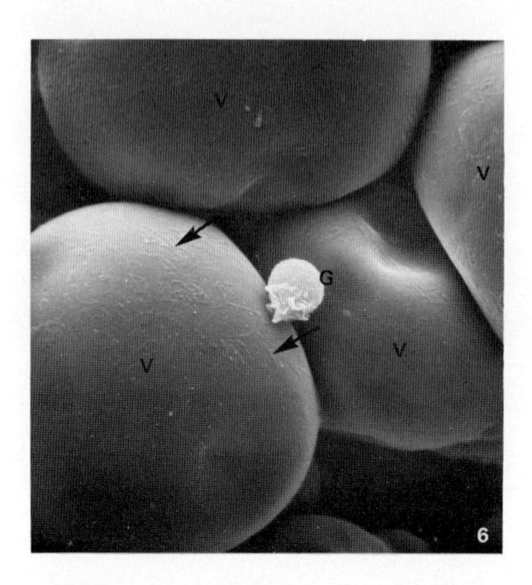


Fig. 6. Scanning electron micrograph of an area of the monolayer with 4 giant fat cell vacuoles (V) and a single granulocyte (G). The enormous difference in magnitude between the fat vacuoles and granulocyte is readily apparent. On the surface of the fat vacuoles the mitochondria in the thin layer of cytoplasm stand out as VERMIFORM ridges (arrowed). Mag. × 575

cells themselves may contain all sizes of fat droplets, varying from less than 1 $\mu$ to 30 $\mu$ to 40 $\mu$ within a single profile (Fig. 4). The droplets themselves are bounded by a single membrane indicating cellular synthesis of the lipid rather than phagocytic ingestion. In view of the large size difference between individual droplets, it is reasonable to assume that the larger ones are formed by sequential coalescence of smaller vacuoles. Sectioned profiles of these cells show a nucleus of approximately 5 $\mu$ in diameter, with a smooth profile, no segmentation and little condensed chromatin. The remainder of the cytoplasm has numerous mitochondria (Fig. 4) and these are sometimes visible on the surface of the large vacuoles in the scanning electron microscope preparations where the surrounding veil-like layer of cytoplasm has contracted during presentation to leave the mitochondria slightly proud of the surface (Fig. 6). The large size differential between a single fact vacuole and a complete granulocyte is also shown in Fig. 6. All cells in the monolayer, the mononuclear, "epithelial", and "giant fat" cells may attach directly to the substratum or to each other (Figs. 2, 5, 7 and 8).

Direct attachment of the larger feeder cells is less evident in *in situ* preparations, but this may well be due to the assumption of an overall spherical shape after the accumulation of a large amount of lipid. In cells at an earlier

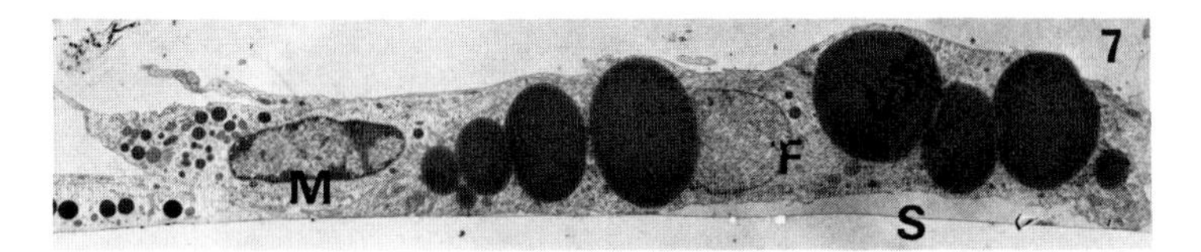


Fig. 7. Section cut at right angles to the growing surface (S), showing a giant fat cell (F) in early stages of fat accumulation while still attached in an epithelial manner to the substratum. A phagocytic mononuclear cell is also attached (M). Mag. × 2550

stage of lipid accumulation however, direct attachment to the substratum and the original morphology of these cells is visible (Fig. 7).

*C. The Role of CSF in Liquid Cultures*

In the agar CFU-C system the presence of colony stimulating factor (CSF) appears to be essential for the development of colonies of granulocytes and macrophages' (Metcalf, 1970). In view also of the *in vivo* evidence supporting a role of CSF as a physiological regulator of granulopoiesis (Moore *et al*, 1974) the relevance of CSF in our liquid culture system was also investigated.

The growth medium removed after various times in culture was centrifuged – and the cell free supernate tested for CSF activity i.e. its ability to stimulate the development of CFU-C from freshly isolated bone marrow cells. A wide

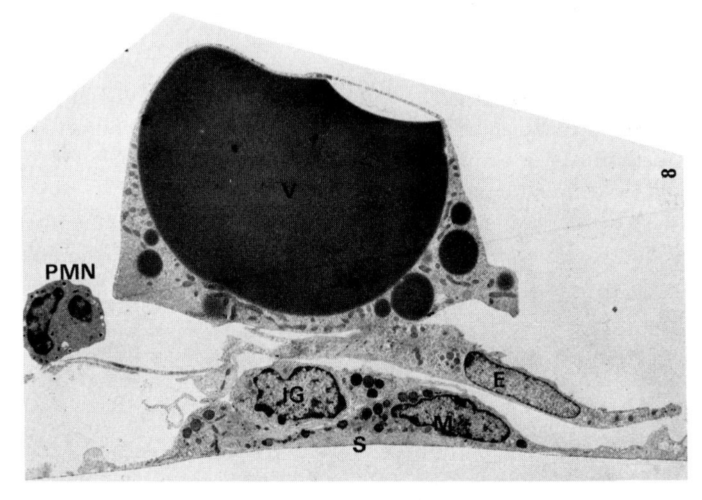


Fig. 8. Section cut at right angles to the growing surface showing the composite nature of the monolayer. A phagocytic mononuclear cell (M) is attached to the substratum (S) and overlying this an epithelial (E) cell. Attached to the surface of the epithelial cell are a mature granulocyte (PMN) and a giant fat cell, with a single larger vacuole (V) and several smaller ones. Between the epithelial cell (E) and the monocyte there is an immature granulocyte (IG). Mag. × 1800

range of concentrations (5–75%) of the conditioned medium obtained from the liquid cultures was tried. It was found that, irrespective of the time in culture, such conditioned medium induced neither cluster nor colony formation in fresh bone marrow cells (nor indeed in the cultured cells), even after heat treatment and/or dialysis (thought to remove molecules with CSF inhibitory activity). Further experiments have also established that neither the suspension cells nor the monolayer cells would stimulate the development of bone marrow CFU-C when used as a feeder layer in the agar system. It seems therefore that in the liquid culture system, CFU-S and CFU-C maintenance as well as granulopoiesis is occurring in the absence of detectable CSF. Experiments are now in progress to determine the effects of addition of CSF, at optimal stimulating levels, to these cultures.

## IV. DISCUSSION

Previous attempts to culture stem cells *in vitro* using the Marbrook type liquid culture system (Summer *et al,* 1972) or agar culture systems (Testa and Lajtha, 1973) have been unsuccessful in that stem cells have been maintained for only a matter of days. However, Dexter and Lajtha (1974) showed that in a proportion of "co-cultures" of thymus and bone marrow cells, proliferation of stem cells and granulopoiesis occurred for several weeks. Subsequently it was demonstrated that stem cell proliferation also occurred in cultures of bone marrow cells only, provided that an adherent population of bone marrow cells was established prior to inoculation of a second (target) bone marrow cell population (Dexter and Lajtha, 1976; Dexter and Testa, 1976). In these cultures there is also production of granulocyte precursor cells (CFU-C) and granulopoietic maturation for several months, although many cultures eventually show an accumulation of blast cells. Horse serum seems to provide the most consistent growth factors for these cultures but, using the same batch of serum, there are considerable variations seen in the extents of stem cell maintenance. This variation appears to be associated with the quality or type of adherent cell layer produced.

There appears to be an association between the long term survival of stem cells and the presence in the monolayer of the 'giant fat' cells, apparently of a lipid nature. It is proposed that these cells may function as nurse or feeder cells, promoting the proliferation of stem cells. These nurse cells are quite distinct from the phagocytic mononuclear cells which make up the large part of the monolayer since the mononuclear cells show a distinct granule population and phagocytic inclusions within the cytoplasm which are absent from the fat cells, and their nuclear morphology is also different, there being rather more condensed peripheral chromatin in the nucleus of the phagocytic cells.

The relevance of these nurse cells *in vivo* is not known. However, Lord *et al,* 1975 have shown that stem cells are not distributed in a homogenous fashion throughout the femoral shaft. Rather the concentration of stem cells was greatest at the endosteal surface and lowest in the center of the femoral bone. It could well be that *in vivo* specific cell populations near the bone surface which may produce factors facilitating maintenance and proliferation of stem cells. If so, the isolation and proliferation of these cells would be of considerable interest.

Of interest was our finding that CSF was not detectable in these cultures, *i.e.* stem cell proliferation, production of granulocyte precursor cells and their subsequent maturation to granulocytes is occurring in the apparent absence of CSF. The explanation for this prolonged granulopoism is difficult to explain since in the soft agar system, colonies of granulocytes do not form in the absence of CSF. Also, in the agar system, *low* CSF levels favour macrophage production and *high* CSF levels favour granulocyte development. The answer to this paradox may lie in a closer investigation of the interaction between the adherent and the non-adherent cell population in the liquid cultures. Certainly, the developing granulocytes are selectively concentrated in particular areas of the culture bottles – areas where there is a close association between giant fat cells, epithelial cells and phagocytic mononuclear cells. Such areas may be providing an *in vitro* microenvironment where proliferation and differentiation of haemopoietic cells occurs.

Possible interaction occurring in such sites between developing granulocytes and the adherent cells need to be further investigated, as well as the effects of addition of CSF and other haemopoietic precursor cell stimulators. However, such an *in vitro* system may enable us to clarify some of the control mechanisms operating on proliferation and differentiation in the haemopoietic system.

## ACKNOWLEDGMENTS

This work was supported by the Medical Research Council and the Cancer Research Campaign. The authors wish to thank Nicola P. J. Higgins and Mr. G. R. Bennion for excellent technical assistance.

## REFERENCES

Bradley, T. R. and Metcalf, D. (1966). *Austr. J. Exp. Biol. Med. Sci.* **44,** 287–300.

Bradley, T. R. and Sumner, M. A. (1968). *Austr. J. Exp. Biol. Med. Sci.* **46,** 607–618.

Dexter, T. M., Allen, T. D., Lajtha, L. G., Schofield, R. and Lord, B. I. (1973). *J. Cell. Physiol.* **82,** 461–474.

Dexter, T. M. and Lajtha, L. G. (1974). *Brit. J. Haematol.* **28,** 525–530.

Dexter, T. M. and Lajtha, L. G. (1976). *In* "Proceedings of VII International Symposium on Comparative Research on Leukaemia and Related Diseases" Karger (Basel).

Dexter, T. M. and Testa, N. G. (1976). *In* "Methods in Cell Biology" (D. M. Prescott, ed.), Vol. 14. Academic Press, N.Y. (In press).

Edwards, G. E., Miller, R. G. and Phillips, R. A. (1970). *J. Immunology.* **105,** 719–729.

Fibach, E., Gerassi, E. and Sachs, L. (1976). *Nature,* **259,** 127–128.

Lord, B. I., Testa, N. G. and Hendry, J. H. (1975). *Blood* **46,** 65–72.

Metcalf, D. (1970). *J. Cell Physiol.,* **76,** 89–100.

Metcalf, D. and Moore, M. A. A. (1971). "Haemopoietic Cells." North Holland Publishing Co., Amsterdam.

Metcalf, D., MacDonald, H. P., Odartchenko, N. and Sordat, L. B. (1975a). *Proc. Nat. Acad. Sci. U.S.A.* **72,** 1744–1748.

Metcalf, D., Warner, N. L., Nossal, G. J. V., Miller, J. F. A. P., Shortman, K. and Rabellion, E. (1975b). *Nature* **255,** 630–632.

Moore, M. A. S., Spitzer, G., Metcalf, D. and Pennington, D. G. (1974). *Brit. J. Haematol,* **27,** 47–55.

Raff, M. C. (1969). *Nature* **224,** 378–379.

Raff, M. C., Sternberg, M. and Taylor, R. B. (1970). *Nature,* **225,** 553–534.

Sredni, B., Kalechman, Y., Michlin, H. and Rozenszajn, L. A. (1976). *Nature* **259,** 130–132.

Stepheson, J. R., Axelrad, A. A., McLeod, D. L. and Shreeve, M. M. (1971). *Proc. Nat. Acad. Sci. U.S.A.* **68,** 1542–1546.

Sumner, M. A., Bradley, T. R., Hodgson, G. S., Cline, M. J., Fry, A. and Sutherland, L. (1972). *Brit. J. Haematol.* **23,** 221–234.

Till, J. E. and McCulloch, E. A. (1961). *Rad. Res.* **14,** 213–222.

Testa, N. G. and Lajtha, L. G. (1973). *Brit. J. Haematol.* **24,** 367–376.

Wu, A. M., Till, J. E., Siminovitch, L. and McCulloch, E. A. (1968). *J. Exp. Med.* **127,** 455–463.

# Aspects of Erythroid Differentiation and Proliferation

John W. Adamson, M.D. and James E. Brown, M.D.

*Hematology Research Laboratory*
*Veterans Administration Hospital*
*and the*
*Division of Hematology*
*Department of Medicine*
*University of Washington School of Medicine*
*Seattle, Washington*

## I. INTRODUCTION

This review deals with the role of erythropoietin in the regulation of mammalian erythropoiesis and has three purposes. The first is to discuss the known characteristics and mode of action of the hormone, erythropoietin, at the biochemical level. The second is to review the evidence that multiple, distinct classes of erythropoietin-responsive progenitors can be identified in culture. The final purpose is to present new information which suggests that the action of erythropoietin on its target cell may be modulated *in vitro* by other hormones and small molecules, providing an additional possible dimension to the regulation of erythropoiesis.

## II. ERYTHROPOIETIN AND ITS ACTION

Erythropoietin is thought to regulate both steady-state and accelerated erythropoiesis in mammals. Its major effects are to promote erythroid differentiation and to initiate hemoglobin synthesis (Krantz and Jacobson, 1970). To an unknown extent, this hormone may also promote limited proliferative self-renewal of the pool(s) of immature erythroid precursors (Reissmann and Samorapoompichit, 1970). In the absence of erythropoietin, these activities are markedly diminished. Erythropoietin production is regulated by the kidney and the hormone circulates in the plasma space (Gordon *et al.*, 1967). Approximately 2–5% of the daily production is lost in the urine (Weintraub *et al.*, 1964).

Erythropoietin is a glycoprotein of approximately 46,000 molecular weight (Goldwasser and Kung, 1972). Only minute quantities of highly purified material have been produced to date due to the inadequate supply of starting material – plasma or urine from anemic animals or man – and the low recovery from purification procedures. Consequently, the hormone is not certifiably pure, and detailed characterization of its structure/function relationships has not been possible. It is believed, however, that about 11% of the molecule is composed of sialic acid. The purpose of the sialic acid moiety is to preserve the activity of the hormone *in vivo*, preventing its rapid removal by the liver from circulation (Morell *et al.*, 1968). Thus, while desialated erythropoietin is ineffective *in vivo*, (Lowy *et al.*, 1960), it has at least normal activity *in vitro* (Goldwasser *et al.*, 1974).

The mechanism by which erythropoietin exerts its effects on either the cytoplasm or nucleus of the target cell is not clear. Because of its size, erythropoietin is not likely to enter its target cell in order to promote differentiation and hemoglobin synthesis. Studies by Chang *et al.* (1974) have suggested that there is a protein-like membrane receptor with which erythropoietin interacts. In their experiments, exposure of rat bone marrow cells to trypsin plus cycloheximide blocked the effect of erythropoietin on RNA synthesis. Exposure of target cells to either agent alone, however, failed to reduce RNA synthesis significantly. These results were felt to demonstrate a protein receptor on the surface of responsive cells which was required for erythropoietin effect.

An additional fact which is consistent with a surface receptor for erythropoietin is the observation that the biological activity of erythropoietin in the incubation medium does not decline *in vitro* even on prolonged culture with large numbers of marrow cells. This would suggest that the hormone is neither degraded by nor incorporated into the target cells in significant amounts (Iscove and Sieber, 1975).

In extending the membrane receptor model of erythropoietin action, Chang and Goldwasser (1973) have proposed that there is a cytoplasmic mediating factor which arises in specific target cells upon exposure to the hormone. In cultures of various tissues of the rat, erythropoietin stimulated the appearance of this mediating factor only in bone marrow cells; but, once formed, the factor promoted RNA synthesis in nuclear preparations from a variety of tissues. The activity of the mediating factor became maximal within one hour of incubation of the cells with erythropoietin, but its appearance was not dependent on new protein synthesis. These workers concluded that the mediating factor is not related to the adenyl cyclase/cyclic AMP mechanism because cyclic AMP does not stimulate hemoglobin synthesis in this system (Goldwasser, 1975). As will be discussed, these results may be an artifact of the assay employed, for more recent data clearly demonstrate a stimulatory effect of cyclic nucleotides on *in vitro* mammalian erythropoiesis.

The early biochemical events in the action of erythropoietin on its responsive cell and the initial alterations in hemoglobin synthesis have been studied in cultures of rodent hematopoietic cells by several groups and have recently been summarized (Marks and Rifkind, 1972; Rifkind and Marks, 1975; Goldwasser, 1975). The sequence of events induced by erythropoietin is broadly outlined in Fig. 1. Following exposure of cells to erythropoietin, the earliest detectable macromolecular event occurs within 15–30 minutes (Krantz and Goldwasser, 1965) and is marked by an increase in the synthesis of a variety of RNA species. This response precedes the rises in DNA and protein synthesis and does not depend on these processes (Gross and Goldwasser, 1970, 1972; Djaldetti *et al.*, 1972; Maniatis *et al.*, 1973). The RNA response, however, is completely inhibited by actinomycin D and, thus, probably results from DNA-dependent RNA polymerase activity (Krantz and Goldwasser, 1965). Further characterization of this early material reveals ribosomal precursor RNA (from 32S to 45S), higher molecular weight RNA (greater than 45S), as well as transfer RNA (4-5S) (Gross and Goldwasser, 1969; Nicol *et al.*, 1972; Maniatis *et al.*, 1973). After 1 hour of culture, processed ribosomal RNA (18S and 28S) appears (Nicol *et al.*, 1972; Maniatis *et al.*, 1973), but not until after 5–10 hours is globin messenger RNA detected, either by its activity in the Krebs ascites cell-free system (Terada *et al.*, 1972) or by complementary DNA hydridization techniques (Ramirez *et al.*, 1975). Thus, the appearance of globin messenger RNA is probably due to transcription of globin genes and not to an increase in the rate of post-transcriptional modification of heterogeneous nuclear RNA.

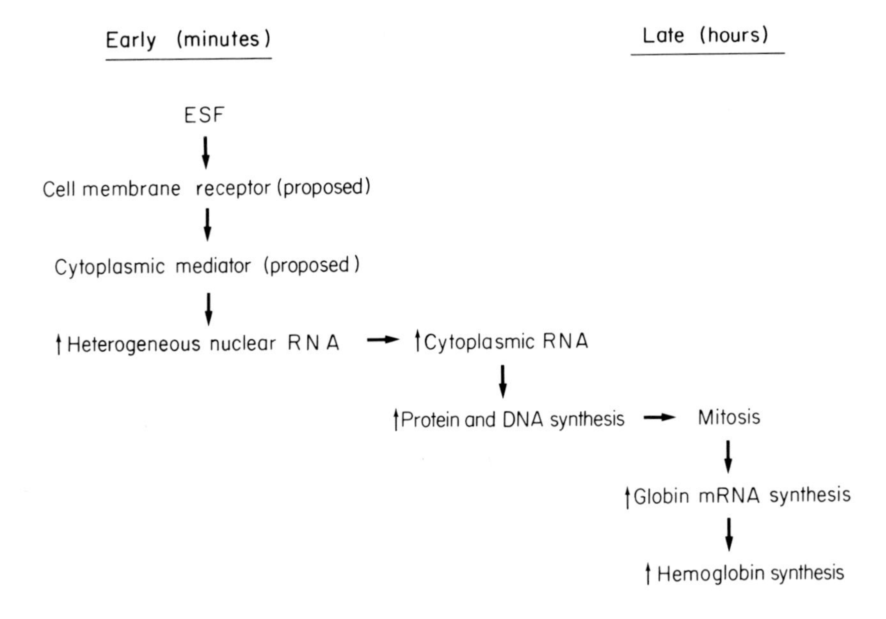


Fig. 1. The sequence of events induced by erythropoietin in erythropoietic tissues.

These earliest transcriptional events resulting from the action of erythropoietin may involve an activation of RNA polymerases (Piantadosi *et al.*, 1976) or an effect on DNA template availability or activity (Spivak, 1976), since these changes appear to be independent of new protein synthesis. The increased number of more mature erythroid cells results in an increment in the total number of globin messenger RNA molecules per culture (Conkie *et al.*, 1975), as well as an accumulation of the number of molecules of globin messenger RNA per cell.

While the initial hormone effects are on RNA synthesis, inhibition of DNA synthesis has a profound effect on subsequent globin synthesis (Gross and Goldwasser, 1970, 1972). The sum of these observations has led to the concept that erythropoietin triggers precursors to undergo proliferation and, at the same time, initiates biochemical events peculiar to the erythroid program which result in globin messenger RNA and globin synthesis. These changes occur at the level of an immature precursor cell and lead to the terminal stages of erythropoietic maturation (Rifkind and Marks, 1975; Fig. 1).

Among the reported effects which have been ascribed to the influence of erythropoietin on already differentiated and maturing elements are the shortening of the intermitotic time of erythroid progenitors and an increase in the amount of hemoglobin synthesized per cell (Stohlman, 1970; Papayannopoulou and Finch, 1972). It is not clear whether these findings reflect a non-specific response of an actively proliferating system or, in fact, indicate a direct but limited action of erythropoietin on maturing erythroid precursors. Recent studies by Glass *et al.* (1975) suggest that heme and hemoglobin synthesis are differentially stimulated by erythropoietin throughout the erythroid maturation sequence. This concept remains to be clarified. It is generally believed, however, that once the erythroid program is expressed, erythropoietin has little further effect on erythroid maturation or the amount of hemoglobin synthesized per cell (Chui *et al.*, 1971).

Considerable understanding of the biochemistry of both erythropoietin and the initial events of erythropoietin stimulation has been provided by such studies. However, the lack of sufficient quantities of pure hormone and the fact that strictly homogeneous populations of target cells are not available limit refined studies of hormone-cell interaction.

## III. CELLS RESPONDING TO ERYTHROPOIETIN

Despite the lack of purity of hormone and target cell, significant progress has been made within the last several years in the functional definition of erythropoietin-responsive cells and their relationship to other hematopoietic progenitors. These advances have been made possible by the development of clonal assays for cells responding *in vitro* to erythropoietin. Until recently, erythropoietin stimulation of cells in culture was monitored by the incorporation of various radioactive tracers into newly synthesized heme or hemoglobin (Krantz *et al.,* 1963). This suspension culture technique permitted the analysis of certain cumulative effects of the hormone but provided limited insight into the nature of the target cell. In 1971, Stephenson and colleagues presented the first quantitative *in vitro* assay for erythropoietin-responsive cells. The technique allowed the growth of clusters of hemoglobin synthesizing cells in a semisolid medium (Fig. 2). The number of colonies or clusters formed was related directly to the concentration of erythropoietin in the medium and the number of cells plated. Because linearity of colony formation was found even at very low cell concentrations, and the regression of this relationship passed through the origin (Fig. 3), it was believed that each colony arises from a single progenitor which is termed the erythroid colony-forming unit or CFU-E. Independent confirmation of the clonal origin of such colonies has recently been provided for cultures of human marrow cells by analyzing the glucose-6-phosphate dehydrogenase (G-6-PD) content of individual erythroid colonies. If marrow cells obtained from a G-6-PD heterozygote are cultured, only one isoenzyme type is found in each colony and the ratio of colonies containing one isoenzyme type to the other conforms to the ratio of isoenzyme activities observed in other mesenchymal tissues (Prchal *et al.,* 1976).

The advantages of a colony forming assay for enumerating erythropoietin-responsive cells are several. First, the expression of colony growth requires both differentiated function and proliferation. Differentiated function here is defined as the capacity of the target cell to respond to a recognizable stimulus by the production of cells containing hemoglobin. Second, because colony numbers are linearly related to cell concentration, the assay can be used to quantitate the size

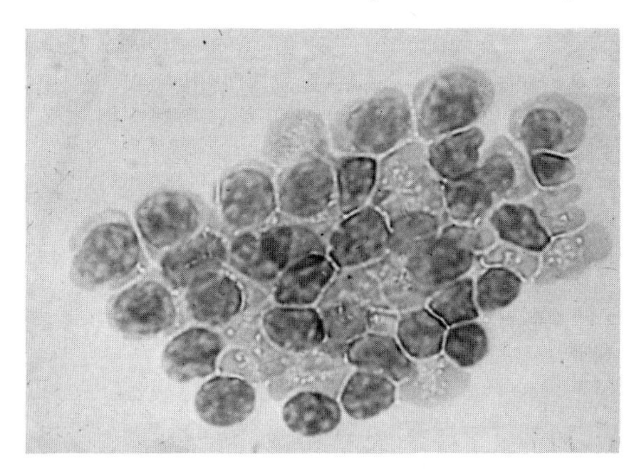

Fig. 2. An erythroid colony cultured from mouse marrow and fixed and stained with benzidine. (Magnification × 800 .)

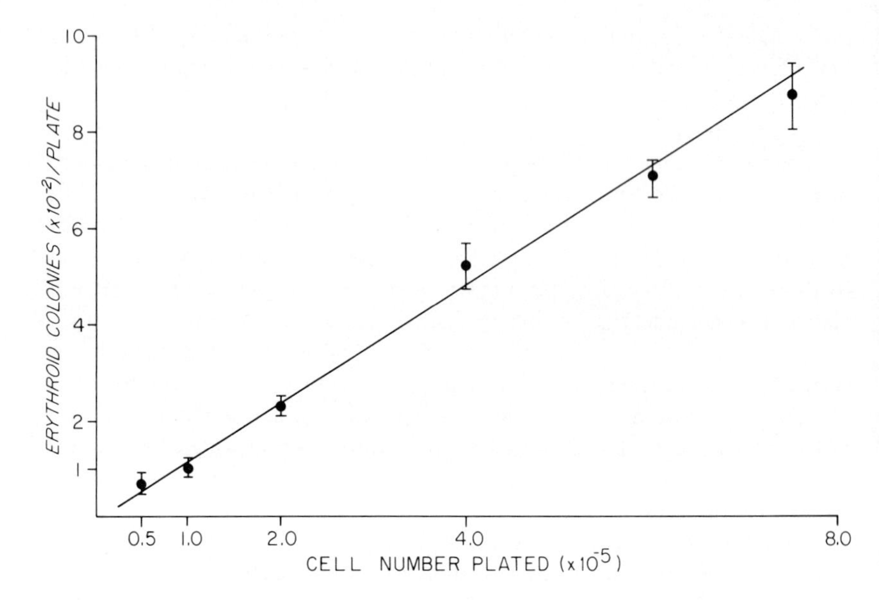


Fig. 3. The relationship between the number of marrow nucleated cells plated and the number of erythroid colonies formed. Values represent the mean ± SEM in this and subsequent figures. The ESF concentration was 0.1 International Reference Preparation (IRP) units/ml.

of the responsive cell population. Finally, an experimental animal may be manipulated physiologically to characterize the response of the cell population to various erythropoietic stimuli.

The originally described colony-forming unit, the CFU-E, has limited growth properties, and is capable of only 4–5 doubling divisions. The presence in culture of erythropoietin is an absolute requirement for colony formation, and size, as well as number of colonies, is related directly to the concentration of the hormone (Cooper *et al.*, 1974). The colonies grow out over a relatively short period of time, reach maximal numbers at about 2 days for mouse marrow and then rapidly decline (McLeod *et al.*, 1974; Fig. 4). The number of CFU-E in normal mice is approximately 1 in 250 marrow cells. The cycle characteristics of these cells reflect the fact that the majority – 75–80% – are in active DNA synthesis, or S phase, as estimated by the tritiated thymidine suicide technique (Singer and Adamson, 1976b).

Because of the limited growth potential and the relative numbers of CFU-E in the marrow, it was considered unlikely that this progenitor is a direct or near descendant of the pluripotent stem cell – or CFU-S – capable of giving rise to macroscopic spleen colonies as assayed by the method of Till and McCulloch (1961). Experiments designed to analyze the degree of relatedness between CFU-E and CFU-S clearly demonstrate that these cells are separated by several steps in differentiation. This analysis was based on the hypothesis that cells

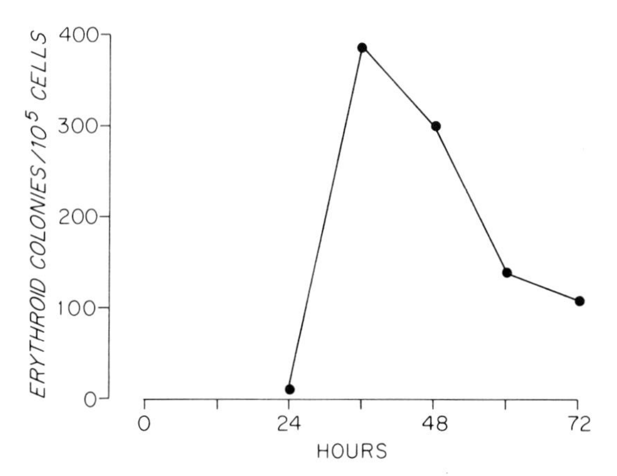


Fig. 4. Time course of appearance of mouse erythroid colonies after initiation of culture. Colonies containing 8 or more cells are counted.

closely related in development will appear in proportional numbers in a growing clone. One example of such a clone would be the individual spleen colony, known to arise from a single cell (Becker *et al.*, 1963). The more randomizing events which occur between a pluripotent cell and a more differentiated element, the greater the disparity between the numbers of these two elements (Fig. 5A). If there are relatively few randomizing events in the transition from a pluripotent to a more differentiated cell, then the numbers of cell types to be compared should be reasonably proportional within a given spleen colony (Fig. 5B). To perform this type of analysis, marrow cells are injected into appropriately prepared recipients to provide exogenous spleen colonies. Individual colonies are then harvested, disaggregated, and their content of various

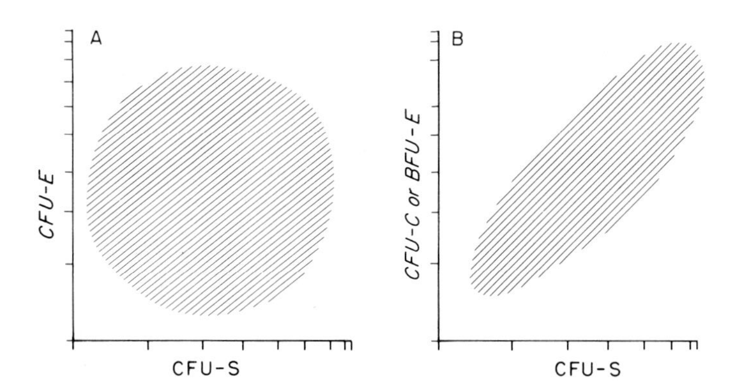


Fig. 5A. The schematic relationship between the numbers of CFU-E and CFU-S present in individual spleen colonies. No correlation is observed.

Fig. 5B. The schematic relationship between the numbers of CFU-C or BFU-E and the numbers of CFU-S present in individual spleen colonies.

colony-forming units determined. As demonstrated by Wu *et al.* (1968), a close relationship was observed within individual colonies between CFU-S and granulocytic colony-forming units (CFU-C). By similar analysis, Gregory *et al.* (1973) clearly demonstrated that such a relationship did not hold for erythroid colony-forming units and CFU-S. In contrast with the positive correlation between CFU-S and CFU-C, the lack of correlation between CFU-E within individual spleen colonies suggests that eryhroid colony-forming units represent a stage of differentiation further removed from CFU-S than the CFU-C. Thus, more randomizing events must occur in the generation of CFU-E.

Functional studies of the regulation of the size of the CFU-E population demonstrate that the numbers of this relatively differentiated cell depend upon the erythropoietic state of the intact animal. Thus, if mice are given a single injection of erythropoietin, or made anemic by bleeding, the numbers of CFU-E per femur and spleen increase 2–3 fold (Gregory *et al.,* 1974; Fig. 6). With hypertransfusion, a manipulation which reduces endogenous erythropoietin to unmeasurable levels, the numbers of CFU-E per organ decline to less than 10% of initial values (Gregory *et al.,* 1973; Fig. 6). In neither instance are significant alterations in the numbers of CFU-C or CFU-S observed. Thus, the size of this population of erythroid colony-forming units is dependent on erythropoietin, as is the growth of these precursors in culture. The finding that the size of the CFU-E population is dependent upon the hormone indicates that the acquisition of sensitivity to erythropoietin occurs before the differentiation step leading to CFU-E (Gregory *et al.,* 1973).

Of additional interest are studies suggesting that erythropoiesis may be finely regulated by altered responsiveness of CFU-E to erythropoietin. Thus, Gregory *et al.* (1974) reported decreased sensitivity to erythropoietin of CFU-E from regenerating marrow as compared to normal marrow. These findings suggest that erythropoiesis may be controlled rather precisely through alterations in surface receptor conformation or density.

Because of the apparent differentiative distance between pluripotent stem cells and CFU-E, it was predicted that intermediate populations of cells might exist which participate in erythropoietic proliferation (Lajtha *et al.,* 1971; Gregory *et al.,* 1973). In 1973, Axelrad demonstrated that normal mouse bone marrow cells, exposed for a prolonged period of time in culture to increased concentrations of erythropoietin, gave rise to large aggregates of hemoglobin synthesizing cells. Because of their appearance and size, these aggregates were referred to as "erythroid bursts" and the cell giving rise to the burst the "burst-forming unit," or BFU-E (Axelrad *et al.,* 1973). While the numbers of such colony forming units were much fewer than those of CFU-E – about $^1/_{10}$ – the proliferative capacity was much greater, with up to 5,000 cells per colony. The properties of BFU-E and the CFU-E, summarized in Table I, are quite different not only in terms of their numbers, but in other parameters as well. First, the erythropoietin requirements necessary for optimal colony formation in culture

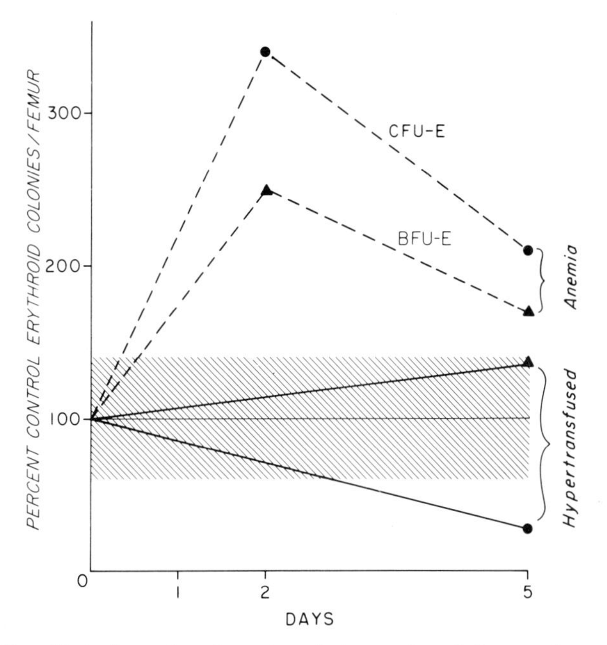


Fig. 6. The effect of anemia or hypertransfusion on the relative numbers of CFU-E (•) and BFU-E (▲) per mouse femur. Hypertransfusion was achieved by the intraperitoneal administration of washed packed homologous red cells (1.0 ml × 2 injections) given at −3 and −2 days of the experiment. Anemia was produced by bleeding of one-half the calculated blood volume of the animal from the retro-orbital plexus on day 0.

are different, approximately 10 times higher for the BFU-E (Iscove and Sieber, 1975). Second, with mouse marrow the time of appearance in culture is significantly different. Whereas the peak number of small erythroid colonies is reached at about 48 hours and then declines rapidly, erythroid bursts appear in optimal numbers between days 7 and 10 in culture. These two classes of erythroid progenitors have different cell cycle characteristics and can be

**Table 1**

*Comparison of Mouse Erythroid Colony Forming Units and Bursts*

| | CFU-E | Bursts |
|---|---|---|
| Frequency | 300–400/$10^5$ cells | 20–40/$10^5$ cells |
| Cells per Colony | 8–64 | Up to 5000 |
| Erythropoietin Requirement | 0.2–0.5 u/ml | 2.5–5.0 u/ml |
| Maximum Growth | 2 days | 7–10 days |
| Percent in S Phase | 75–80% | 40–45% |
| Sedimentation Velocity | 7.5 mm/hr | 4.6 mm/hr |
| Relationship to CFU-S | Relatively distant | Relatively close |
| Hypertransfusion | Reduced number | No effect |

separated physically, largely on the basis of size, by velocity sedimentation analysis (Axelrad *et al.*, 1973; Heath *et al.*, 1976). Using tritiated thymidine suicide kinetics, it has been shown that about 40% of burst-forming units are in DNA synthesis in resting marrow, as opposed to the 75–80% for erythroid colony-forming units. By velocity sedimentation analysis, CFU-E are found to be both larger in number and in size and sediment with a modal velocity of 7.5 mm/hr compared to BFU-E which sediment at about 4.6 mm/hr (Fig. 7).

Recently, Gregory (1976) has analyzed the relatedness of CFU-S to BFU-E and a different pattern was observed than that which had been seen previously for CFU-E. These studies demonstrate a close proportional relationship between the numbers of CFU-S and BFU-E in individual spleen colonies (Fig. 5B). Thus, in a manner analogous to the studies of Wu *et al.* (1968) relating CFU-C to CFU-S, it appears that the burst-forming unit is removed by fewer randomizing events from the pluripotent stem cell than is the CFU-E.

Finally, information has been gained recently about the regulation of the number of BFU-E. Unlike CFU-E, which depend upon erythropoietin for *in vivo* maintenance of their numbers, the presence of BFU-E does not require erythropoietin. Consequently, hypertransfusion has no demonstrable effect on the number of burst-forming units in the femurs of appropriately manipulated animals (Fig. 6). However, in response to an anemic stimulus, the population of burst-forming units increases in size (Fig. 6), and many become activated into cycle as demonstrated by tritiated thymidine suicide values which rise to 60% or

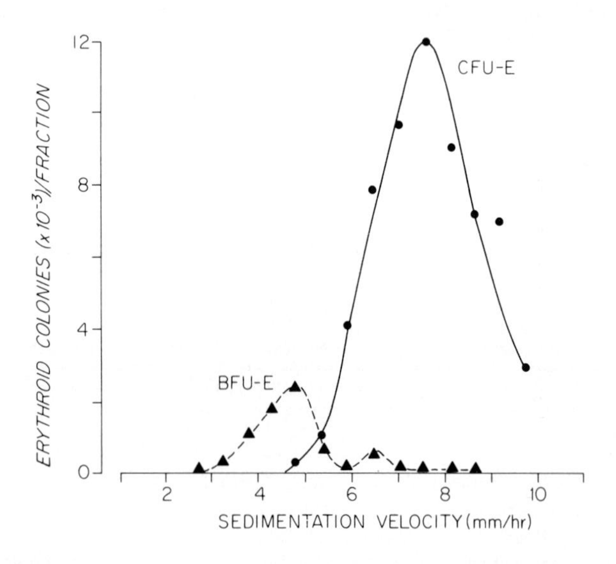


Fig. 7. The velocity sedimentation gradient profile of murine marrow CFU-E (●) and BFU-E (▲).

more (Adamson, unpublished observations). A portion of these observations confirm those recently published by Axelrad and colleagues (Axelrad *et al.*, 1973; Heath *et al.* 1976). These results are at some variance with those of Hara and Ogawa (1976, 1977) and Iscove (1976) who found no increase in the number of murine BFU-E in response to acute anemia. These differences may be based on species of animal chosen or degree of erythropoietic stress imposed. However, Iscove (1976) has demonstrated an increase in BFU-E and their activation into cycle in animals recovering from radiation-induced marrow damage. Thus, similar to granulocytic progenitors, the number of erythroid bursts may increase and the cycle characteristics of the responsive cells may change as a result of physiologic stimulation. While BFU-E exist in small numbers in normal marrow, apparently independently of erythropoietin, they bear an important marker of differentiation – the capacity to respond to erythropoietin by proliferation *in vitro*.

More recently, a third population of erythropoietin-responsive cells has been postulated by Gregory (1976) to lie intermediate in both number, growth characteristics, and sensitivity to erythropoietin between the BFU-E and CFU-E. Such complexities in the properties and interrelationships of the erythropoietin-responsive populations have been unrecognized previously.

The dissection of the various erythropoietin-responsive cell classes provides a tool for the study of differentiation control of expression of genetic information. For example, it may be possible to localize the capability of expressing certain genetic information to distinct levels of differentiation as marked by specific populations of erythropoietin-responsive cells. This concept has already been demonstrated, in part, for the *in vitro* activation of the structural gene which regulates the synthesis of a novel globin chain – $\beta^c$ – in cells from certain sheep and goats. The "switching on" of $\beta^c$ globin synthesis occurs in direct response to erythropoietin *in vivo* and was also demonstrated in suspension culture by Adamson and Stamatoyannopoulos, (1973). The capacity for this "switch" is not found in all erythoid progenitors and has been localized physically to erythroid colony-forming units (Anderson *et al.*, 1975; Barker *et al.*, 1975). However, no distinction was made between the various stages of differentiation of cells capable of erythroid colony formation. Thus, it may be that the ability to synthesize $\beta^c$ globin is confined to progeny of earlier colony-forming units and is lost in more differentiated colony-forming units.

Recent studies of factors influencing human gamma chain synthesis in erythroid colonies suggests that this problem also could be analyzed in terms of the differentiative level at which modulation of fetal hemoglobin synthesis may be achieved (Papayannopoulou *et al.* 1976). If such an experimental approach is applicable to the analysis of individual colonies, the findings would provide information on the state of differentiation of the colony-forming population as well as confirm, by an independent marker, the proposed classification. The interpretation of these observations arises logically from the clonal origin of such erythroid colonies and demonstrates how these techniques can provide the opportunity to study specific information programmed into an individual cell and then amplified by *in vitro* growth.

As a result of these considerations, a hierarchy of erythroid progenitors may be put forth (Fig. 8). The earliest differentiation step, whose control mechanism is unknown, is represented by the transition from the pluripotent stem cell to the erythroid burst-forming unit (McCulloch *et al.,* 1974). This transition does not appear to be dependent upon erythropoietin, for if erythropoietin were the primary regulator, hypertransfusion should suppress the number of these cells; such suppression is not seen. However, the numbers of BFU-E may increase in response to hematopoietic stress. Although the burst-forming unit is capable of responding to erythropoietin *in vitro*, it cannot be rigorously demonstrated that the increase in its numbers *in vivo* is in response to erythropoietin. The transition between pluripotent stem cell and BFU-E may be regulated by control factors, operating over short distances, which arise in response to changes in compartment size. Finally, the possibility of limited self-replication by an early

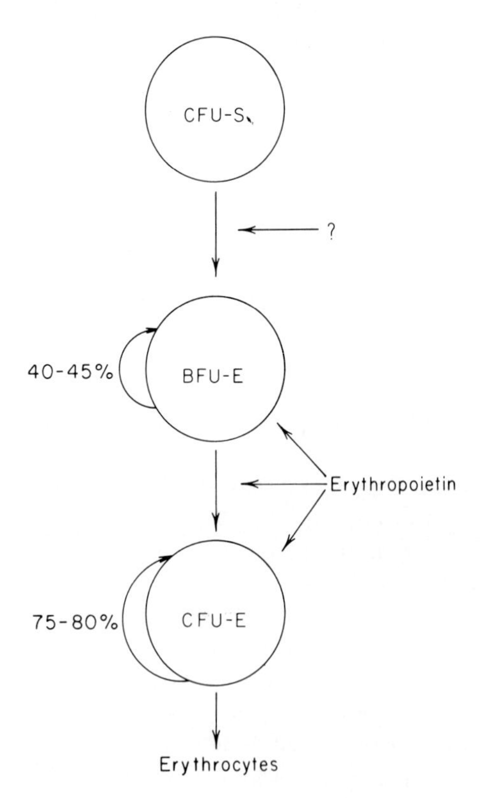


Fig. 8. A proposed hierarchy of erythroid progenitors.

erythroid population, such as BFU-E, suggests a degree of elasticity in the amplification steps involved in erythropoietic regulation (Reissmann and Samorapoompichit, 1970; Lajtha *et al.*, 1971).

The next differentiation steps, leading to the CFU-E, appear to be under the control of erythropoietin. Hypertransfusion, which suppresses endogenous levels of the hormone, reduces the numbers of CFU-E. Erythropoietic stresses, such as hypoxia or anemia which increase endogenous erythropoietin, as well as the injection of erythropoietin itself, increase the number of CFU-E. Furthermore, possible alterations in sensitivity to erythropoietin which may accompany different states of erythropoietic demand suggest yet another type of modulatory control (Gregory *et al.*, 1974; Till *et al.*, 1975). The pattern emerging from these analyses implies that decreasing proliferative capacity of erythroid progenitors is associated with progressively increasing sensitivity to erythropoietin as cells move from a position close to the pluripotent stem cell through several differentiation steps (Gregory, 1976).

## IV. MODULATORS OF ERYTHROPOIESIS

The concept of receptor modification or modulation of hormone action can be extended to regulators other than erythropoietin. In keeping with previous studies of the possible importance of small molecules in processes regulating cell differentiation and proliferation (Holley, 1975; Tomkins, 1975), we have studied the effects of selected classes of compounds on erythroid colony formation. Because the growth and function of a number of cell systems are regulated by mechanisms linked to adenyl cyclase (Jost and Rickenberg, 1971; Sutherland, 1972), the possibility that such mechanisms could influence erythropoiesis has been examined. A number of previous studies have alleged both positive and negative influences of cyclic nucleotides and related compounds on hematopoiesis. Byron (1971; 1975) has clearly demonstrated that both cyclic nucleotides and compounds linked to cell function through adenyl cyclase are able to activate DNA synthesis in mouse pluripotent stem cells. In contrast, the work by Krantz and co-workers has indicated that cyclic adenosine nucleotides have no positive effect on *in vitro* hemoglobin synthesis (Graber *et al.*, 1972). However, these studies may have been limited both by the species and assay system selected (Brown and Adamson, 1977a).

To determine whether cyclic nucleotides could influence erythropoiesis directly, we have examined the effect on erythroid colony growth of a number of compounds known to participate in the adenyl cyclase system. These compounds have included various cyclic nucleotides, phosphodiesterase inhibitors, and adenyl cyclase stimulators. Marrow cells from a variety of sources were examined, and canine and human marrow cells were found to be most responsive. While incapable of initiating erythroid colony formation by itself, dibutyryl cyclic AMP significantly enhanced colony numbers, and the effect was reproduced by the phosphodiesterase inhibitor, RO-20-1724 (Sheppard and Wiggan, 1971). Maximal effects with each compound were seen at a concentration of $10^{-5}$ M. To determine the specificity of the response, cyclic AMP and its mono- and dibutyryl derivatives were examined along with other non-cyclic adenosine derivatives, including adenosine, AMP, ADP, and ATP. Enhanced erythroid colony formation was specifically related to adenosine cyclic nucleotides (Fig. 9). In addition, cyclic guanosine monophosphate and cyclic cytosine monophosphate, two other cyclic nucleotides reported to play a positive role in cell proliferation, were inactive.

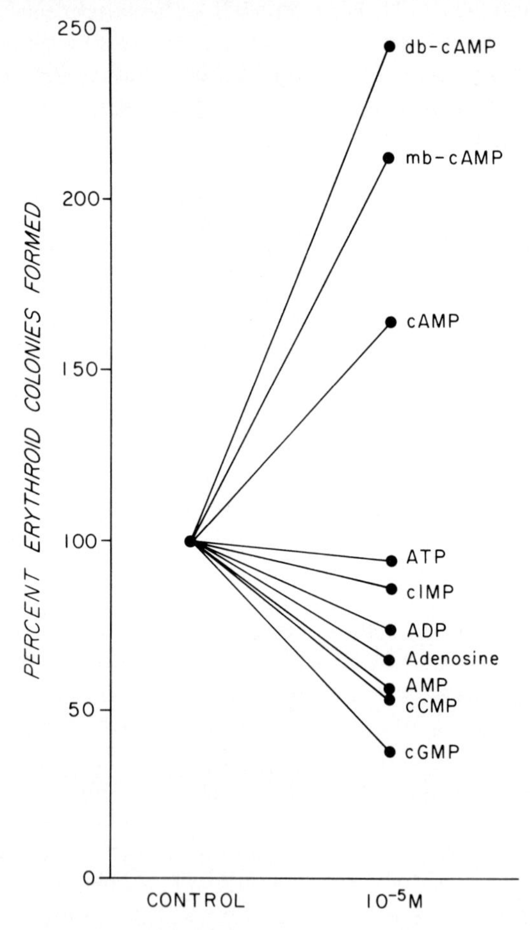


Fig. 9. The influence of cyclic nucleotides and adenosine congeners on erythroid colony formation in cultures of canine marrow. Compounds tested included dibutyryl and monobutyryl cyclic AMP (db-cAMP and mb-cAMP, respectively), cyclic inosine monophosphate (cIMP), cyclic cytosine monophosphate (cCMP) and cyclic guanosine monophosphate (cGMP).

Further support for the concept that cyclic nucleotides influence erythropoiesis was obtained from studies of adenyl cyclase stimulators. Purified cholera enterotoxin, which interacts with surface receptors to activate membrane-bound adenyl cyclase (Greenough *et al.*, 1970), markedly enhanced erythroid colony growth. Adrenergic agonists also bind to surface receptors and thereby stimulate adenyl cyclase in a wide variety of tissues (Sutherland, 1972). Membrane receptors for adrenergic compounds are classified by their binding properties as $\alpha$, $\beta_1$, and $\beta_2$ receptors (Ahlquist, 1948; Lands *et al.*, 1967). When isoproterenol, an agonist which interacts with both types of $\beta$ receptors (Lands *et al.*, 1967) was added to cultures, enhanced erythroid colony growth was seen (Fig. 10). Optimal effects with the agonist were observed at $10^{-7}$M. The isoproterenol-induced colony growth was completely inhibited, however, by propranolol, a $\beta$-adrenergic blocker, at $10^{-8}$M (Fig. 11). In contrast, propranolol does not

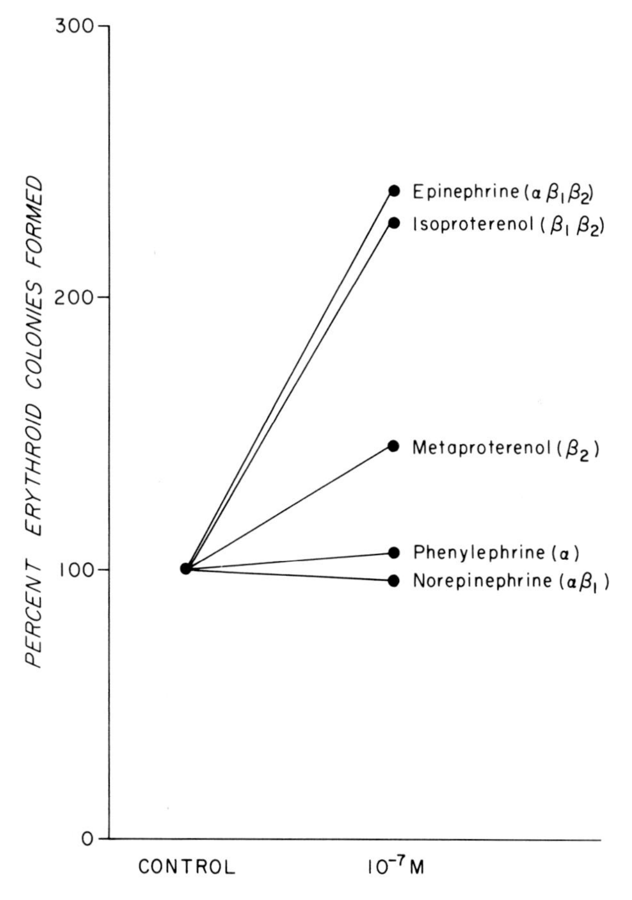


Fig. 10. The effect of various catecholamines on canine erythroid colony numbers. The relative receptor specificity of each agent is indicated in parenthesis.

interfere with erythropoietin-dependent colony growth. Since the blocking effect of propranolol is due to competitive binding to $\beta$ receptors, it would seem that the receptors for erythropoietin and $\beta$ agonists are distinct from one another. When the blocking effect of propranolol was tested in cultures containing dibutyryl cyclic AMP, the phosphodiesterase inhibitor RO-20-1724, or cholera enterotoxin, no significant inhibition of colony growth occurred. Thus, these other agents probably act either at cytoplasmic sites or at cell surface locations which are separate from the $\beta$ receptor. Subspecificity of the $\beta$ receptor was determined by testing compounds having different relative affinities for the subtypes of receptors. Neither the $\alpha$ agonist, phenylephrine, nor the primarily $\alpha$ and $\beta_1$ agonist, norepinephrine, enhanced colony growth. In contrast, agents with $\beta_2$ effects, such as epinephrine and isoproterenol, and the relatively selective $\beta_2$ agonist, metaproterenol, all enhanced erythroid colony formation (Fig. 10). The selectivity of the $\beta$ receptor was confirmed by testing equimolar concentrations of various adrenergic blocking agents. The $\alpha$ blocker,

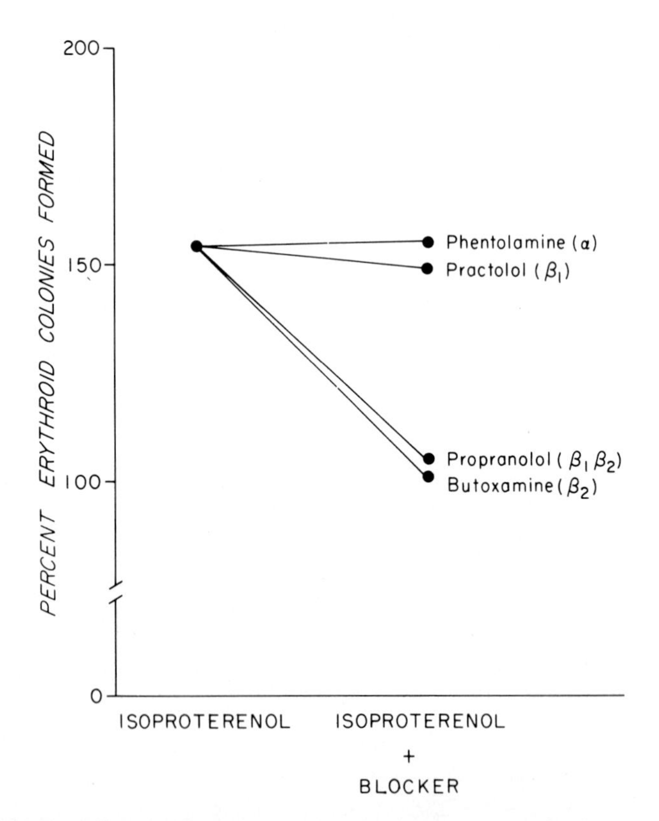


Fig. 11. The influence of equimolar amounts of various catecholamine antagonists on $10^{-7}$ M isoproterenol-enhanced canine erythroid colony formation. The relative specificity of each antagonist is indicated in parentheses.

phentolamine, and $\beta_1$ blocker, practolol, failed to inhibit isoproterenol-induced colony growth (Fig. 11). The $\beta_2$ blocker, butoxamine, however, was as effective as propranolol in inhibiting the formation of these colonies. Once again, no inhibitory effect on erythropoietin-dependent colony growth was seen. Thus, the adrenergic receptor is distinct from that for erythropoietin and appears to have $\beta_2$ specificity (Brown and Adamson, 1977b).

Since erythroid colonies require the presence of erythropoietin for their growth, it is possible that such compounds and mechanisms linked to adenyl cyclase and cyclic AMP play a modulatory role in the *in vitro* response to erythropoietin. These findings suggest a mechanism by which other hormones or regulatory molecules may modulate erythropoiesis.

More recently, work by others has demonstrated that leukocyte conditioned medium (Aye, 1975), selected steroids (Golde *et al.,* 1976b; Singer and Adamson, 1976a) and thyroid hormones (Golde *et al.,* 1976a) are capable of enhancing erythroid colony formation. Again, these agents do not initiate colony formation by themselves but, rather, act in a permissive or modulatory fashion. The eventual meaning of these observations for *in vivo* regulation of erythropoiesis remains to be defined. Nevertheless, these studies demonstrate that other intracellular events may have a profound effect on the expression of a differentiating system in response to hormonal regulation.

In summary, the picture of the regulation of erythropoietic differentiation and proliferation is becoming increasingly detailed and complex. It is a system in which erythropoietic development may be amplified or muted by a number of factors. The system appears to have a great deal of elasticity, and its eventual expression will depend upon the proliferative potential of the various classes of responsive cells as well as possible modulation by other factors of the interaction between erythropoietin and its target cells.

## ACKNOWLEDEMENTS

The authors would like to acknowledge Dr. Connie Gregory for providing experimental results prior to publication, Dr. Richard Rifkind for generously reviewing portions of the manuscript, and Ms. Nancy Lin and Mrs. Christine Reichgott for expert technical assistance. This work was supported in part by grant HL-06242 and AM-19410 of the NIH, DHEW, and designated research funds of the Veterans Administration. Dr. J. E. Brown is a Research and Education Associate of the Veterans Administration and Dr. J. W. Adamson is the recipient of a Research Career Development Award (AM-70222) of the NIAMDD, NIH. A portion of this review was completed while Dr. Adamson was supported by a Faculty Scholar Award for 1976–1977 from the Josiah Macy, Jr. Foundation.

## REFERENCES

Adamson, J. W. and Stamatoyannopoulos, G. (1973). *Science* **180,** 310–312.

Ahlquist, R. P. (1948). *Am. J. Physiol.* **153,** 586–600.

Anderson, W. F., Barker, J. E., Elson, N. A., Merrick, W. C., Steggles, A. W., Wilson, G. N., Kantor, J. A. and Nienhuis, A. W. (1975). *J. Cell. Physiol.* **85,** 477–494.

Axelrad, A. A., McLeod, D. L., Shreeve, M. M. and Heath, D. S. (1973). *In* "Hemopoiesis in Culture" (W. A. Robinson, ed.), pp. 226–234. HEW (NIH) publication 74–205.
Aye, M. T. (1975). *Blood* **46,** 1022 (a).
Barker, J. E., Anderson, W. F. Nienhuis, A. W. (1975). *J. Cell Biol.* **64,** 515–527.
Becker, A. J., McCulloch, E. A. and Till, J. E. (1963). *Nature* **197,** 452–454.
Brown, J. E. and Adamson, J. W. (1977a). *Brit. J. Haemat.* **35,** 193–208.
Brown, J. E. and Adamson, J. W. (1977b). *J. Clin. Invest.* in press.
Byron, J. W. (1971). *Nature* **234,** 39–40.
Byron, J. W. (1972). *Exp. Cell Res.* **71,** 228–232.
Byron, J. W. (1975). *Exp. Hematol.* **3,** 44–53.
Chang, S. C.-S. and Goldwasser, E. (1973). *Develop. Biol.* **34,** 246–254.
Chang, S. C.-S., Sikkema, D. and Goldwasser, E. (1974). *Biochem. Biophys. Res. Comm.* **57,** 399–405.
Chui, D. H. K., Djaldetti, M., Marks, P. A. and Rifkind, R. A. (1971). *J. Cell Biol.* **51,** 585–595.
Conkie, D., Kleiman, L., Harrison, P. R. and Paul, J. (1975). *Exp. Cell Res.* **93,** 315–324.
Cooper, M. C., Levey, J., Cantor, L. N., Marks, P. A. and Rifkind, R. A. (1974). *Proc. Nat. Acad. Sci. U.S.A.* **71,** 1677–1680.
Djaldetti, M., Preisler, H., Marks, P. A. and Rifkind, R. A. (1972). *J. Biol. Chem.* **247,** 731–735.
Glass, J., Lavidor, L. M. and Robinson, S. H. (1975). *J. Cell Biol.* **65,** 298–308.
Golde, D. W., Bersch, N., Chopra, I. J. and Cline, M. J. (1976a). *Clin. Res.* **24,** 309 (a).
Golde, D. W., Bersch, N. and Cline, M. J. (1976b). *J. Clin. Invest.* **57,** 57–62.
Goldwasser, E. (1975). *Fed. Proc.* **34,** 2285–2292.
Goldwasser, E. and Kung, C. K.-H. (1972). *J. Biol Chem.* **247,** 5159–5160.
Goldwasser, E., Kung, C. K.-H. and Eliason, J. (1974). *J. Biol. Chem.* **249,** 4202–4206.
Gordon, A. S., Cooper, G. W. and Zanjani, E. D. (1967). *Semin. Hematol.* **4,** 337–359.
Graber, S. E., Carillo, M. and Krantz, S. B. (1972). *Proc. Soc. Exp. Biol. Med.* **141,** 206–210.
Greenough, W. B. III, Pierce, N. F. and Vaughan, M. (1970). *J. Infect. Dis.* **121,** Suppl., S 111–113.
Gregory, C. J. (1976). *J. Cell Physiol.* **89,** 289–301.
Gregory, C. J., McCulloch, E. A. and Till, J. E. (1973). *J. Cell. Physiol.* **81,** 411–420.
Gregory, C. J., Tepperman, A. D., McCulloch, E. A. and Till, J. E. (1974). *J. Cell. Physiol.* **84,** 1–12.
Gross, M. and Goldwasser, E. (1969). *Biochem.* **8,** 1795–1805.
Gross, M. and Goldwasser, E. (1970). *J. Biol. Chem.* **245,** 1632–1636.
Gross, M. and Goldwasser, E. (1972). *Biochim. Biophys. Acta* **287,** 514–519.
Hara, H. and Ogawa, M. (1976). *Amer. J. Hematol.* **1,** 453–458.
Hara, H. and Ogawa, M. (1977). *Exp. Hematol.* **5,** 141–148.
Heath, D. S., Axelrad, A. A., McLeod, D. L. and Shreeve, M. M. (1976). *Blood* **47,**777–792.
Holly, R. W. (1975). *Nature* **258,** 487–490.
Iscove, N. N. (1976). XVI International Congress of Hematology, Kyoto, Japan. Abstracts of Symposia; page 12.
Iscove, N. N. and Sieber, F. (1975). *Exp. Hematol.* **3,** 32–43.
Jost, J. P. and Rickenberg, H. V. (1971). *Ann. Rev. Biochem.* **40,** 741–774.
Krantz, S. B., Gallien-Lartigue, O. and Goldwasser, E. (1963). *J. Biol. Chem.* **238,** 4085–4090.
Krantz, S. B. and Goldwasser, E. (1965). *Biochim. Biophys. Acta* **103,** 325–332.
Krantz, S. B. and Jacobson, L. O. (1970). "Erythropoietin and the Regulation of Erythropoiesis". The University of Chicago Press, Chicago, IL.

Lands, A. M., Arnold, A., McAuliff, J. P., Luduena, F. P. and Brown, T. G. Jr. (1967). *Nature* **214**, 597–598.

Lajtha, L. G., Gilbert, C. W. and Guzman, E. (1971). *Brit. J. Haemat.* **20**, 343–354.

Lowy, P. H., Keighley, G. and Borsook, H. (1960). *Nature* **185**, 102–103.

Maniatis, G. M., Rifkind, R. A., Bank, A. and Marks, P. A. (1973). *Proc. Nat. Acad. Sci. U.S.A.* **70**, 3189–3194.

Marks, P. A. and Rifkind, R. A. (1972). *Science* **175**, 955–961.

McLeod, D. L., Shreeve, M. M. and Axelrad, A. A. (1974). *Blood* **44**, 517–534.

McCulloch, E. A., Mak, T. W., Price, G. B. and Till, J. E. (1974). *Biochim. Biophys. Acta* **355**, 260–299.

Morell, A. G., Irvine, R. A., Sternlieb, I. and Scheinberg H. (1968). *J. Biol. Chem.* **243**, 155–159.

Nicol, A. G., Conkie, D., Lanyon, W. G., Drewienkiewicz, C. E., Williamson, R. and Paul, J. (1972). *Biochim. Biophys. Acta* **277**, 342–353.

Papayannopoulou, T., Brice, M. and Stamatoyannopoulos, G. (1976). *Proc. Nat. Acad. Sci. U.S.A.* **73**, 2033–2037.

Papayannopoulou, T. and Finch, C. A. (1972). *J. Clin. Invest.* **51**, 1179–1185.

Piantadosi, C. A., Dickerman, H. W. and Spivak, J. L. (1976). *J. Clin. Invest.* **57**, 20–26.

Prchal, J., Adamson, J. W., Fialkow, P. J. and Steinmann, L. (1976). *J. Cell. Physiol.* **89**, 489–492.

Ramirez, F., Gambino, R., Maniatis, G. M., Rifkind, R. A., Marks, P. A. and Bank, A. (1975). *J. Biol. Chem.* **250**, 6054–6058.

Reissmann, K. R. and Samorapoompichit, S. (1970). *Blood* **36**, 287–296.

Rifkind, R. A. and Marks, P. A. (1975). *Blood Cells* **1**, 417–428.

Sheppard, H. and Wiggan, G. (1971). *Biochem. Parmacol.* **20**, 2128–2130.

Singer, J. W. and Adamson, J. W. (1976a). *J. Cell. Physiol.* **88**, 127–134.

Singer, J. W. and Adamson, J. W. (1976b). *Blood.* **48**, 855–864.

Spivak, J. L. (1976). *Blood* **47**, 581–592.

Stephenson, J. R. and Axelrad, A. A. (1971). *Blood* **37**, 417–427.

Stephenson, J. R., Axelrad, A. A., McLeod, D. L. and Shreeve, M. M. (1971). *Proc. Nat. Acad. Sci. U.S.A.* **68**, 1542–1546.

Stohlman, F. (1970). *In* "Regulation of Hematopoiesis" (A. S. Gordon, ed.), Vol. 1, pp. 317–326. Appleton, Century, Crofts, New York, NY.

Sutherland, E. W. (1972). *Science* **172**, 401–408.

Terada, M., Cantor, L., Metafora, S., Rifkind, R. A., Bank, A. and Marks, P. A. (1972). *Proc. Nat. Acad. Sci. U.S.A.* **69**, 3575–3579.

Till, J. E. and McCulloch, E. A. (1961). *Radiat. Res.* **14**, 213–222.

Till, J. E., Price, G. B., Mak, T. W. and McCulloch, E. A. (1975). *Fed. Proc.* **34**, 2279–2284.

Tomkins, G. M. (1975). *Science* **189**, 760–763.

Weintraub, A. H., Gordon, A. S., Becher, E. L., Camiscoli, J. F. and Contrera, J. F. (1964). *Am. J. Physiol.* **207**, 523–529.

Wu, A. M., Siminovitch, L., Till, J. E. and McCulloch, E. A. (1968). *Proc. Nat. Acad. Sci. U.S.A.* **59**, 1209–1215.

# The Role of Factors in Differentiation of Non-Erythroid Cells

Malcolm A. S. Moore

*Sloan-Kettering Institute for Cancer Research*

## I. INTRODUCTION

The hemopoietic system, exhibiting as it does a high potential for proliferation and, of course, differentiation, is ideally suited for studies on the characteristics of both specific regulatory macromolecules and on less specific modulating agents. Progress in this area has been facilitated by the development of specific and quantitative *in vitro* assays for various committed cells. Semisolid cloning systems have been developed for detection of granulocyte-macrophage committed stem cells (CFU-c) as well as eosinophil, megakaryocyte and erythroid precursors (Bradley and Metcalf, 1966; Metcalf *et al.*, 1974; Metcalf *et al.*, 1975a; Stephenson *et al.*, 1971). Functional subpopulations of more differentiated hemopoietic cell types such as macrophages, T lymphocytes and B lymphocytes may also be cloned in agar culture with the latter lymphocyte assays requiring the presence of an appropriate mitogenic stimulus (Lin and Stewart, 1966; Fibach *et al.*, 1976, Metcalf *et al.*, 1975b).

Granulocyte-macrophage colony formation in soft agar or methylcellulose exhibits an absolute requirement for what has been termed colony stimulating factor (CSF) or colony stimulating activity (CSA). The existence of CSF was first demonstrated in studies on the growth of mouse bone marrow and spleen cells, in which colonies developed provided a feeder layer of cells was present (Bradley and Metcalf, 1966). Subsequent work showed that the feeder layer cells need not be capable of proliferating and indeed could be substituted for by medium harvested from liquid cultures of feeder layer cells. Other studies showed that small volumes of mouse or human serum or of human urine could serve as a suitable stimulus for *in vitro* colony formation (Metcalf and Moore, 1971).

The action of CSF is complex and is not simply an inducer of colony formation, since it is required continuously for proliferation and/or differentiation throughout the period of culture. Cell separation procedures and single cell manipulation studies have shown that the CFU-c in primates is a transitional lymphocyte-like mononuclear cell comprising 0.1–0.5% of the marrow-nucleated cell population (Moore *et al.*, 1972). The majority of CFU-c appear to be bi-potential generating colonies containing both neutrophilic granulocytes and monocyte-macrophages. In murine bone marrow minor populations of CFU-c appear to be restricted to neutrophilic differentiation or to eosinophilic differentiation and, in the latter case, an eosinophil colony stimulating activity elaborated by mitogen-stimulated lymphocytes appears to be a prerequisite for colony development (Metcalf *et al.*, 1975a).

While the action of CSF appears to be restricted to the granulocyte-macrophage series, it is not limited to stimulation of immature myeloid precursors. Macrophages present in thioglycollate-stimulated mouse peritoneal exudate, when cultured in agar for 2–4 weeks in the presence of CSF, commence proliferation to form colonies of 100–200 macrophages (Lin and Stewart, 1973; Stanley *et al.*, 1976). The peritoneal macrophage CFU (CFU-PM) can be distinguished from the CFU-c on the basis of its adherence properties, incidence (approximately 5% of activated peritoneal macrophages) and tissue distribution. In addition, CFU-PM do not commence proliferation for 10–14 days in contrast to the much earlier proliferation seen with CFU-c. This lag period is CSF-independent and cannot be shortened by increasing the CSF concentration. Both CFU-c and CFU-PM show a sigmoidal dose-responsiveness to CSF; however, a 7-fold greater concentration of stimulus is required to produce a plateau incidence of CFU-PM than of CFU-c (Kurland and Moore, 1976). In this context, CSF can be considered as a macrophage growth factor (MGF), and it appears that the L-cell-conditioned medium factor described by Defendi, which activates 30–50% of macrophages into DNA synthesis, is identical to L-cell-derived CSF (Mauel and Defendi, 1971; Stanley *et al.*, 1976).

## II. SOURCES OF CSF

CSF was originally detected as a product elaborated by certain cells *in vitro*, e.g., neonatal mouse kidney cells, lymphoma cells or fibroblasts. Subsequent studies showed that CSF was extractable from a wide variety of murine organs of which lung, kidney, submaxillary gland, heart and pregnant mouse uterus were particularly rich sources (Sheridan and Stanley, 1971; Metcalf *et al.*, 1974). The intravenous injection of endotoxin promotes a rapid increase of CSF extractable from various tissues, particularly the lung, and causes a dramatic increase in serum CSF levels which are maximum within 3–6 hours (Sheridan and Metcalf, 1973, 1974). CSF molecules extracted from the different organs, with and without endotoxin stimulation, are heterogenous with respect to charge as revealed by electrophoretic mobility and by binding to calcium phosphate gels.

In contrast to the ubiquity of murine sources of CSF, human sources of CSF active against human CFU-c are limited. Adaptation of the culture system to support human granulocytic colony formation involved the use of a feeder layer containing 1 $\times$ $10^6$ normal human peripheral blood leukocytes (Pike and Robinson, 1970). Cell separation using buoyant density and adherence techniques identified the blood monocyte as the source of CSF in the leukocyte feeder layers (Moore and Williams, 1972; Moore *et al.*, 1973). Subsequent studies have shown that the phagocytic mononuclear cell population including the fixed tissue macrophages of spleen and marrow, as well as Kupffer cells, alveolar macrophages and pleural and peritoneal macrophages are a major source of human CSF (Cline *et al.*, 1974; Moore, 1974). The presence of CSF producing cells in hemopoietic tissue accounts for the phenomenon of spontaneous colony formation observed when marrow cells from a variety of species are cultured at high cell concentrations in the absence of exogenous CSF (Moore and Williams, 1972; Moore *et al.*, 1973). The threshold for spontaneous colony formation may be as low as $10^4$ cells per ml (monkey, guinea pig), $10^5$ cells per ml (human), or as high as $10^6$ cells per ml in the case of mouse marrow culture, which shows the least evidence of spontaneous colony formation. The degree of spontaneous colony formation may reflect the relative incidence and activity of CSF producing cells in the marrow of various species or, alternatively, variation in the incidence of hemopoietic cells (e.g., granulocytes), which can suppress CSF production. In this context, mouse bone marrow-conditioned medium is a relatively poor source of CSF, whereas bone devoid of hemopoietic cells is a very active source (Chan and Metcalf, 1972). This suggests that stromal cells may be active in local production of CSF or that the osteoclasts, components of the phagocytic mononuclear series, may be responsible. Mitogen-stimulated human lymphocytes also produce human active CSF as do human embryonic kidney cells (Ruscetti and Chervenick, 1975).

The species specificity of CSF is such that human CSF can stimulate colony formation by bone marrow cells from most mammalian species. The reverse is not, however, true since human marrow cultures are not stimulated by sources of CSF from species other than human or monkey (Moore *et al.*, 1973). It is unfortunate that the most readily obtainable and highly purified CSF prepared from human urine does not stimulate human colony formation despite retaining full activity in stimulating mouse colony formation. This rather paradoxical observation may indicate that some alteration of the CSF molecule has occurred upon excretion with loss of activity or possibly some cofactor is absent which may be required for human colony stimulation. Despite the functional differences between human leukocyte and human urine CSF, antisera can be prepared against the latter, which can inhibit human leukocyte CSF (Metcalf, 1974).

## III. BIOCHEMICAL CHARACTERIZATION OF CSF

CSFs vary with respect to size, and mouse CSFs have apparent sedimentation coefficients ranging from 2.0–6.0 S (Sheridan and Stanley, 1971).

The size appears to be dependent not only on source, but also according to whether a tissue injury stimulus, such as endotoxin, has been used. In the hours following endotoxin injection, CSF levels increase, but the CSF molecular size appears to fall (Sheridan and Metcalf, 1974). The approximately 2.0 S CSF present in endotoxin-treated mouse lung-conditioned medium probably represents the most extreme example of this phenomenon. This latter 20,000 molecular weight factor has been extensively purified to a point where it has $1 \times 10^{10}$ units of CSF per ml of protein (1 unit being the amount of CSF required to stimulate the formation of 1 colony in standard cultures of 75,000 C57BL marrow) (Sheridan and Metcalf, 1974). Highly purified CSFs have also been obtained from L-cell-conditioned medium and human urine, being respectively, 60,000 and 45,000 molecular weight sialic acid-containing glycoproteins (Table I) (Stanley *et al.,* 1976; Stanley and Metcalf, 1972). Four molecular species of CSF have been characterized in human leukocyte-conditioned medium, 3 being nondialyzable with apparent molecular weights of 15,000, 35,000 and 100,000 as determined by gel filtration and sucrose gradient centrifugation (Price *et al.,* 1975). These CSFs could also be isolated from leukocyte cell membranes and were active against both mouse and human CFU-c. A fourth species of CSF active only on human marrow was found to be less than 1,300 molecular weight on the basis of gel filtration on Sephadex G25 (Price *et al.,* 1973). This low molecular weight CSF was trypsin-sensitive and extractable in organic solvents. Although the true identity of this CSF is not yet known, it is possible that it consists of one or more peptides having cyclic structures and which are known to be soluble in organic solvents.

Gel filtration of medium conditioned by human monocytes or PHA-stimulated lymphocytes has revealed two main regions of CSF activity; one with an apparent molecular weight of 25,000–35,000, which is active against both

**TABLE 1**

*Characteristics of Various CSF's*

| Source of CSF | Apparent Mol. Wt. | Activity Mouse | Activity Human | Reference |
|---|---|---|---|---|
| Human Leukocyte CM | 1,300 | – | + | Price *et al.*, 1973 |
| Human Leukocyte CM | 15,000 | + | + | Price *et al.*, 1975 |
| Human Leukocyte CM | 35,000 | + | + | Price *et al.*, 1975 |
| Human Leukocyte CM | 100,000 | + | + | Price *et al.*, 1975 |
| Human Urine | 45,000 | + | – | Stanley & Metcalf, 1972 |
| Endotoxin Mouse Lung CM | 20,000 | + | – | Sheridan & Metcalf, 1974 |
| L-Cell CM | 60,000 | + | – | Stanley *et al.*, 1976 |
| Human Macrophage CM | 35,000 | + | + | Shah *et al.*, 1976 |
| Human Macrophage CM | 150,000 | + | – | Shah *et al.*, 1976 |
| PHA-stimulated Human Lymphocyte CM | 25,000 | + | + | Shah *et al.*, 1976 |
| PHA-stimulated Human Lymphocyte CM | 150,000 | + | – | Shah *et al.*, 1976 |

mouse and human marrow, and a CSF of 150,000 apparent molecular weight, which is active against mouse bone marrow only (Shah *et al.,* 1976). This high molecular weight CSF has functional similarities to human urine CSF, which also has an apparent high molecular weight on Sephadex G150 when unpurified urine concentrates are used. The gel filtration characteristics of these CSFs suggest that they are probably bound to other proteins or associated with a nonprotein component such as lipid or carbohydrate.

The heterogeneity of the various CSFs, even those that have been highly purified, suggests that they may be a family of factors based on multiples of a monomeric subunit with additional heterogeneity superimposed by variations in sialic acid content. An alternative possibility is that a small molecule, such as the 1,300 molecular weight CSF, may be bound to a variety of glycoproteins present in the cell membrane.

## IV. REGULATION OF CSF PRODUCTION

The role of CSF in recruiting monocytes and macrophages from the CFU-c compartment, as well as in promoting local macrophage proliferation, indicates that a positive feedback control is operative based on CSF production by phagocytic mononuclear cells. The release of CSF by macrophages may be augmented by bacterial endotoxins, suggesting a mechanism whereby the presence of bacterial products can cause an increase in granulocyte and monocyte production by the bone marrow (Cline *et al.,* 1974). Indeed, the macrophage may act as a surveillance cell capable of mobilizing host defense against gram-negative bacteria. The direct production of CSF by activated immunocytes and the enhancement of macrophage CSF production by lymphokines also provides a mechanism for increasing granulopoiesis and macrophage production during certain immunological reactions.

Evidence that CSF is in fact an *in vivo* regulator of granulopoiesis and monocytopoiesis at present remains indirect; however, clear relationships between CSF levels and granulocyte production have been reported in mice injected with endotoxin and in dogs and man with cyclic neutropenia (Quesenberry *et al.,* 1972; Moore *et al.,* 1974; Dale *et al.,* 1972). It is probable, however, that the action of CSF on the CFU-c or fixed tissue macrophages may be a very local event, possibly even requiring the direct interaction of target cells in the hemopoietic tissue with the CSF producing cells. The inability to detect CSF in medium of long-term cultures of mouse bone marrow, in which sustained granulopoiesis is occurring, supports this view (Dexter *et al.,* 1976).

The production of CSF by monocytes and macrophages appears to be closely regulated by mechanisms designed to limit inappropriate overproduction of the factor. Mouse peritoneal macrophages in culture release CSF during the first 24 hours of incubation and if the medium is not changed, little further CSF production occurs (Moore and Kurland, 1976). In contrast, daily medium replacement promotes continuing, incremental production of CSF over many weeks, suggesting that CSF secretion is modulated by CSF in the external mileau or that some inhibitor of CSF production accumulates in the medium. Parallel studies with human monocytes have shown that in the first week of culture two types of CSF are produced, the main activity being a factor of apparent molecular weight 35,000 active against human and mouse CFU-c, and a high molecular weight factor active only against mouse CFU-c (Shah *et al.*, 1976). Media collected from monocyte-macrophage cultures at weekly intervals showed a rapid loss of human active CSF with increasing production of mouse active CSF, which was the sole CSF type secreted after the third week.

Mature granulocytes have been implicated in mediating a mitotic inhibitory negative feedback in which granulocyte chalone inhibits CFU-c proliferation (Rytomaa, 1973). Recent studies indicate a more indirect mechanism by which mature granulocytes may inhibit granulopoiesis. Removal of mature granulocytes from human marrow by density and adherence procedures produced a significant enhancement of spontaneous colony formation, whereas readdition of intact granulocytes, granulocyte extracts or granulocyte-conditioned medium suppressed colony formation (Broxmeyer *et al.*, 1976). Further analysis of this inhibition revealed that mature granulocytes contain and release a labile factor (colony inhibitory activity – CIA), which suppresses the synthesis or release of CSF by monocytes and macrophages (Broxmeyer *et al.*, 1977). CIA was not species-specific nor cytotoxic and had no inhibitory influence on CFU-c stimulated by an exogenous source of CSF. Such an inhibitory mechanism whereby CSF production is controlled by mature granulocyte turnover or population size may place an important role in maintaining steady-state granulocyte production, but can presumably be overridden in situations where a marked granulocytosis occurs; for example, following bacterial infection or when extensive macrophage proliferation occurs associated with an immune or inflammatory response. In this context, granulocyte-mediated inhibition of macrophage CSF production is not observed if the macrophages are activated with endotoxin and CIA does not inhibit mitogen-stimulated lymphocyte CSF production.

## V. THE ROLE OF PROSTAGLANDINS IN MODULATION OF HEMOPOIESIS

While attention has mainly been focused on specific inhibitory mechanisms involved in regulation of hemopoiesis, a further level of control modulation involves the role of E-type prostaglandins (PGE, $PGE_2$), a ubiquitous class of molecules which affect the function of a diverse number of organ systems via activation of adenylate cyclase. Prostaglandin $E_1$ at concentrations as low as $10^{-9}$–$10^{-10}$ M inhibits the proliferation of both CFU-c and peritoneal macrophage colonies; however, in both systems increasing the CSF concentration can partially counteract this inhibition, indicating an antagonistic interaction between a specific regulatory macromolecule and prostaglandin (Moore and Kurland, 1976; Kurland and Moore, 1976). Metcalf *et al.* (1975b) recently reported that colonies of B lymphocytes can be cloned from normal mouse spleen and lymph node using a modification of the soft agar culture system containing 2-mercaptoethanol. Besides the obligatory requirement for the reducing agent, a mitogen which can be extracted from bacto-agar is a specific requirement for B lymphocyte clonal development (Kincade *et al.*, 1976). The cell responsible for B lymphocyte colony formation (CFU-BL) was extremely sensitive to PGE since 50% inhibition of cloning was seen at PGE concentrations of $10^{-8}$–$10^{-9}$ M, and significant inhibition was seen with $10^{-10}$ M. In contrast to the inhibition seen with CFU-c and CFU-PM, addition of high concentrations of CSF did not protect against the PGE inhibition of CFU-BL (Kurland and Moore, 1976).

The prostaglandins are produced by many cells of bone marrow origin; however, the macrophage, particularly in the activated state, represents a major hemopoietic source. While we have observed *in vitro* inhibition of cell proliferation at PGE concentrations within the range reported in the serum, it is more likely that the regulatory role of PGE in hemopoiesis is mediated locally, for example, within the marrow environment or in sites of local inflammation.

To balance and counteract continuing proliferation and recruitment of macrophages following enhanced CSF production by activated macrophages, it is probable that increasing macrophage prostaglandin synthesis would serve to limit this positive feedback. The ability of PGE to inhibit lymphoid proliferation further suggests that PGE production by activated macrophages may function to limit continuing macrophage activation by limiting lymphokine production. Indeed, it has been proposed that a defect in such a prostaglandin-mediated control system may be involved in the pathogenesis of rheumatoid arthritis (Morley, 1974).

Many of the conflicting observations concerning the capacity of macrophages to influence the proliferation of both lymphoid and myeloid cells may be placed in better perspective by recognition of the capacity of these cells to elaborate a spectrum of regulatory molecules which may have mutually antagonistic actions.

## ACKNOWLEDGMENTS

Supported by NCI grants CA-08748, CA-17353, CA-17085 and the Gar Reichman Foundation.

## REFERENCES

Bradley, T. R. and Metcalf, D. (1966). *Aust. J. Exp. Biol. Med. Sci*, **44**, 287–300.
Broxmeyer, H. E., Baker, F. L. and Galbraith, P. R. (1976). *Blood* **47**, 389–402.
Broxmeyer, H. E., Moore, M. A. S. and Ralph, P. (1977). *Exp. Hematol.*, **5**, 87–93.

Chan, S. H. and Metcalf, D. (1972). *Blood* **40,** 646–649.
Cline, M. J., Rothman, B. and Golde, D. W. (1974). *J. Cell. Physiol.* **84,** 193–196.
Dale, D. C., Alling, D. W. and Wolff, S. M. (1972). *J. Clin. Invest.* **51,** 2197–2204.
Dexter, T. M., Allen, T. D. and Lajtha, L. G. (1977), in press, this volume.
Fibach, E., Gerassi, E. and Sachs, L. (1976). *Nature* **259,** 127–128.
Kincade, P., Ralph, P. and Moore, M. A. S. (1976). *J. Exp. Med.* **143,** 1265–1270.
Kurland, J. and Moore, M. A. S. (1976). *Exp. Hematol.,* in press.
Lin, H. and Stewart, C. C. (1973). *Nature New Biol.* **243,** 176–179.
Mauel, J. and Defendi, V. (1971). *Exp. Cell Res.* **65,** 377–379.
Metcalf, D. (1974). *Exp. Hematol.* **2,** 157–173.
Metcalf, D. and Moore, M. A. S. (1971). "Haemopoietic Cells." North Holland Publishing Co., Amsterdam.
Metcalf, D., MacDonald, H. P., Odartchenko, N. and Sordat, L. B. (1975a). *Proc. Nat. Acad. Sci. U.S.A.* **72,** 1744–1748.
Metcalf, D., Warner, N. L., Nossal, G. J. V., Miller, J. F. A. P., Shortman, K. and Rabellino, E. (1975b). *Nature* **255,** 630–632.
Metcalf, D., Parker, J., Chester, H. M. and Kincade, P. W. (1974). *J. Cell. Physiol.* **84,** 275–290.
Moore, M. A. S. (1974). *In* "Advances in Acute Leukaemia" (F. J. Cleton, D. Crowther, J. S. Malpas, eds.), pp. 161–227. ASP-Biological and Medical Press, Amsterdam.
Moore, M. A. S., Kurland, J. I. (1976). *In* "Progress in Differentiation Research" (N. Muller-Berat, ed.), pp. 483–492. North-Holland Publishing Co., Amsterdam.
Moore, M. A. S. and Williams, N. (1972). *J. Cell. Physiol.* **80,** 195–206.
Moore, M. A. S., Williams, N. and Metcalf, D. (1972). *J. Cell. Physiol.* **79,** 283–292.
Moore, M. A. S., Williams N. and Metcalf, D. (1973). *J. Natl. Cancer Inst.* **50,** 591–602.
Moore, M. A. S., Spitzer, G., Metcalf, D. and Penington, D. (1974). *Brit. J. Haematol.* **27,** 47–55.
Morley, J. (1974). *Prostaglandins* **8,** 315–320,
Pike, B. and Robinson, W. A. (1970). *J. Cell. Physiol.* **76,** 77–84.
Price, G. B., McCulloch, E. A. and Till, J. E. (1973). *Blood* **42,** 341–348.
Price, G. B., McCulloch, E. A. and Till, J. E. (1975). *Exp. Hematol.* **3,** 227–229.
Quesenberry, P., Morley, A. and Stohlman, F., Jr. (1972). *N. Engl. J. Med.* **286,** 227–232.
Ruscetti, F. W. and Chervenick, P. A. (1975). *J. Immunol.* **114,** 1513–1516.
Rytomaa, T. (1973). *Brit. J. Haematol.* **24,** 141–146.
Shah, R., Caporale, L. and Moore, M. A. S. (1976), *Blood* (in press).
Sheridan, J. W. and Metcalf, D. (1973). *J. Cell. Physiol.* **81,** 11–24.
Sheridan, J. W. and Metcalf, D. (1974). *Proc. Soc. Exp. Biol. Med.* **146,** 218–221.
Sheridan, J. W. and Stanley, E. R. (1971). *J. Cell. Physiol.* **78,** 451–460.
Stanley, E. R. and Metcalf, D. (1972). *In* "Cell Differentiation," Proc. 1st Int. Conf. (R. Harris, and R. Viza, eds.), pp. 272–276. Munksgaard, Copenhagen.
Stanley, E. R., Cifone, M., Heard, P. M. and Defendi, V. (1976). *J. Exp. Med.* **143,** 631–647.
Stephenson, J. R., Axelrad, A. A., McLeod, D. L. and Shreeve, M. M. (1971). *Proc. Nat. Acad. Sci. U.S.A.* **68,** 1542–1546.

# V. Cell Interactions in the Immune System

# Cell Interactions in Cell Mediated Immunity

Michael Feldman

*Department of Cell Biology*
*The Weizmann Institute of Science*
*Rehovot, Israel.*

## I. STIMULATOR CELL-LYMPHOCYTE INTERACTIONS

The induction of cell-mediated immunity (CMI) against cell surface antigens results in the differentiation of antigen-specific effector T (thymus-derived) lymphocytes. *In vivo*, such cell populations mediate the rejection of tissue transplants. *In vivo*, they manifest target cell lysis. We investigated the *in vitro* induction of CMI, following a primary exposure of lymphocytes to cell surface antigens. These studies are reviewed in Feldman *et al.*, 1972. I shall briefly discuss here some aspects of cell interactions associated with both the induction and the functional manifestation of effector cytotoxic T-lymphocytes (CTL). The experimental system we used involved sensitization of lymphocytes by culturing them on monolayers of fibroblasts. At first, we investigated xenogeneic systems in which rat lymphocytes are sensitized in culture on monolayers of mouse fibroblasts. Following 5 days of culture on the sensitizing monolayers, lympohcytes appear capable of lysing target cells syngeneic with the sensitizing fibroblasts (Feldman *et al.*, 1972). The specificity of the lytic reaction could be attributed to a state of diversity of the rat T-lymphocytes with regard to their receptors for the mouse cell surface antigens. We tested whether indeed diversity of T-lymphocytes is the basis for the immunological specificity of CTL (Lonai *et al.*, 1972; Wekerle *et al.*, 1972). Rat lymphocytes were seeded on monolayers of mouse fibroblasts of a given H-2 phenotype. Following 30–120 minutes, a small proportion of the lymphocytes adhered to the fibroblast monolayer. To test whether the initial adherence of lymphocytes to the fibroblast cell surfaces represents recognition, i.e., specific binding to the fibroblast antigens via the lymphocyte cell receptors for antigens, we separated the non-adherent from the adherent lymphocytes and transferred the non-adherent cells to other monolayers, either syngeneic or allogeneic to the "adsorbing" monolayers. We found that: (a) the adherent lymphocytes cultured for 5 days differentiated to specific highly cytotoxic T-cells; (b) the non-adherent lymphocytes transferred to a monolayer syngeneic to the first "immunoadsorbent" monolayer manifested a significant depletion of reactivity; and (c) the non-adherent lymphocytes transferred to a monolayer allogeneic to the "immunoadsorbent" developed a high level of specific cytotoxicity. The depletion, via the cellular immunoadsorbent, of immunospecific lymphocytes indicated that: (a) the adherence or binding of lymphocytes to the fibroblast monolayers represents the process of antigen recognition; and (b) the T-lymphocytes constitute a diverse population of cells with regard to their cell receptors for antigen (Lonai *et al.*, 1972).

Previous studies indicated that antigen recognition by B-lymphocytes does take place at low temperatures and is energy independent (Wigzell and Makela, 1970; Walters and Wigzell, 1970). We found, however, that the specific binding of T-lymphocytes to the cell surface antigens does not take place at 0–4° C. Since recognition of antigen by T-cells is temperature dependent we tested whether it requires metabolic energy. We found that metabolic inhibitors prevented the specific binding of lymphocytes to fibroblast monolayers (Wekerle *et al.*, 1972). To further analyze the implication of metabolic energy, we tested whether the removal of neuraminic acid will alter the temperature dependence of recognition. We found that following neuraminidase treatment, specific binding by lymphocytes did take place at 4° C. Whether or not the energy requirement is related to the process of exposure of otherwise cryptic cell surface receptors, which were exposed following neuraminidase treatment, is a possibility which merits further investigation (Wekerle *et al.*, 1972).

Is antigen recognition, i.e., specific binding, which is necessary for triggering immunospecific T-lymphocytes to differentiate to "killer" cells, also sufficient for the trigger effect? To test this, we used fibroblast monolayers pretreated with glutaraldehyde (Wekerle *et al.*, 1974). Such treatment, resulting in cross-linking of protein moieties on the cell surface, did not alter the antigenicity of the monolayers. This was deduced from the capacity of the glutaraldehyde-treated cells to specifically adsorb allo-antibody. In accordance, we found that T-lymphocytes did adhere specifically to the cell surface antigens of the glutaraldehyde-treated monolayers. Thus, recognition of antigen on glutaraldehyde-treated cells did take place. These adherent cells, however, did not differentiate to CTL. Thus, recognition of antigen by T-cells is necessary, yet not sufficient, for signalling lymphocyte development to effector cells. What is the process which, following recognition, triggers the differentiation of the lymphocytes to killers? This is still an open question. One possibility is that triggering of the lymphocytes requires the intramembrane migration leading to concentration ("capping") of antigen-cell receptor complexes. The fixation of the antigen on the fibroblast membrane prevents the antigen receptor complexes from lateral movement. Without such capping the intracellular signal from the lymphocyte receptor site leading to gene activation may not be generated (Wekerle *et al.*, 1974).

## II. MACROPHAGE-LYMPHOCYTE INTERACTIONS

Studies on the induction of antibody production indicated that macrophages play a determining role in presenting the antigen to the lymphocytes. We therefore approached the question of whether the induction of effector T-lymphocytes *in vitro* also involves interactions between macrophages and lymphocytes. Rat lymphocytes from which the adherent (macrophage-containing) subpopulation was depleted were cultured on monolayers of mouse fibroblasts. Such macrophage-depleted lymphocyte populations did not produce effector CTL. When peritoneal macrophages were added to depleted lymphocyte populations, reactivity was regained. It appeared, therefore, that macrophages are involved in the *in vitro* induction of cell-mediated immunity against surface antigens (Lonai and Feldman, 1971). Their role, however, could not be inferred from these results. To test whether macrophages might be involved in presenting cell surface antigens to T-cells we tested tumor associated antigens. Experimenting with a C57Bl transplantable carcinoma (the 3LL Lewis lung carcinoma), we demonstrated that syngeneic lymphocytes cultured on monolayers of the tumor cells developed tumor-specific CTL (Schechter *et al.,* 1976). We then tested whether macrophages, fed with membrane-containing cell-free extract of the 3LL tumor, will be capable of signalling the production of anti-tumor CTL. Lymphocytes cultured on such monolayers developed within 5 days a high level of CTL (Treves *et al.,* 1976a). In more recent experiments we extended this approach to antigens associated with the radiation leukemia virus (RadLV). Here again, macrophages fed with viral preparations sensitized lymphocytes, and this resulted in the differentiation of CTL capable of affecting targets possessing the viral antigens (Treves *et al.,* 1976b). A direct exposure *in vitro* of the lymphocytes themselves to the viral preparations did not result in the differentiation of effector cells. These results indicate that antigen-fed macrophages can effectively trigger lymphocyte sensitization. They do not, however, necessarily imply that graft reactions *n vivo* are triggered by macrophages presenting the transplantation antigens to the T-lymphocytes, although this may indeed be the case. Whatever the *in vivo* mechanisms may be, it appears that the macrophage-lymphocyte interaction plays a role in signalling sensitization of T-lymphocytes.

## III. CELL INTERACTIONS OF T-LYMPHOCYTES

Are the T-lymphocytes which have initially recognized the antigen differentiating themselves to CTL? The alternative situation would involve interactions between T-cells, whereby lymphocytes which had recognized antigen would signal other T-cells differentiate to actual CTL. In our laboratory, Dr. I. R. Cohen investigated this question (Cohen and Livnat, 1976). He sensitized mouse lymphocytes against an allogeneic monolayer of fibroblasts, then injected the lymphocytes into the hind foot pad of a syngeneic mouse. This resulted in a dramatic enlargement of the draining lymph nodes, associated with the accumulation, within these lymph nodes, of effector T-lymphocytes capable of specific lysis of target cells. *In vitro* sensitized lymphocytes could not generate effector cells if injected into recipients which had been depleted of their T-cell population by total body irradiation. On the other hand, lymphocytes exposed after sensitization to 2000r could generate the effector response when injected into normal recipients. It was, therefore, deduced that one type of lymphocytes, initiator T-lymphocytes (ITL), recruit from the recipient population of T-lymphocytes cells which differentiate to effector CTL (Treves and Cohen, 1973). Thus, an interaction between initiator T-lymphocytes (ITL) and recruited T-lymphocytes (RTL) takes place, in which the ITL constitute the afferent phase of the immune reaction. Following sensitization, they migrate to the lymph nodes and there they signal another type of T-lymphocytes, the RTL, to differentiate to effector CTL. The latter, then, constitute the efferent phase of the reaction. Sensitization *in vitro* for 6 hours at 37° C, but not at 4° C, was sufficient to convert normal T-lymphocytes to ITL.

Do the ITL function merely by presenting the antigen to the RTL? In approaching this question it was observed that trypsinization of the ITL prevented their capacity to recruit (Livnat and Cohen, 1975). This could have been interpreted to indicate that trypsin removed antigen from the ITL cell surface. Yet, 4 hours after trypsinization the ITL regained, in the absence of antigen, the capacity to recruit effectors. One could argue that the intracellular antigens were exteriorized after trypsinization. Yet, cycloheximide, applied after trypsinization, prevented the reappearance of recruiting potency. Although exposure of internalized antigen may depend on protein synthesis, it seems equally possible that ITL-RTL interaction may not be based simply on antigen presentation. The RTL were found to be circulating T-cells, pre-programmed to react against the cells which triggered the ITL. They seemed to be trapped by the ITL in the draining lymph nodes (Cohen and Livnat, 1976).

## IV. EFFECTOR-TARGET CELL INTERACTIONS

The final phase of cell interactions following the induction of cell-mediated immunity against cell surface antigens involves the effector-target cell interaction, whereby the target is lysed. Obviously, this should comprise two distinct stages: (a) recognition, i.e., specific binding of CTL via their receptors for antigen to target cells; and (b) the actual lytic process. To analyze the effector phase, G. Berke of our laboratory used Balb/c-sensitized peritoneal lymphocytes obtained following an intraperitoneal allograft rejection of a C57Bl-originated EL-4 leukemia (Berke and Amos, 1973). To analyze the process of recognition as distinct from the terminal lytic phase, sensitized lymphocytes were mixed with EL-4 cells, centrifuged and resuspended at room temperature. Conjugates of lymphocyte-target cells were formed. The anti-C57Bl sensitized Balb/c lymphocytes formed conjugates also when mixed with other cells of C57Bl origin. They did not, however, produce conjugates with $C_3H$ or Babl/c targets. Thus, conjugation seems to represent recognition of strain-specific cell surface antigens of the targets by the CTL. Conjugation requires divalent cations such as $Mg^{2+}$ and $Ca^{2+}$, and the inhibitory effects of cytochalasin B and of colchicine suggested the involvement of microfilaments and microtubuli (Plant *et al.*, 1973; Stulting *et al.*, 1973). The maximum number of lymphocytes conjugated to target cells is obtained at 21–22° C. When the temperature is raised to 37° C, lysis of the target cells takes place. Thus, recognition and lysis have two distinct ranges of temperature-dependence, a phenomenon which led to an analysis of the factors controlling each of these phases (Berke and Gabison, 1975; Berke *et al.*, 1975; Zagury *et al.*, 1975). Although lysis had a higher temperature optimum than binding, it is the latter which was found to depend on metabolic energy; metabolic inhibitors inhibited conjugation. They did not affect the maintenance of existing conjugates, nor did they prevent the lytic pahse (Berke and Gabison, 1975). Thus, the lytic process itself, although temperature-dependent, is independent of metabolic energy. The interaction between CTL and target cells is prevented by pretreating the targets with allo-antibodies. Thus, the target allo-antigens seem to be involved. Yet, solubilized antigen did not inhibit effector-target interactions (Brunner *et al.*, 1968; Berke and Amos, 1973). Berke has therefore suggested that the organized pattern of the target cell membrane plays a role in the binding process. Following binding, aggregated membrane antigens of the target cells cap at the site of contact interface of CTL target conjugates (Berke and Fishelson, 1975). Such a polar localization of target cell antigens seems to be determined by the initial binding process. Whether it is related to the stabilization of the bond between the two cells or whether it determines the following lytic phase is still an open question.

In this particular system of EL-4 targets, every bound lymphocyte is an actual killer (Berke *et al.*, 1975; Zagury *et al.*, 1975). Furthermore, an effector cell which has lysed one target cell can bind to and kill another. Experimenting with individual conjugates led to a study of the identity of the killer lymphocyte and of its ultrastructural properties (Zagury *et al.*, 1975). The analysis of the contact area between the effector and the EL-4 target indicated that these regions are confined to surfaces of the microvilli of both cells. The cytoplasm at the contact sites (300–1500 A°) contained microfilaments, but no morphological manifestations of junctions were observed. This, obviously, does not rule out the possibility of molecular transfer between effector and target. Yet, neither $^{51}Cr$ labeling of either the CTL or the target nor fluorescein probes gave any indication of intercellular exchange (Kalina and Berke, 1976). The molecular events which take place at the membranal site of contact-interaction between the two cells leading to target cells lysis is a subject of current investigation.

## REFERENCES

Berke, G. and Amos, B. (1973). *Transpl. Rev.* **17,** 1971. 71–107.

Berke, G. and Fishelson, Z. (1975). *J. Exp. Med.* **142,** 1011–1016.

Berke, G., and Gabison, D. (1975). *Eur. J. Imm.* **5,** 671–675.

Berke, G., Gabison, D., and Feldman, M. (1975). *Eur. J. Imm.* **5,** 813–818.

Brunner, K. T., Mauel, J., Cerottini, J. C., and Chapius, B. (1968). *Immunology* **14,** 181–196.

Cohen, I. R., and Livnat, S. (1976). *Transpl. Rev.* **29,** 24–58.

Feldman, M., Cohen, I. R., and Wekerle, H. (1972). *Transpl. Rev.* **12,** 57–90.

Kalina, M., and Berke, G. (1976). *Cell. Immunol.* **25,** 41–51.

Livnat, S., and Cohen, I. R. (1975). *Eur. J. Imm.* **5,** 357–360.

Lonai, P., and Feldman, M. (1971). *Immunology* **21,** 861–867.

Lonai, P., Wekerle, H., and Feldman, M. (1972). *Nature New Biol.* **235,**235–236.

Plaut, M., Lichtenstein, L. M., and Henney, C. S. (1973). *J. Imm.* **110,** 771–780.

Schechter, B., Treves, A. J., and Feldman, M. (1976). *J. Nat. Cancer Inst.* **56** 975–979.

Stulting, R. D., Berke, G., and Heimstra, K. (1973). *Transpl.* **16,** 684–686.

Treves, A. J., and Cohen, I. R. (1973). *J. Nat. Cancer Inst.* **51,** 1919–1925.

Treves, A. J., Feldman, M., and Kaplan, H. S. *J. Nat. Cancer Inst.,* in press.

Treves, A. J., Schechter, B., Cohen, I. R., and Feldman, M. (1976). *J. Imm.* **116,**(4) 1059–1064.

Walters, C. S., and Wigzell, H. (1970). *J. Exp. Med.* **132,** 1233–1249.

Wekerle, H., Kolsch, E., and Feldman, M. (1974). *Eur. J. Imm.* **4,** 246- 250.

Wekerle, H., Lonai, P., and Feldman, M. (1972). *Proc. Nat. Acad. Sci.* **69,** 1620–1624.

Wigzell, H., and Makela, O. (1970). *J. Exp. Med.* **132,** 110–126.

Zagury, D., Bernard, J., Thierness, N., Feldman, M., and Berke, G. (1975). *Eur. J. Imm.* **5,** 818–822.

# Lymphocyte Differentiations Induced by Thymopoietin, Bursopoietin and Ubiquitin

Gideon Goldstein*

*Memorial Sloan-Kettering Cancer Center*
*New York, NY 10021*

## I. INTRODUCTION

The ancestry of lymphocytes, which comprise the immunological system of vertebrates, can be traced back to pluripotential stem cells of the hemopoietic tissues. These pluripotential stem cells, which also give rise to other cells such as red blood cells, granulocytes, monocytes and platelets, presumably give rise to a lymphoid precursor which is common to the two main lines of lymphocyte development, thymus-dependent lymphocytes (T cells) and thymus-independent or bone marrow derived lymphocytes (B cells). Thus the various differentiated forms of lymphocytes which comprise the vertebrate immune system are related to each other by sequential differentiation along the various differentiative pathways or by occupancy of parallel branches of differentiation in the genealogical tree of lymphocyte development. In this respect the vertebrate immune system represents, in microcosm, an entire metazoan organism, which has developed its various differentiated cells from a single cell, the zygote. With lymphocytes differentiation from self sustaining precursors continues throughout the life of the organism.

Developmental biologists are all familiar with the concept of induction of differentiation; namely that differentiation of cells in the developing embryo can be regulated by molecules secreted by adjacent tissues, there are many examples of such inductions in the literature (Holtfreter, 1934). The principles are clear; the inducing molecules trigger differentiation in tissues already committed to such differentiation. Isolation of embryonic inducing agents has been hindered by the phenomenon of non-specific induction, which is the induction of differentiation by extracts of inappropriate tissues which would not be expected to regulate this differentiation under physiological circumstances.

I shall describe the serendipitous isolation of a specific inducing molecule from thymus, the polypeptide thymopoietin, and studies on the differentiation

*Present Address: Ortho Pharmacetial Corp., Paritan, NJ 08869

of prothymocyte-to-thymocyte that is induced by thymopoietin. Furthermore I shall describe the isolation of a polypeptide which is ubiquitous in tissue extracts, hence its name ubiquitin, and show that ubiquitin can non-specifically induce a number of differentiative events. Finally, I shall describe a method whereby non-specific induction by ubiquitin can be prevented in assay systems to permit the identification of tissue-specific inducing agents in tissue extracts and show how this method was used to identify bursopoietin, an inducing agent in the avian bursa of Fabricius that induces probursocyte-to bursocyte differentiation.

## II. THE T CELL INDUCTION ASSAY

Thymocytes of the mouse can be identified with some precision by the molecules they carry on their surface membranes. These are identified by alloantisera using the complement mediated cytotoxicity test (Boyse et al. 1970). Using such methods on thymocytes from appropriate strains, the phenotype of the cortical thymocyte can be listed as $TL^+Thy\text{-}1^+Ly\text{-}1^+Ly\text{-}2/3^+Ly\text{-}5^+G_{IX}^+$. Komoro and Boyse (1973) first showed that when murine spleen or bone marrow cells were fractionated by flotation on discontinuous bovine serum albumin gradients the cells of the lighter layers, which lacked thymocyte surface molecules, could be induced to display such molecules by incubation at 37°C for two hours with an inducing agent. It was quickly established that thymus extracts were effective inducing agents, but that a host of other non-thymus-related substances, including extracts of other tissues, were also effective in inducing prothymocyte-to-thymocyte differentiation *in vitro* (Scheid et al. 1973). These findings, which are reminiscent of the findings with induction in experimental embryology, were explained by our subsequent studies of molecular mechanisms which have established the following points:

(i) Induction is rapid with appearance of cell surface molecules after 1½ to 2 hours of exposure to the inducing agent (Komuro and Boyse 1973).

(ii) Active metabolism is required because induction does not proceed at 4°C (Komuro and Boyse 1973).

(iii) Induction involves implementation of genetic information in the induced cell and display of antigens is dependent on the genetic background of the mouse from which the precursors were obtained (Komuro and Boyse 1973).

(iv) Only brief exposure to the polypeptide inducing agents thymopoietin or ubiquitin (see below) is required. The inducing polypeptides can be washed away after 10 minutes exposure yet differentiation proceeds with appearance of surface markers at 1½ to 20 hours (Scheid et al. 1976).

(v) The action of inducing agents is mediated by cyclic AMP. This was shown by enhancement of induction by theophylline, which raises intracellular cyclic AMP levels, and inhibition of induction by imidazole, which depresses intracellular cyclic AMP levels (Scheid et al. 1975). Additionally, dibutyryl cyclic AMP or even cyclic AMP itself were effective inducing agents and induction was inhibited by 8-bromocyclic GMP (Scheid et al. 1975, Scheid et al. 1976).

(vi) The cyclic AMP signal was only required during the first 30 minutes of induction because imidazole was only effective in inhibiting induction when added during this period. After 30 minutes the addition of imidazole did not prevent the appearance of T cell markers at 1½ to 2 hours (Scheid et al. 1976).

(vii) Studies with drugs inhibiting macromolecular synthesis showed that induction required transcription of mRNA (it was inhibited by actinomycin D, camptothecin and cordycepin) and translation of mRNA to protein (it was inhibited by cycloheximide and puromycin) but did not require replication of DNA (it was not inhibited by hydroxyurea or cytosine arabinoside) (Storrie et al. 1976).

## III. THE DUAL INDUCTION ASSAY

Other lymphocyte differentiative events can be induced *in vitro* in a similar manner to prothymocyte-to-thymocyte differentiation. Induction *in vitro* of a B cell differentiation, the conversion of B cells lacking complement receptors ($CR^-$) to $CR^+$ B cells, represents differentiation in an entirely separate cell line yet the induction process appears similar to T cell differentiation in that it is triggered by non-specific agents which elevate intracellular cyclic AMP; the cell surface marker CR appears after 1½ to 2 hours (Goldstein et al. 1975). Induction of $CR^-$ to $CR^+$ B cell differentiation was tested in parallel with induction of prothymocyte-to-thymocyte differentiation to constitute the dual induction assay, which was used to monitor the inductive specificity of inducing agents. A specific inducing agent should only induce the appropriate differentiative event, specificity being determined by the presence of specific receptors on the precursor cell. By contrast agents which non-specifically elevate intracellular cyclic AMP would be expected to induce differentiation in both systems. These expectations were fulfilled (see below).

## IV. THYMOPOIETIN

Thymopoietin is a polypeptide inducing hormone of the thymus. It was isolated serendipitously in studies that were related not to its effect in inducing T cell differentiation but to a presumably secondary effect on neuromuscular transmission (Goldstein 1974). In the disease myasthenia gravis impaired neuromuscular transmission is consistently related to disease of the thymus (Goldstein 1966). Thymopoietin was identified in experimental studies of the pathogenesis of myasthenia gravis as a polypeptide hormone of the thymus that impairs neuromuscular transmission (Goldstein, 1974, Hofmann 1969). This neuromuscular effect of thymopoietin, which is not mimicked by any other agents, was used to isolate thymopoietin. Purified thymopoietin was shown to be a specific T cell inducing agent in the dual induction assay (Goldstein et al.

1975) and it was effective in concentrations below 1 ng/ml (Basch and Goldstein 1974). Thus in retrospect it appears that the neuromuscular properties of thymopoietin are secondary and that its primary function in the body is to induce prothymocyte-to-thymocyte differentiation in the thymus.

The complete 49 amino acid sequence of thymopoietin has been determined (Schlesinger and Goldstein 1975a) and a synthetic tridecapeptide corresponding residues 29–41 of thymopoietin was synthesized by solid phase techniques (Schlesinger et al. 1975a). This tridecapeptide was shown to have the biological properties of the parent molecule in that it induced selective T cell differentiation *in vitro* in the dual induction assay (Schlesinger et al. 1975a) and also affected neuromuscular transmission (Goldstein and Schlesinger 1975).

## V. UBIQUITIN

During the isolation of thymopoietin a major polypeptide was identified in thymus extracts; it had a molecular weight of 8451, somewhat greater than that of thymopoietin (5562) (Goldstein et al. 1975). This polypeptide was present in far greater amounts than thymopoietin in thymus extracts but could also be identified in tissues other than thymus. It was purified and a radioimmunoassay for it was applied to extracts of tissues from various sources. Ubiquitin, as we named this molecule, was detected in extracts of mammalian, avian and prevertebrate tissues and also in yeasts, bacteria and higher plants (Goldstein et al. 1975). When we studied this extraordinary polypeptide in the dual induction assay we found that it induced both T cell and B cell differentiation and that, unlike thymopoietin, induction by ubiquitin was inhibited by the $\beta$-adrenergic inhibitory drug propranolol (Goldstein et al. 1975). We infer therefore that ubiquitin causes an elevation of intracellular cyclic AMP by interacting with $\beta$-adrenergic receptors on precursor cells. We do not know the normal function of ubiquitin but it is clear from its effects that it could well account for the inappropriate induction found in many systems using tissue extracts if these inductions are mediated by elevation of intracellular cyclic AMP in committed percursor cells. The complete 74 amino acid sequence of ubiquitin has been determined (Schlesinger et al. 1975b; Schlesinger and Goldstein 1975b) and a synthetic hexadecapeptide corresponding to the last 16 residues was synthesized and shown to have the biological properties of the parent molecule.

## VI. APPROACH TO DETECTING SPECIFIC INDUCING AGENTS IN TISSUE EXTRACTS

Ubiquitin is present in virtually all tissue extracts and theoretically it could initiate non-specifically any differentiative event in which the committed precursor cell had $\beta$-adrenergic receptors on its surface and was triggered to differentiate by an elevation of intracellular cyclic AMP. Ubiquitin could therefore not only account for inductions by inappropriate tissues but could also mask the inductive effects of a specific inducing agent in tissue extracts. We formulated the following approach to searching for specific inducing agents in new systems:

(i) Can we demonstrate induction with appropriate concentrations of ubiquitin?

(ii) Can we demonstrate induction by cyclic AMP in the presence of a high concentration of ubiquitin? Ubiquitin at higher concentrations is self-inhibitory and this inhibition appears to lie at the level of the ubiquitin receptor because high concentrations of ubiquitin do not prevent induction by, for example, thymopoietin in the T cell induction assay.

(iii) Can we detect specific induction by tissue extracts "spiked" with 100 μg/ml ubiquitin: This concentration of ubiquitin, which is non-toxic, would inhibit induction by ubiquitin in the extract but permit induction by any specific inducing substance in the extract.

This approach is exemplified by our detection of a specific probursocyte-to-bursocyte inducing agent in extracts of the avian bursa of Fabricius (Brand et al. 1976).

## VII. BURSOPOIETIN

Birds have a thymus in which differentiation of T cells proceeds, and they also have an organ, the bursa of Fabricius, in which differentiation of B cells proceeds. In this respect birds differ from mammals in which the analog of the bursa of Fabricius has yet to be determined. Gilmour et al. (1976) developed two alloantisera between inbred strains of chickens which are specific for (a) an antigen present on thymocytes and T cells (Th-1 antigen) and (b) an antigen present on bursocytes and B cells (Bu-1 antigen). Using these alloantisera as a marker for T cell and B cell differentiation we developed *in vitro* induction assays in the chicken analogous to the dual induction assay in the mouse, and used the approach outlined above to detect the inducing agent bursopoietin in extracts of chicken bursa (Brand et al. 1976).

(i) Bone marrow cells from newly hatched chickens were separated by flotation on discontinuous gradients of bovine serum albumin. Cells in the lighter layers, which lacked Th-1 or Bu-1 antigens on their surfaces, were incubated for two hours with ubiquitin at various concentrations. Ubiquitin induced two populations of cells to appear, one bearing Bu-1 antigens on their surfaces, the other bearing Th-1 antigens. High doses of ubiquitin were inhibitory.

(ii) Differentiation of $Bu\text{-}1^{+}$ and $Th\text{-}1^{+}$ cells was induced by cyclic AMP and this induction proceeded unhindered in the presence of high concentrations of ubiquitin.

(iii) Studies of induction by extracts of chicken tissues to which a high concentration of ubiquitin had been added gave the following results:

(a) Bursal extracts induced differentiation of both B cells and T cells but on dilution specific B cell differentiation was obtained.

(b) Extracts of chicken thymus induced specific T cell differentiation.

(c) Extracts of inappropriate tissues such as spleen, muscle, liver and kidney did not induce T cell or B cell differentiation.

We interpret these results as demonstrating the presence of thymopoietin in thymus extracts and a lymphocyte differentiating hormone bursopoietin in bursal extracts. Bursopoietin appears to have inductive specificity for B cells at lower concentrations. This assay is being used to monitor the purification of bursopoietin.

I believe that ubiquitin in tissue extracts may have contributed to the non-specificity of some embryonic induction systems and that a similar approach may be fruitful in elucidating other inductive systems operative in embryological development.

## ACKNOWLEDGMENTS

Supported by U.S. Public Health Service Grants CA-08748, AI-12487, CA-17085 and Contract CB-53868 from the National Cancer Institute.

## REFERENCES

Basch, R. S. and Goldstein, G. (1974). *Proc. Nat. Acad. Sci. U.S.A.* **71**, 1474–1478.

Boyse, E. A., Hubbard, L., Stockert, E. and Lamm, M. D. (1970). *Transplantation* **10**, 446–449.

Brand, A., Gilmour, D. and Goldstein, G. (1976). *Science* **193**, 319–321.

Gilmour, D. G., Brand, A., Connelly, N. and Stone, H. A. (1976). *Immunogenetics* (in press).

Goldstein, G. (1966). *Lancet* **2**, 1164–1167.

Goldstein, G. (1974). *Nature* **247**, 11–14.

Goldstein, G. and Hofmann, W. W. (1969). *Clin. Exp. Immunol.* **4**, 181–189.

Goldstein, G. and Schlesinger, D. H. (1975). *Lancet* **2**, 256–259.

Goldstein, G., Scheid, M., Hammerling, U., Boyse, E. A., Schlesinger, D. H. and Niall, H. D. (1975). *Proc. Nat. Acad. Sci. U.S.A.* **72**, 11–15.

Holtfreter, J. (1934). Wilhelm Roux' *Arch. Entwicklungsmech. Organismen* **133**, 367.

Komuro, K. and Boyse, E. A. (1973). *Lancet* **1**, 740–743.

Scheid, M. P., Hoffman, M. K., Komuro, K., Hammerling, U., Abbott, J., Boyse, E. A., Cohen, G. H., Hooper, J. A., Schulof, R. S. and Goldstein, A. L. (1973). *J. Exp. Med.* **138**, 1027–1032.

Scheid, M. P., Goldstein, G., Hammerling, U. and Boyse, E. A. (1975). *Ann. N.Y. Acad. Sci.* **249**, 531–538.

Scheid, M. P., Goldstein, G. and Boyse, E. A. (1976). (Manuscript in preparation).

Schlesinger, D. H. and Goldstein, G. (1975a). *Cell* **5**, 361–365.

Schlesinger, D. H. and Goldstein, G. (1975b). *Nature* **255**, 423–424.

Schlesinger, D. H., Goldstein, G., Scheid, M. P. and Boyse, E. A. (1975a). *Cell* **5**, 367–370.

Schlesinger, D. H., Goldstein, G. and Niall, H. D. (1975b) *Biochem.* **14**, 2214–2218.

Storrie, B., Goldstein, G., Boyse, E. A. and Hammerling, U. (1976). *J. Immunol.* **116**, 1358.

# VI. Cell Interactions in Organogenesis

# An Analysis of Pancreatic Development: Role of Mesenchymal Factor and Other Extracellular Factors

William J. Rutter, Raymond L. Pictet,
John D. Harding, John M. Chirgwin,
Raymond J. MacDonald and Alan E. Przybyla

*Department of Biochemistry and Biophysics*
*University of California*
*San Francisco, California 94143*

## I. INTRODUCTION

The development of the pancreas *in vitro* closely approximates its development *in vivo*. Thus, it is an attractive system for the analysis of terminal differentiation. The mature pancreas has two distinctive histological types, the exocrine (acinar cells and ducts) and endocrine (the islets of Langerhans) components which can be characterized by their distinctive morphology and by the specific proteins they produce (Table I). The predominant cells are the acinar cells (Fig. 1) that contain the set of digestive enzymes (or their precursors) that are sequestered within the zymogen granules. The duct cells (Fig. 2) produce the fluid of the pancreatic juice and line the tubules connecting the acinar structures with the gut. Randomly interspersed within the exocrine structure of the pancreas are the islets containing the endocrine cells. Each produces a hormone that is stored in a characteristic secretory granule. The predominant cells in the islets are the insulin-containing B cells; the A cells contain glucagon and the relatively rare D cells contain somatostatin, that may act to suppress secretion of the two other hormones. Since each of the secretory granules has a characteristic structure, the A, B and D cells can be recognized morphologically (Fig. 3). In addition to these epithelial cells, the mature pancreas contains a relatively small proportion of mesenchymal (mesodermal, connective tissue) cells as well as nerves and blood vessels.

**TABLE 1**

*Epithelial Pancreatic Cell Types*

| Cell Type | | Product/Function |
|---|---|---|
| Exocrine | Acinar | Digestive enzymes (zymogens) |
| | Duct | Transport of zymogen to the gut |
| Endocrine | A | Glucagon |
| | B | Insulin |
| | D | Somatostatin |

## II. THE DEVELOPING PANCREAS CONTAINS THE INFORMATION REQUIRED FOR MORPHOGENESIS AND CYTODIFFERENTIATION

The pancreas appears at 20–25 somites (approximately 11th day of gestation in the rat) as a diverticulum comprising a few hundred epithelial cells. Nine to ten days later the pancreas has about 7 million cells and the morphogenesis and cytodifferentiation of both exocrine and endocrine components are completed. If the region of the gut forming the pancreas is explanted at 20–25 somites and cultured for 9 days in a nutrient medium devoid of protein or any hormone, both exocrine and endocrine cells differentiate normally, although growth is impaired. As illustrated in Figures 4, 5 & 6, these cultured pancreases contain acinar cells with typical zymogen granules and clusters of endocrine cells that are comprised of B cells, as *in vivo*, containing $\beta$ granules. As implied by these secretory granules, these explants also contain high levels of amylase, a major exocrine enzyme, and insulin (Table II). The great similarity in morphology and concentration of the cell specific

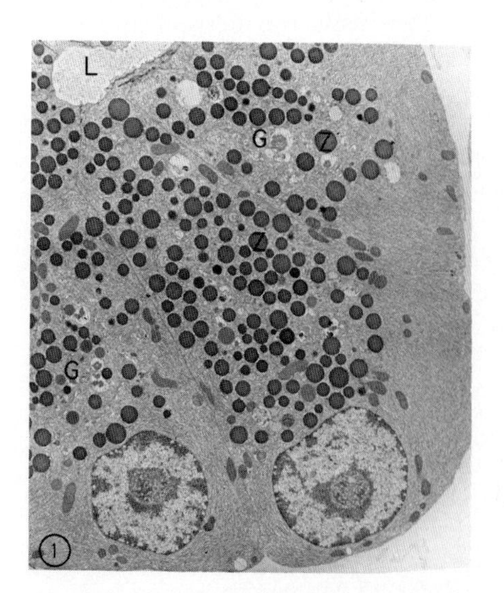


Fig. 1. Acinar cells. The acinar cells have a pyrimidal shape and the organelles are clearly polarized within the cells. The nucleus is located in the basal portion with most of the rough endoplasmic reticulum (RER). The Golgi apparatus (G) and the zymogen granules (Z), are located between the nucleus and the lumen (1) into which the exocrine enzymes are secreted. ×3000

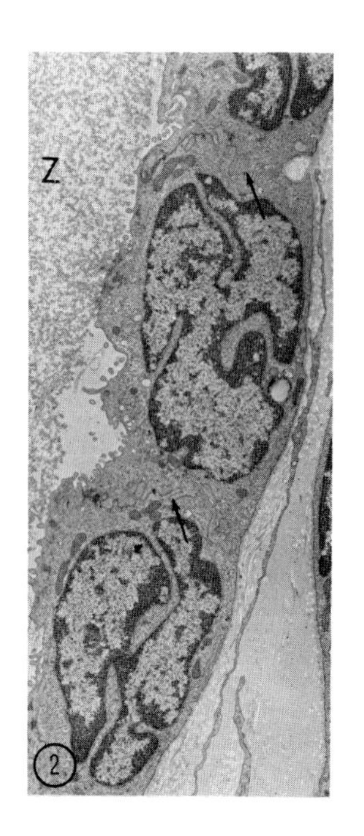


Fig. 2. Duct cells. The duct cells are cuboidal in shape, have basal and apical surfaces, but otherwise have no distinctive intracellular morphological features. The lateral cell membrane shows many intercalated digitations (arrows) found in cells active in fluid transport. The lumen contains the zymogens (Z) secreted by the acinar cells. × 6300

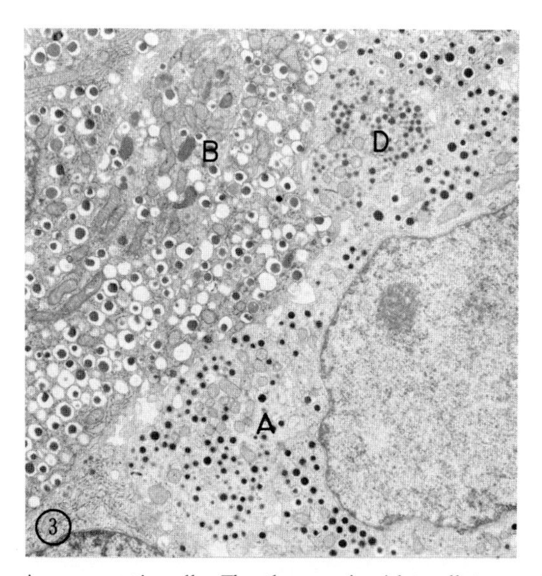


Fig. 3. Endocrine pancreatic cells. The three major islet cell types, A, glucagon, B, insulin and D, somatostatin, are distinguished mostly on the basis of the appearance of their storage granules. In B cells the insulin-containing $\beta$ granules loosely fit their vesicles and in addition may have a polygonal shape. In A cells the space between the glucagon-containing $\alpha$ granules and their membrane vesicle is smaller. In the D cells the somatostain-containing $\delta$ granules completely fill the vesicles. ×5280. From Pictet & Rutter, 1972

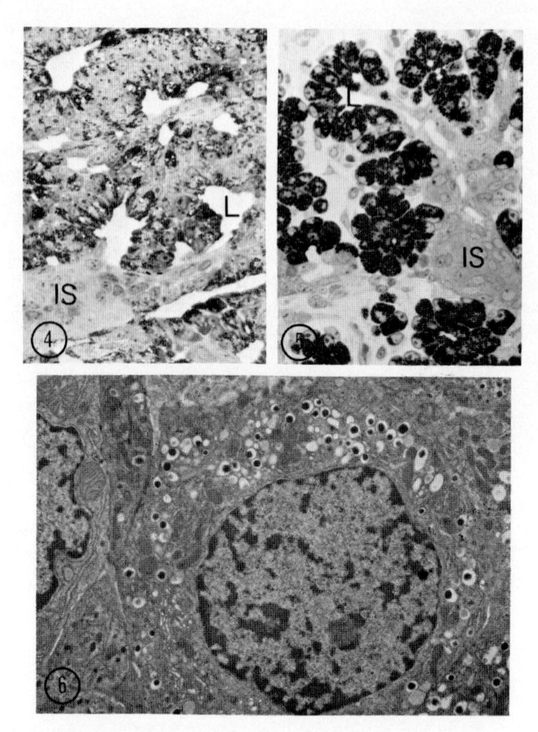


Figs. 4, 5 and 6. Differentiation of pancreatic rudiments in a defined medium. 20–25 somite (11th day) pancreas-forming gut region cultivated for 9 days in the absence of protein and hormones (Fig. 4), develops as *in vivo*. At the corresponding stage of development (20th day of embryonic life) (Fig. 5) the acinar cells surround the lumen (1) and contain zymogen granules. There are also clusters of cells corresponding to the islets (is) which (by electron microscopic analysis) show the typical pancreatic endocrine granules (Fig. 6). *In vivo* the acinar cells contain more zymogen granules and the various components are more individualized. Fig. 4 & 5, × 293; Fig. 6, × 5400

proteins in the *in vivo* and *in vitro* systems demonstrates that hormones or exogenous factors supplied by remote embryonic or maternal tissues are not required for normal expression of the developmental program of the pancreas during this period. Development is also not dependent on innervation.

## III. ORIGIN OF THE ENDOCRINE PANCREAS

It is generally accepted that the acinar and duct cells are derived from the gut endodermal cells; however, the origin of endocrine cells has been the subject of controversy. Whereas a similar endodermal origin for these cells has been favored, Pearse has proposed that the endocrine cells of the pancreas and the endocrine cells of the digestive tract are, like the adrenal medulla, derived from

**TABLE 2**

*The Pancreas Develops in a Defined Medium Lacking Protein or Any Hormone*

| | *In Vivo* | | *In Vitro* |
|---|---|---|---|
| Gestational Age (days) | 14 | 20 | 20 |
| Amylase (ng/μg total tissue protein) | 0.010 | 130 | 26 |
| Insulin (ng/μg total tissue protein) | 0.002 | .59 | 2.9 |

Day 11 rudiments at the onset of pancreatic organogenesis were excised and cultured for 9 days. The medium was not supplemented with a protein source. After 9 days, amylase and insulin levels were measured. These concentrations are over a thousand fold compared with that measured *in vivo* at day 14 of development.

the neural crest (Pearse, 1969; Pearse & Polak, 1971; Pearse *et al.*, 1973). This intriguing theory is based on the observation that these cells, like neural crest cells, have the capacity to concentrate and decarboxylate amine precursors (APUD cells) and therefore can be recognized by a histological test. We have tested the Pearse hypothesis by removing the whole ectoderm including the neural fold of <3 somite embryos (9th day of gestation), prior to the formation of the neural crest, and allowing *in vitro* development for 11 days (equivalent to the remaining period of embryonic development) (Pictet *et al.*, 1976). As shown in Figure 7, typical acinar and endocrine B cells containing secretory granules were evident in the pancreas. Sufficient levels of insulin were also detected to account for a few hundred differentiated cells, a reasonable number for the size of the tissue. These experimental results suggest that the pancreatic endocrine cells are not derived from the neural crest per se. This conclusion is reinforced by the finding of Le Douarin and colleagues (Le Douarin & Teillet, 1973) and of Andrew (1976) that heterografts of quail neural tube including the neural crest, when grafted to neuroectoderm deprived chicken embryos did not lead to the presence of quail cells in the islets of Langerhans or among the gut epithelial cells (though such cells were found among the calcitonin producing cells (Polak *et al.*, 1974). The migration of the quail cells was not impaired since they were found in the nervous ganglia of the digestive tract. We conclude that the neural crest is not the source of the endocrine cells of the pancreas. The experimental

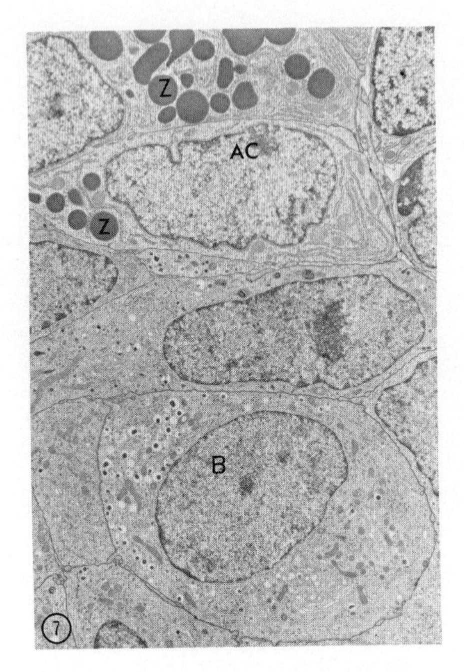


Fig. 7. Endocrine B cells differentiate *in vitro* in embryos deprived of neural crest. The neuroectoderm was removed from rat embryos having 2 to 4 somite pairs (9th day of development). The remaining parts of the embryo were cultured 11 days. Both acinar cells (Ac) containing zymogen granules (Z) and B cells (B) containing insulin granules are present in the pancreas. See also Table II. ×4150

results, however, do not preclude earlier migration from the ectoderm, nor do they require that the cells are derived from the endoderm itself.

## IV. DIFFERENTIATION OF THE A CELL: GLUCAGON AND DEVELOPMENT

The glucagon producing endocrine A cells differentiate concomitantly with the formation of the pancreatic epithelial bud at approximately 20–25 somites (11th day of embryonic development in the rat); and the glucagon content of the pancreas reaches its highest level with 24 hours (Rall *et al.*, 1973). At that time the A cells are clustered in 3 to 5 small islets and represent 5–10% of the cells in the epithelial bud (Pictet & Rutter, 1972). The A cells may be the first endocrine cells to differentiate. Glucagon is known to increase intracellular cyclic AMP by stimulating adenylcylase. Since cyclic AMP has been implicated as a regulatory intermediary in a number of developmental processes (Deshpande & Siddiqui, 1976; Wahn *et al.*, 1976), it is possible that glucagon plays a developmental role.

## V. MORPHOGENESIS AND CYTODIFFERENTIATION

In the 2–3 day period after the formation of the pancreatic bud (12–14) days of gestation), there is rapid morphogenesis with the formation of acinar structures but no evidence of cytodifferentiation of acinar or endocrine B cells.

The typical secretory granules found in B cells and the zymogen granules of acinar cells begin to appear 2 to 3 days later concomitant with the enlargement of the Golgi apparatus and the development of arrays of endoplasmic reticulum.

We have previously shown that there is a biphasic accumulation of exocrine proteins (Rutter *et al.*, 1967a,b, 1968) and of insulin (Clark & Rutter, 1972; Rall *et al.*, 1973) in the pancreas. During the embryonic period from 12 to 14 days the levels of exocrine enzymes and insulin are very low and relatively constant; thus the rate of their accumulation parallels the rate of cell proliferation. Between 14 and 15 days, however, the synthetic rates of insulin and the exocrine proteins increase dramatically, and zymogen granules begin to appear within the cells. The differentiated level of exocrine proteins, 3 to 4 orders of magnitude above the 12–14 day levels, is reached at about 20 days, just prior to birth (21st day). These observations suggest there is an intermediate period in the process of development between the appearance of the pancreas and its products and overt cytodifferentiation. We have termed this period the protodifferentiated state (Rutter *et al.*, 1967a,b; Rutter *et al.*, 1968). It can therefore be inferred that there are at least two differentiative transitions, the first results in the formation of the pancreatic diverticulum and the capacity to form pancreas specific structures and synthesize low levels of pancreas specific products, the second involves the amplification of the specialized synthetic processes.

The process of macromolecular synthesis during the secondary transition has been studied at the translational level (Kemp *et al.*, 1972). The rate of synthesis of two of the principle exocrine enzymes, chymotrypsin and amylase, during this period, accounts quantitatively for the rate of their accumulation; there is no evidence for turnover. Similar observations have been made for the synthesis of proinsulin and the accumulation of insulin (Rall, 1975). The exocrine enzymes accumulate in a non-coordinate fashion; there is about a two and a half day delay between the accumulation profiles of the first (amylase) and the last (carboxypeptidase B) enzyme. Thus the genes for at least some of the exocrine proteins are regulated separately. However some of the accumulation profiles appear congruent so that coupled regulation is feasible (but not proved).

We have also begun an analysis of the pattern of accumulation of specific pancreatic transcripts. For these studied we have employed a cDNA "probe" prepared against poly(A+) pancreatic RNAs. This approach seemed feasible since over 90% of the cells in the adult pancreas are acinar cells, and at any one time the secretory proteins represent 50–90% of the total protein synthesized. Thus the mRNA coding for these proteins should be a dominant proportion of the poly(A+) pool. A major obstacle to these experiments however is the presence of large amounts of RNAse (one of the secretory proteins of the rat pancreas) that rapidly degrades mRNA during isolation. This problem was obviated by developing a method that destroys RNAse activity. Intact, translatable poly(A+)

mRNA can now be routinely prepared from pancreases rich in RNAse. Translation of the poly(A+) RNA in an *in vitro* protein synthesizing system produces a set of proteins that strikingly resembles, both in number and relative size, the secretory proteins isolated from zymogen granules. The polypeptides synthesized are, however, slightly larger than the secretory proteins isolated from the zymogen granules and may be precursors. Many other secretory proteins are synthesized as precursors. Blobel and his associates (Blobel & Sabatini, 1971; Devillers-Thiery *et al.*, 1975) have presented evidence suggesting that a "signal peptide" present in the precursor is involved in the attachment of the ribosomes to the membrane of the endoplasmic reticulum and in the transport of the polypeptide into the cisternal space.

A complementary DNA was prepared from the poly(A+) mRNA using reverse transcriptase and oligo(dT) as a primer. As shown in Figure 8, total pancreas RNA hybridizes with pancreatic cDNA at higher $R_0t$ values than does pancreas poly(A+) RNA. An evaluation of the $R_0t_{1/2}$ values of the reactions indicates that the poly(A+) RNA is approximately 50-fold enriched in the sequences complementary to the DNA probe. Thus this mRNA fraction represents approximately 2% of the total RNA. Tests of the specificity of the probe have been made by hybridization with RNA extracted from other tissues. Approximately 10% of the probe hybridizes to these RNAs at one thousand fold higher RNA concentrations than total pancreas RNA. These sequences are apparently common to all the cells tested since combinations of the RNA fractions from these tissues do not yield an increased degree of hybridization to the probe. The specificity of the probe is further illustrated by *in situ* hybridization with RNA of individual cells in pancreatic tissue. As shown in Figure 9, the acinar cells are heavily labeled; the basal part of the cells, which is enriched in the rough endoplasmic reticulum, contains most of the bound cDNA. The islets show little reactivity and the connective tissue cells even less; the labeling index is at least one hundred fold higher in the acinar cells than in the mesenchymal cells.

The kinetics of hybridization of a cDNA with the RNA from which it is prepared is a function of the number of components, their relative concentrations and their sequence length. For example, rabbit α-globin cDNA, composed of approximately 620 nucleotides (Gould and Hamlyn, 1973; Maniatis *et al.*, 1976), hybridizes with rabbit α-globin mRNA (after correction for the salt concentration used in our experiments (20)) between 2 log units of $R_0t$ with a $R_0t_{1/2}$ of $3.8 \times 10^{-4}$ mol·s/l. Under the same conditions, the hybridization of the pancreatic cDNA is both displaced to higher $R_0t$ values and extends over a wider $R_0t$ range than expected for a single component. At least two populations are easily resolved by a computer analysis of the hybridization curve of the pancreas cDNA with adult pancreas poly(A+) RNA. The first represents approximately half of the cDNA, and has an observed $R_0t_{1/2}$ of $5.1 \times 10^{-3}$ mol·s/l correspond-

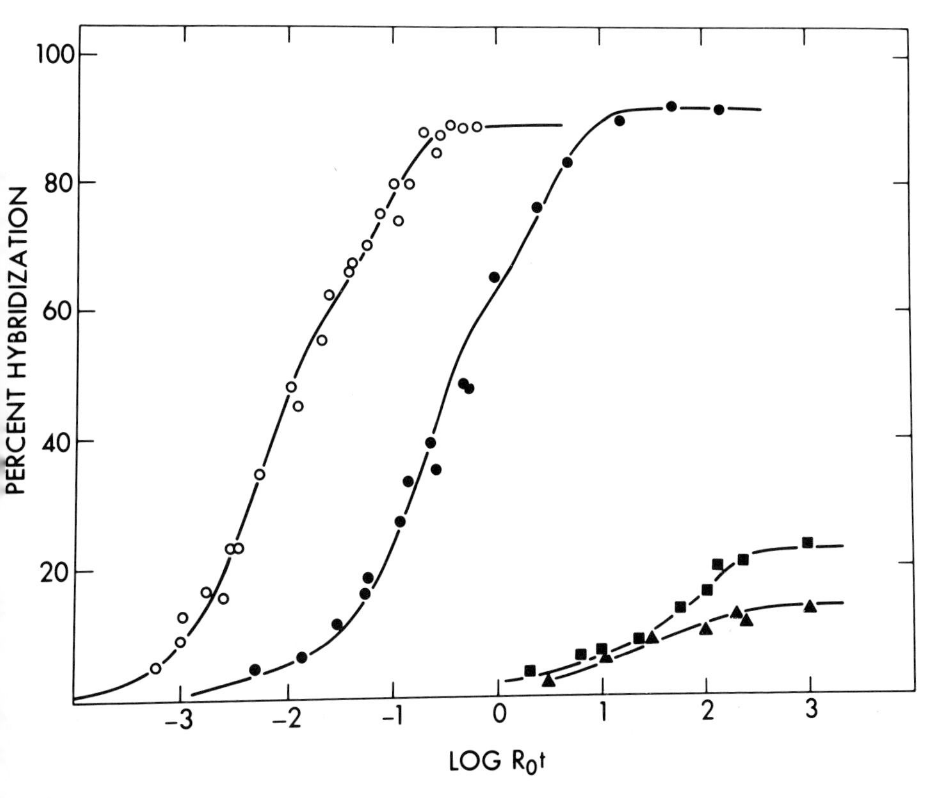


Fig. 8. Specificity of rat pancreatic cDNA probe: Hybridization of adult rat pancreas cDNA with RNA prepared from rat tissue RNAs. cDNA was synthesized from adult rat pancreas oligo(dT) bound RNA. The extent of hybridization was measured after Sl nuclease digestion of unhybridized nucleic acid. 2.4% of the cDNA is resistant to Sl nuclease in the absence of hybridization; this background is subtracted. Hybridization with pancreas oligo(dT) bound RNA, ○ ○. The two component curve was fit to the data points by computer analysis. Hybridization with total adult rat pancreas RNA, ● ●. The curve fit to the open circles has been shifted on the abscissa to fit these data points. Hybridization with total liver RNA, ■ ■. Total brain RNA, ▲ ▲.

ing, by comparison with rabbit α-globin RNA, to a complexity of 1600 nucleotides. This suggests there is a high concentration of one or two mRNA species. The second population comprising the remaining ∿40% of the cDNA has an observed $R_0t_{1/2}$ of $7.4 \times 10^{-2}$ mol·s/l, corresponding to a complexity of approximately 20,000 nucleotides. This complexity is in the range of values expected for the population of mRNA molecules coding for the set of exocrine proteins synthesized from poly(A+) RNA *in vitro*, and found in secretory granules and in the pancreatic juice.

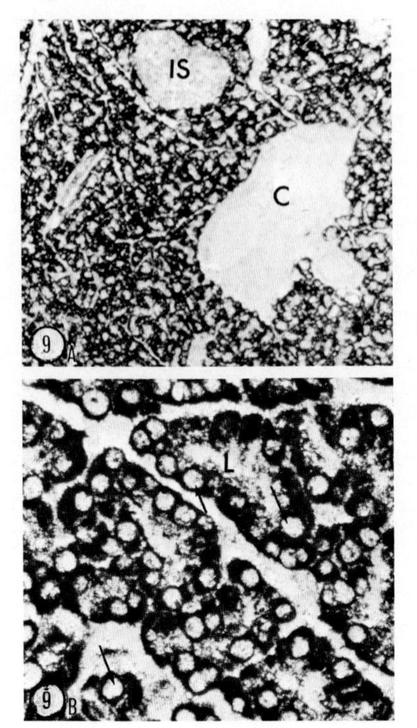


Figs. 9a and 9b. In situ hybridization of the pancreatic cDNA probe with acinar cell RNA. Frozen sections of adult rat pancreases were fixed, incubated with pancreatic cDNA and prepared for autoradiography. Figure 9a shows that the cDNA hybridizes predominantly to the acinar tissue. The endocrine cells of the islets of Langerhans (IS) show little hybridization. The connective tissue (C) cells, blood vessels and fibroblasts serve as an internal control for the cell specificity of the hybridization. Figure 9b shows at a higher magnification that hybridization is mostly restricted to the basal and perinuclear region of the acinar cells where most of the rough endoplasmic reticulum is located (see Fig. 1). Nuclei (arrows) and lumen (l) of the acini are relatively unlabelled. 9a, × 60 ; 9b, ×300

This cDNA probe has been used to measure the relative concentrations of complementary transcripts during the course of embryological development (Fig. 10). The kinetics of the hybridization gives a qualitative indication of the number and relative concentration of the mRNA components. As shown in Figure 10, the relative concentration of transcripts complementary to pancreas cDNA changes dramatically during the course of development. The hybridization curves of embryonic RNA with the probe shift toward the position of the adult curve on the $R_0t$ scale as development proceeds. The hybridization curves are however complex, indicating that all the components do not change in a coordinate fashion. This is compatible with the varying accumulation profiles of individual pancreas specific proteins. Both qualitative and quantitative changes are detected during the course of development. The apparent absence of RNA species complementary to some of the cDNA at early developmental times may be due to the fact that the analyses were not carried out at high enough $R_0t$ values, because of limitation in the amount of available RNA. In the 18 day curve, an additional component is detected at high $R_0t$ values. Quantitative comparisons of the $R_0t$ profiles show an approximate 10-fold (1 log unit)

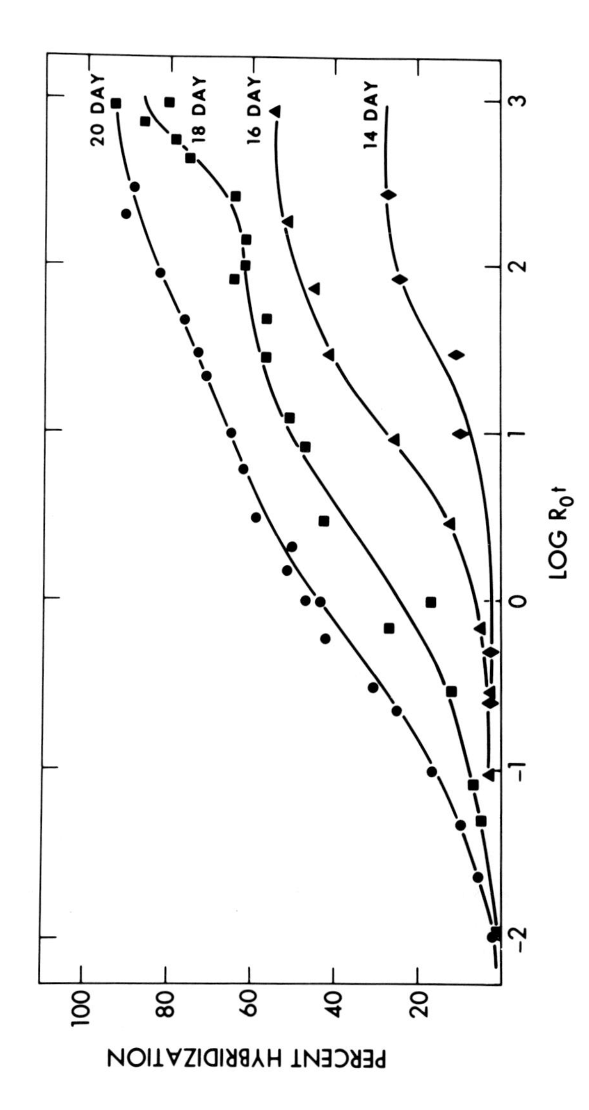


Fig. 10. Hybridization of adult rat pancreas cDNA with embryonic rat pancreas RNA. Total RNA was extracted from embryonic pancreas of the indicated gestational age and hybridized with adult pancreas cDNA as in Figure 8. Hybridization with 20 day embryonic pancreas RNA, ●—●; 18 day RNA, ■—■; 16 day RNA, ▲—▲; 14 day RNA ◆—◆

increase in the concentration of RNA transcripts complementary to the pancreas cDNA during each two days of development except for the 18–20 day period in which there is a somewhat smaller increment; this corresponds to an approximate 500-fold increase in specific pancreas transcripts occurring between 14 days of development and adulthood. There are therefore dramatic changes in the relative concentrations of pancreas transcripts hybridizing with pancreas cDNA during the developmental period. It is not possible in these experiments, however, to discern which species are involved.

There is a significantly higher proportion of the probe which hybridizes to the RNA from protodifferentiated pancreas (14 days) than to RNA from other tissues. Thus there are pancreas specific RNA transcripts present during the protodifferentiated state. It cannot be ascertained from these experiments whether all pancreas specific transcripts are present, because only limited quantities of RNA were available for hybridization; a probe for each species will be needed to answer this question. In the meantime our interpretations are only tentative: (1) The primary transition involves some change in the chromosomes that results in the capability of producing pancreatic specific transcripts. (2) The secondary transition is associated with enhanced transcription of a certain set of genes. (3) The qualitative similarity in the profile of accumulation of pancreas specific transcripts and that of the synthetic rate of the polypeptides suggests that gene expression during this period is controlled at least in part at the transcriptional level.

## VI. MESENCHYMAL FACTOR CONTROLS NORMAL PANCREATIC DEVELOPMENT

Interactions between mesenchymal and epithelial cells are prominently involved in morphogenesis and terminal differentiation of many organ systems (Grobstein, 1967; Wolff, 1968). As in other systems, pancreatic epithelial anlagen separated from their mesenchymal components fail to develop, but co-culture of the recombined tissues leads to normal development. A variety of mesenchymal tissues can complement pancreatic development; in most other tissues, homologous mesenchyme is required for normal development. The mesenchyme can exert its "inducing" effect on pancreatic epithelial rudiments across a Millipore filter, but the intensity of the response is inversely proportional to the pore size and the filter thickness. An ultrastructural analysis suggested that there is no direct contact between epithelial and mesenchymal cells (Kallman & Grobstein, 1964). It was thus concluded that the interaction is mediated by molecules traversing the intercellular space. More recently Saxen and his coworkers (Lehtonen *et al.*, 1975; Wartiovaara *et al.*, 1974) have studied the cellular interactions in kidney formation. It was observed in this system that the inductive response is roughly proportional to the degree of cellular contact. These experiments have again raised the question of whether direct cell contact is required for effective tissue interactions.

We have previously shown that an extract of embryos rich in mesenchymal tissues can replace the mesenchymal requirement for pancreas development (Ronzio & Rutter, 1973; Pictet *et al.,* 1975a,b). This mesenchymal factor (MF) is neither species nor tissue specific. Indeed, an extract from chicken embryos replaces the mesenchymal requirement for rat pancreatic development. In the absence of mesenchyme or MF pancreatic epithelial primordia do not proliferate. In the presence of MF, there is active cell proliferation and later acinar cell differentiation as *in vivo*. MF activity can be assayed by measuring the increased incorporation of [$^3$H]thymidine into DNA. MF activity is found largely in the membrane fractions of cells but can be solubilized in solutions of high ionic strength. In crude preparations, the solubility of MF depends on the calcium concentration; MF is precipitated by less than 2 mM calcium, thus well within the physiological range. Thus it is possible that the diffusion of MF between cells is limited or regulated by the calcium concentration. MF has been partially purified. It is trypsin sensitive and hence appears proteinaceous. Its activity is destroyed by oxidation with sodium periodate and is not reactivated upon reduction by sodium borohydride. Thus a carbohydrate moiety is apparently required; however N-acetyl glucosamine, N-acetyl galactosamine, N-acetyl mannosamine and α-methylglucoside at high concentrations do not competitively inhibit MF activity. MF has two component activities (Filosa *et al.,* 1975). One component is apparently more labile and is destroyed by periodate. This component can be replaced by dibutyryl cyclic AMP or 8-OH cyclic AMP, analogues of cyclic AMP that enter cells. However these derivatives are not sufficient for activity. Another component is also required. This second component is not replaced by cyclic GMP or its derivatives, or by manipulation of $Ca^{++}$ ions or by $Ca^{++}$ ionophores (Filosa *et al.,* 1975; Pictet *et al.,* 1975b).

MF apparently acts on the epithelial cell surface (Levine *et al.,* 1973). Purified MF can be bound to cyanogen bromide activated Sepharose beads (MF-Sepharose). Pancreatic epithelial buds adhere to the MF-Sepharose beads at their basal surface (that normally interacts with the mesenchymal cells). The result is that the pancreatic primordia is everted and the apical surface, that normally faces the lumen, is now in contact with the culture medium. Rapid incorporation of DNA precursors into DNA occurs in cells directly attached to the MF-Sepharose beads (Fig. 11) and later acinar cells containing zymogen granules are observed. These experimental results suggest that the mesenchymal-epithelial interaction is mediated by a molecular species that acts at the level of the cell membrane. Infusion of specific components of the mesenchymal cell by junctional complexes does not appear necessary. On the other hand, close proximity of the cells for effective transfer of materials may be efficacious since MF may not be freely soluble. Our results, then, provide a possible explanation for the observations of Saxen and his colleagues. Depending upon the concentration and the relative solubility of the mediating molecules, close proximity or direct contact of mesenchymal and epithelial cells may be required. Since multiple components may be involved, it is also possible that a particular config-

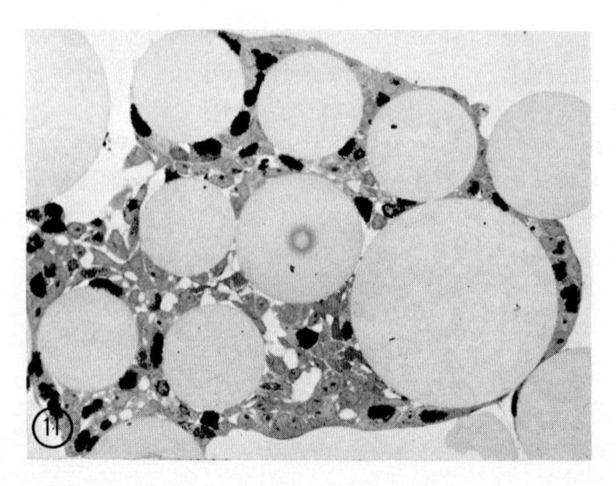

Fig. 11. Pancreatic epithelial cells bind to MF immobilized on Sepharose beads. 12 day epithelia were incubated for 24 hr with MF immobilized on Sepharose beads. During the last 6 hr of the incubation [$^3$H] thymidine was added to the medium. The tissue was fixed and processed for autoradiography. During the culture period the cells have spread on the surface of the beads. The high labeling index demonstrates a stimulation of DNA synthesis by the immobilized MF. If albumin is bound to Sepharose beads there is no binding nor stimulation of DNA synthesis. From: Levine *et al.*, 1973

uration (mosaic) of these molecules is involved at the receptor site; thus direct contact may be required for activity at low concentrations. The role of the basal lamina which normally separates the mesenchymal and epithelial cells should be considered. The mediating molecules may have to traverse this barrier to interact with the epithelial cell; alternatively the basal lamina may be severed in regions so that cell contact and induction can occur. Such selective interruptions in the basal lamina and possible cell contacts have been described (e.g., Mathan *et al.*, 1972). The deterioration of the basal lamina that is observed during active proliferation and development could allow the more effective transit of these molecules. Further experimentation is required to elucidate the details of the proximal regulation involved in epithelial-mesenchymal interactions.

The mesenchymal factor influences the pattern of differentiation in the pancreatic rudiment. Pancreatic epithelial rudiments cultivated for 7 to 8 days in the presence of mesenchymal factor differentiate as *in vivo*. A large majority (greater than 80%) of the cells are differentiated acinar cells. 5% are endocrine, mostly B, cells. The remainder are duct-like cells (Pictet et al., 1975a). When cultivated in the presence of albumin instead of mesenchymal factor, the cultures accumulate very few acinar cells (Fig. 12) but, as shown by electron microscopy, typical $\alpha$ granules are present in the cytoplasm of these cells. Under these circumstances there is no detectable cell proliferation. Thus the majority of the cells in the epithelial primordia have the competence to differentiate into an endocrine A cell. Addition of MF to such rudiments results in a reversal of this endocrine/exocrine cell ratio; within a few days, a large number of differentiated acinar cells containing zymogen granules (Fig. 13) can be

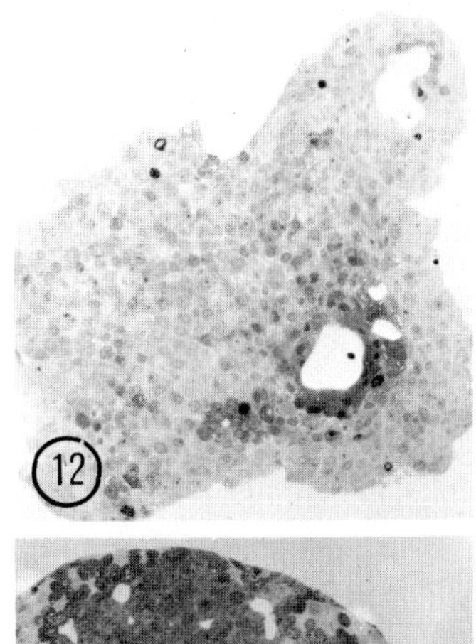


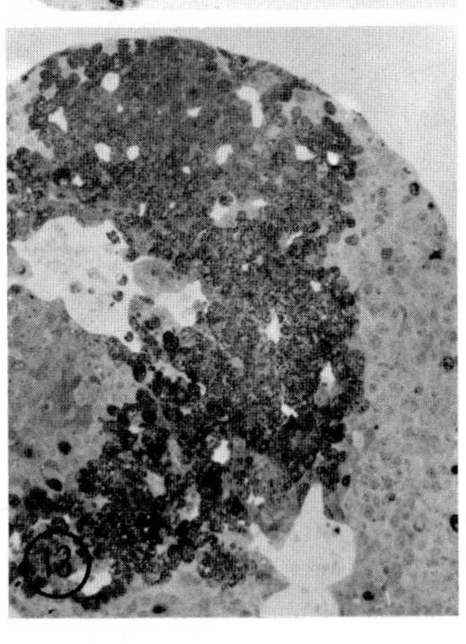


Figs. 12 and 13. The change in the pattern of differentiation occurring in the absence of MF is reversible. Figure 12 is a section of a 12 day pancreatic epithelial bud grown in albumin containing medium for 14 days. Only occasionally acinar cells containing zymogen granules found. Electron microscopic observations show the presence of endocrine granules in most of these cells. If after 7 days in such a culture medium, corresponding to the differentiation period *in vivo*, MF is added to the medium, many acinar cells containing zymogen granules are present 7 days later. Fig. 12 and 13, ×128. (From: Pictet, R. and Rutter, W. J. (1977). Proc. Sigrid Juselius Symp. on "Cell Interactions in Differentiation", Helsinki, Finland, Aug. 1976, in press.)

observed. Thus the pattern of abnormal differentiation is not due to a preferential loss of acinar precursor cells during culture. Cultivation in the presence of MF inactivated at least 90% by periodate treatment, also results in a rudiment containing a lower proportion of acinar cells (approximately 20%), and up to 60% endocrine A cells (Pictet *et al.*, 1975a). In both cases, the duct cell population (the cells that contain no secretory granules) is approximately 15% of the population.

## VII. THE MODULATION OF DEVELOPMENT BY HORMONES AND OTHER EXTRINSIC FACTORS

### *A. Glucocorticoids*

In common with their effects on gene expression in many embryonic tissues (Reif-Lehrer, 1968; Moog, 1971; Moscona, 1971; Eisen *et al.*, 1973; Sugimoto *et al.*, 1974; Topper *et al.*, 1974), glucocorticoids have been shown to influence pancreatic development. Kulka and his colleagues (Cohen *et al.*, 1972) have demonstrated that the potent synthetic glucocorticoid, dexamethasone, substantially stimulates the accumulation of pancreas specific proteins such as chymotrypsinogen in the chicken. It was not determined, however, whether this effect was primarily exerted on the synthesis of the protein or on its secretion. We have carried out an analysis of the effects of glucocorticoids on the development of the rat pancreas, in order to ascertain whether there is a direct requirement of glucocorticoids for the secondary differentiative transition.

Dexamethasone at physiologically active levels exerts two specific effects. It enhances the synthesis of certain pancreas specific proteins and inhibits cell proliferation. Pancreases cultured in dexamethasone during the period of differentiation contain half as many cells as normal but the average cell contains twice as much protein. This increase reflects a higher proportion of secretory proteins. In pancreatic rudiments cultured *in vitro* in the absence of dexamethasone, specific proteins represent approximately 40% of total cell protein; in those cultured in the presence of dexamethasone, the percentage of specific proteins increases to over 70%. Dexamethasone exerts its effects largely on the synthesis of amylase and on procarboxypeptidase B (see Table III). The glucocorticoid effects are evident at the transcriptional level. The relative concentration of pancreatic transcripts which react with the specific pancreatic cDNA probe increases about two orders of magnitude when the rudiments are cultured with dexamethasone. The simplest interpretation of the data is that dexamethasone stimulates the accumulation of certain pancreas specific messenger RNAs (mostly amylase, procarboxypeptidase B), and thus enhances the rate of their synthesis. Whether the effect of dexamethasone is on the rate of synthesis of the transcripts or on some post-transcriptional event is unknown. The dexamethasone effect, however, does not appear to influence the differentiative process per se. Dexamethasone increases the rate of accumulation, but does not result in precocious development of pancreas cells. Furthermore, as stated previously, acinar and B cells differentiate in the absence of glucocorticoids in the medium.

Dexamethasone decreases the relative concentration of insulin by about 3-fold. This effect seems to be partially due to a decrease of insulin synthesis per cell and also to a decrease in the proportion of differentiated B cells in the population. The effect is probably not due to an inhibition of cell proliferation per se, since the inhibitory effect of dexamethasone on cell proliferation is detectable only after two to three days of culture with dexamethasone. At this time, in contrast to accumulation of exocrine proteins, normal insulin accumulation per rudiment is independent of cell proliferation.

The results of these experiments suggest that glucocorticoids modulate the development of the pancreas by affecting cell proliferation and the expression of certain acinar genes, and by inhibiting B cell differentiation. Glucocorticoids are not a requirement for the differentiation process itself, but play a modulatory role.

**TABLE 3**

*Modulation of Growth and Accumulation of Pancreatic Specific Products by Glucocorticosteroids In Vitro*

| | – Dex | + Dex |
|---|---|---|
| μg | | |
| DNA/pancreas | | |
| DNA/pancreas | 6 | 3 |
| protein/pancreas | 150 | 150 |
| μg/DNA | | |
| Amylase | 2.8 | 21.5 |
| Chymotrypsinogen | 6.5 | 13 |
| Procarboxypeptidase A | 0.5 | 1.2 |
| Procarboxypeptidase B | 0.025 | 0.35 |
| Lipase | 0.03 | 0.05 |
| Insulin | 0.072 | 0.031 |

Day 13 pancreatic rudiments were grown for a period of 7 days in the presence of $10^{-7}$ M dexamethasone. Total protein, enzyme and hormone levels are expressed as μg/μg DNA. Units of enzyme activity have been converted to absolute levels knowing the turnover numbers of the various enzymes. (For procarboxypeptidase B and lipase activity pure bovine enzyme has been used (Sanders (1970)).

*B. The Effect of Nutritional Components on Specific Pancreatic Gene Expression*

The rate of growth and the relative content of pancreas specific products are influenced by the nutritional quality, in particular the amino acid content, of the medium. Pancreatic rudiments can differentiate in Eagle's basal medium with the formation of approximately the normal proportion of zymogen granule-containing acinar cells. On the other hand if 6-fold higher levels of amino acids are added, the rate of proliferation of the cells increases and the proportion of pancreas specific proteins increases even more (see Table IV). In the presence of these high levels of amino acids, there is an increase in protein per cell which is quantitatively due to the synthesis of pancreas specific proteins. Thus, it appears that in nutritionally-deprived media the synthesis of cell specific proteins is selectively decreased. This implies that differentiative functions can be regulated independent of the general physiological functions of the cells.

**TABLE 4**

*Amino Acid Levels in the Nutrient Medium Affect Growth and Differentiation of the Embryonic Rat Pancreas**

| | Low Amino Acids (BME) | High Amino Acids (6 × BNE) |
|---|---|---|
| μg protein/pancreas | 50 | 150 |
| μg DNA/pancreas | 3 | 6 |
| protein/DNA ratio | 17 | 25 |
| Exocrine cells | | |
| % total protein | | |
| Chymotrypsinogen | 13 | 25 |
| Amylase | 6 | 12 |
| Endocrine B cells | | |
| Insulin | | |
| (ng hormone/μg DNA) | 28 | 74 |

*Day 13 embryonic pancreases plus 7 days in culture.

## VIII. EMBRYOLOGICAL PRECURSORS TO PANCREAS CELLS: A PLURIPOTENTIAL STEM CELL?

The various differentiated cells in the pancreas can arise from independent or common precursor cells. It has frequently been presumed that the cell origins are independent since the various pancreatic cells are sometimes physically separated; for example, endocrine A and D cells are found in the gut in addition to the pancreas, moreover in certain fish, endocrine B cells are segregated from the exocrine gland. Finally the endocrine and acinar cells differentiate asynchronously. None of the above facts however eliminate a common cell of origin but they do place constraints on the regulation of the differentiation of any putative precursor cell. There are observations which are consistent with but do not prove a common precursor cell for all of the pancreatic cell lines.

A. The precursors to the endocrine islets are originally part of the exocrine epithelial matrix (Pictet & Rutter, 1972). Throughout development, the increase in the number and size of islets occurs by the escape of endocrine cells from the growing exocrine digitations, the endocrine and exocrine cell populations are however never completely separated (Fig. 14). Even in the adult pancreas where the majority of endocrine cells are associated in individual islets, there are many individual or small groups of endocrine cells closely juxtaposed to the duct cells. At the periphery of the larger islets, duct cells and a few acinar cells occasionally can be observed in contact with the endocrine cells (Fig. 15). This close relationship suggests that both acinar and endocrine precursor cells are

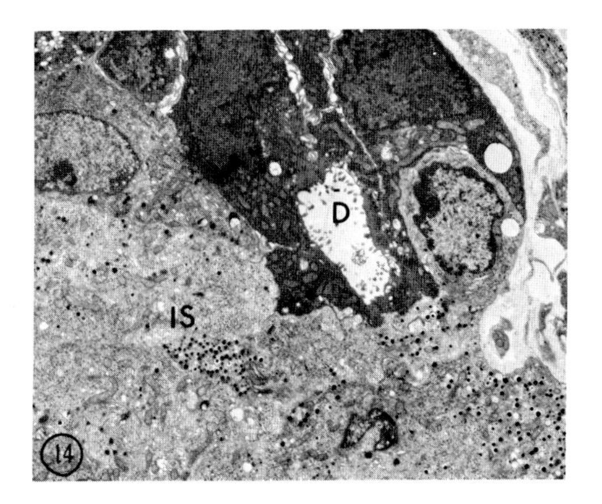


Fig. 14. Fetal rat pancreas. At the end of fetal development the exocrine and endocrine tissue are still in close association. This is demonstrated by direct continuity between endocrine and exocrine tissues. This figure shows the periphery of an islet (IS) which contains differentiated (granule-containing) endocrine cells in direct contact with a duct (D). ×2580

integrated among the duct cells. This integration also explains the random localization of islets throughout the adult pancreas rather than their segregation into localized areas.

B. The proportion of the endocrine cells can be manipulated substantially by cultivation in the absence of mesenchymal factor as described previously. Indeed a large proportion of endocrine A cells observed in rudiments cultivated in the absence of MF suggest that many of the cells are competent to form A cells. This developmental plasticity is somehow modulated by MF.

C. The plasticity of pancreatic development is further indicated by studies with the thymidine analogue, 5-bromodeoxyuridine (BrdU). This compound is known to block cytodifferentiation in many systems (for review see Wilt & Anderson, 1972; Rutter *et al.*, 1973), and to induce virus expression in others (Gerber,

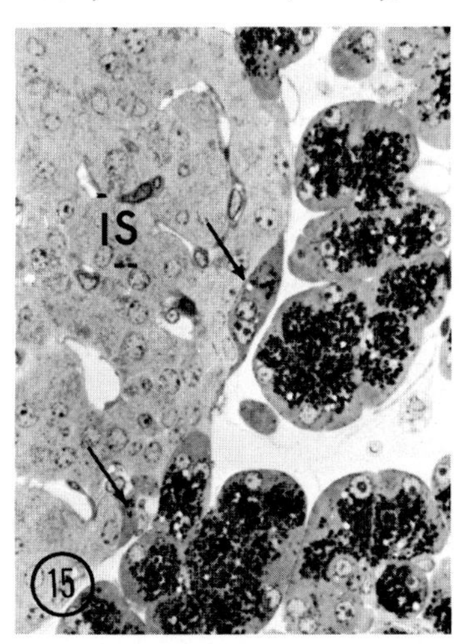


Fig. 15. Adult rat pancreatic islet. In the adult, in contrast to the embryo, the exocrine and endocrine components are individualized. However acinar cells containing zymogen granules (arrows) are often located at the periphery of the islets (IS). ×320

1972; Hampar *et al.*, 1972). In the pancreas, addition of BrdU at low concentrations blocks the accumulation of exocrine specific products without significant alteration of the synthesis of other macromolecules and the proliferation of the cells (Walther *et al.*, 1973). The cells of pancreatic rudiments explanted during the protodifferentiated stage and grown in the presence of BrdU do not undergo acinar or B cell cytodifferentiation (i.e., they do not accumulate secretory products). Instead after 6 or 7 days, the majority of the cells surround large vacuoles filled with fluid (Fig. 16) and resemble duct cells (Fig. 17). There is also a substantial increase in the specific activity of alkaline phosphatase and carbonic anhydrase, two enzymes that are associated with the duct cell population. Specific histochemical analysis shows that these enzyme activities are found in a majority of cells surrounding the vacuoles (Githens *et al.*, 1976). These results suggest that the incorporation of BrdU results in an enhanced formation of duct cells instead of acinar cells. The effects of BrdU on endocrine B cell and acinar cells are not simultaneous. The differentiation of acinar cells is blocked by the addition of BrdU from day 14, while endocrine B cell differentiation proceeds unimpaired. If however the BrdU is added 24 to 36 hours earlier, neither endocrine B cells nor insulin accumulate. Thus there is a sequential period of B and acinar cell differentiation. Differentiated B cells rarely proliferate, rather the increased number of B cells occurring between 16 and 20 days (when 99% of the total insulin is synthesized), occurs by continuous addition of new cells from precursors located within the exocrine matrix.

## IX. STEM CELLS AND THE MECHANISM OF ACTION OF THE MESENCHYMAL FACTOR

During embryonic development differentiated cells in the pancreas accumulate over several days by recruitment of an increasing number of undifferentiated cells and not by the active proliferation of a few early differentiated ones. Thus there must be a pool of dividing precursor cells that are analogous to stem cells. The perpetuation of even a small number of stem cells could secure the maintenance of the differentiated cell population throughout the life of the organism. Whether the five differentiated epithelial cell types distributed in the pancreas arise from a single pluripotential stem cell or from independent stem cells is unknown. The mechanism of regulation of the expression of the various pancreatic phenotypes is also unknown. The alteration in the proportion of the differentiated cells by mesenchyme or MF implies that the mesenchyme plays a basic role in this regulation. It may regulate specifically the expression of acinar function by acting directly on pluripotential cells to induce a differentiative pathway or its effect may be a secondary consequence of the stimulation of proliferation of acinar cell precursors. The determination of the developmental lineage of these cell types and the role of MF is basic to an understanding of pancreas development. The principles may be applicable to organogenesis in general.

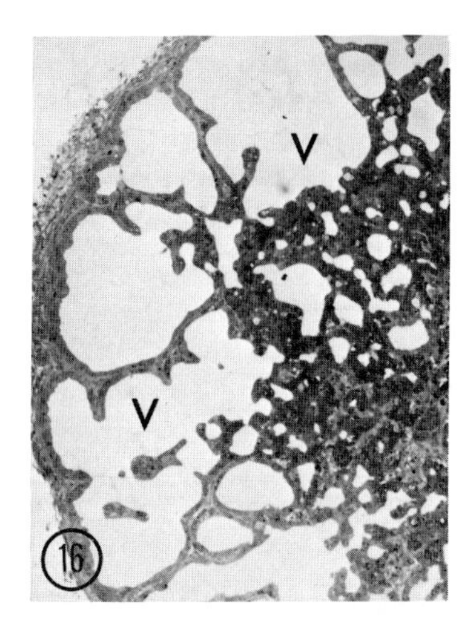


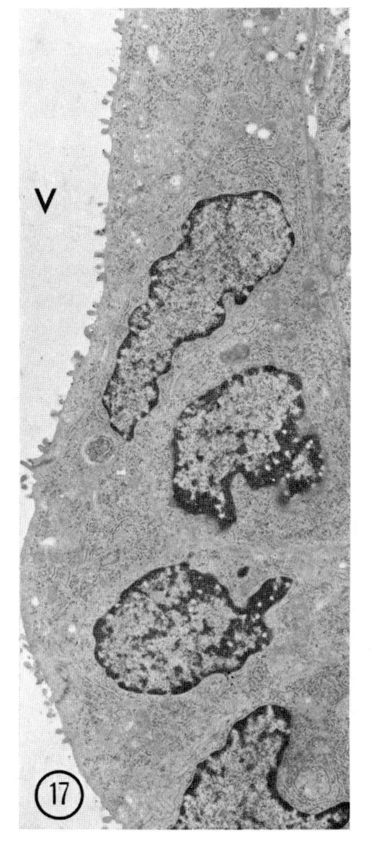


Figs. 16 and 17. BrdU alters the developmental program of the pancreas. Pancreases were explanted at day 14 and culturea ror 6 days in the presence of $2 \times 10^{-5}$ M BrdU. Figure 16 shows that in contrast to normal pancreases, there are large vacuoles (V) surrounded by cells which do not contain zymogen granules. Figure 17 is an electron micrograph of cells which surround the vacuoles (V). They are larger and contain more rough endoplasmic reticulum than the cells of adult or fetal (Githens *et al.*, 1976) pancreatic ducts. They have cuboidal instead of the pyramidal shape of the acinar cells. Furthermore they do not contain zymogen granules nor empty storage vesicles; thus synthesis of product and package membranes are coordinate. Fig. 16, ×64; Fig. 17, ×4960

## ACKNOWLEDGMENTS

This research was supported by grants from the National Science Foundation (BMS72-02222), Juveniles Diabetes Foundation, National Foundation – March of Dimes and a NIH Genetics Center grant (GM 19527). R. P. is a recipient of a National Institutes of Health Career Development Award; J. D. H. is a NIH Postdoctoral Fellow (GM05374); J. C. is a NIH Postdoctoral Fellow (GM05385); R. J. M. is a American Cancer Society Fellow (PF 992) and A. E. P. is a Helen Hay Whitney Foundation Fellow. We would like to thank William Nikovits for preparing tissue sections for in situ hybridization and Jennifer Meek for assistance with embryo dissection.


## REFERENCES


Andrew, A. (1976). IRCS Medical Science Cell and Membrane Biology; Developmental Biology and Medicine **4,** 27.

Blobel, G. and Sabatini, D. (1971). *In* "Biomembranes" (L. A. Manson, ed.), Vol. 2, pp. 193–195. Plenum Press, New York.

Britten, R. J., Graham, D. E. and Neufeld, B. R. (1974). Methods in Enzymol. **29E,** 363–418.

Clark, W. R. and Rutter, W. J. (1972). *Develop. Biol.* **29,** 468–481.

Cohen, A., Heller, H. and Kulka, R. G. (1972). *Develop. Biol.* **29,** 293–306.

Deshpande, A. R. and Siddiqui, M. A. Q. (1976). *Nature* **263,** 588–591.

Devillers-Thiery, A., Kindt, T., Scheele, G. and Blobel, G. (1975). *Proc. Nat. Acad. Sci. U.S.A.* **72,** 5016–5020.

Eisen, H. J., Goldfine, I. D. and Glinsman, W. H. (1973). *Proc. Nat. Acad. Sci. U.S.A.* **70,** 3454–3457.

Filosa, S., Pictet, R. and Rutter, W. J. (1975). *Nature* **257,** 702–705.

Gerber, P. (1972). *Proc. Nat. Acad. Sci. U.S.A.* **69,** 83–85.

Githens, S., Pictet, R., Phelps, P. and Rutter, W. J. (1976). *J. Cell Biol.* **71,** 341–356.

Gould, H. J. and Hamlyn, P. H. (1973). *FEBS Letts.* **30,** 301–304.

Grobstein, C. (1967). *Natl. Cancer Inst. Monogr.* **26,** 279–282.

Hampar, B., Derge, J. G., Martos, L. M. and Walker, J. L. (1972). *Proc. Nat. Acad. Sci. U.S.A.* **69,** 78–82.

Kallman, F. and Grobstein, C. (1964). *J. Cell Biol.* **20,** 399–413.

Kemp, J. D., Walther, B. T. and Rutter, W. J. (1972). *J. Biol. Chem.* **247,** 3941–3952.

Le Douarin, N. and Teillet, M. A. (1973). *J. Embryol. Exp. Morph.* **30,** 31–48.

Lehtonen, E., Wartiovaara, J., Nordling, S. and Saxen, L. (1975). *J. Embryol. Exp. Morph.* **33,** 187–203.

Levine, S., Pictet, R. and Rutter, W. J. (1973). *Nature* **246,** 49–52.

Maniatis, T., Kee, A., Efstradiadis, A. and Kafatos, F. (1976). *Cell* **8,** 163–182.

Mathan, M., Hermos, J. A. and Trier, J. S. (1972). *J. Cell Biol.* **52,** 577–588.

Moog, F. (1971). *In* "Hormones in Development" (M. Hamburgh and E. J. W. Barrington, eds.), pp. 143–160. Appleton-Century-Crofts, New York.

Moscona, A. A. (1971). *In* "Hormones in Development" (M. Hamburgh and E. J. W. Barrington, eds.), pp 169–189. Appleton-Century-Crofts, New York.

Pearse, A. G. E. (1969). *J. Histochem. Cytochem.* **17,** 303–313.

Pearse, A. G. E. and Polak, J. M. (1971). *Gut* **12,** 783.

Pearse, A. G. E., Polak, J. M. and Heath, C. M. (1973). *Diabetologia* **9,** 120.

Pictet, R. and Rutter, W. J. (1972). *In* "Handbook of Physiology, Section 7: Endocrinology" (D. F. Steiner and N. Freinkel, eds.), Vol. 1, pp. 25–66. Williams and Wilkins, Baltimore, Md.

Pictet, R., Rall, L., de Gasparo, M. and Rutter, W. J. (1975a). *In* "Early Diabetes in Early Life" (R. A. Camerini-Davalos and H. S. Cole, eds.). pp. 25–39. Academic Press, New York.

Pictet, R., Filosa, S., Phelps, P. and Rutter, W. J. (1975b). *In* "Extracellular Matrix Influences on Gene Expression" (R. Greulich and H. C. Slavkin, eds.), pp. 531–540. Academic Press, New York.

Pictet, R., Rall, L., Phelps, P. and Rutter, W. J. (1976). *Science* **191,** 191–192.

Polak, J. M., Pearse, A. G. E., Le Lievre, C., Fontaine, J. and Le Douarin, N. (1974). *Histochemistry* **40,** 209–214.

Rall, L. B., Pictet, R. L., Williams, R. H. and Rutter, W. J. (1973). *Proc. Nat. Acad. Sci. U.S.A.* **70,** 3478–3482.

Rall, L. B. (1975). Ph.D. Dissertation, University of Washington, Seattle.

Reif-Lehrer, L. (1968). *Biochem. Biophys. Res. Commun.* **33,** 984–989.

Ronzio, R. A. & Rutter, W. J. (1973). *Develop. Biol.* **30,** 307–320.

Rutter, W. J., Ball, W., Bradshaw, W., Clark, W. R. & Sanders, T. G. (1967). *In* "Secretory Mechanisms in Salivary Glands" (L. H. Schneyer and C. A. Schneyer, eds.), pp. 238–253. Academic Press, New York.

Rutter, W. J., Ball, W. D., Bradshaw, W. S., Clark, W. R. and Sanders, T. G. (1967). *In* "Experimental Biology and Medicine," Vol. 1, pp. 110–124. S. Kargar, Basal/New York.

Rutter, W. J., Clark, W. R., Kemp, J. D., Bradshaw, W. S., Sanders, T. G. and Ball, W. D. (1968). *In* "Epithelial-Mesenchymal Interactions" (R. Fleischmajer and R. E. Billingham, eds.), pp. 114–131. Williams and Wilkins, Baltimore.

Rutter, W. J., Pictet, R. L. and Morris, P. W. (1973). *Ann. Rev. Biochem.* **42,** 601–646.

Sanders, T. G. (1970). Ph.D. Dissertation, University of Illinois, Urbana.

Sugimoto, M., Tajima, A., Kojima, A. and Endo, H. (1974). *Develop. Biol.* **39,** 295–307.

Topper, Y. J. & Vonderhaar, B. K. (1974). *In* "Control of Proliferation in Animal Cells" (I. B. Clarkson and R. Baserga, eds.), pp. 843–852. Cold Spring Harbor Laboratory, Cold Spring Harbor, New York.

Wahn, H. L., Taylor, J. D. and Tchen, T. T. (1976). *Develop. Biol.* **49,** 470–478.

Walther, B. T., Pictet, R. L., David, J. D. and Rutter, W. J. (1974). *J. Biol. Chem.* **249,** 1953–1964.

Wartiovaara, J., Nordling, S., Lehtonen, E. and Saxen, L. (1974). *J. Embryol. Exp. Morph.* **31,** 667–682.

Wilt, F. and Anderson, M. (1972). *Develop. Biol.* **28,** 443–447.

Wolff, E. (1968). *Curr. Top. Dev. Biol.* **3,** 65–94.

# VII. Factors Effecting Differentiation in Lower Eukaryotes

# Action of a Morphogenetic Substance from Hydra

H. Chica Schaller

*European Molecular Biology Laboratory, HEIDELBERG*

## I. INTRODUCTION

From the freshwater coelenterate hydra a morphogenetic substance can be isolated, the head activator, which stimulates head and bud formation in the animal. This head activator is a peptide of approximately 1000 daltons (Schaller, 1973). In the animal the low-molecular-weight head activator is present in an inactive, structure-bound form. It is produced by nerve cells and stored there in neurosecretory granules (Schaller and Gierer, 1973). A peptide with very similar or identical structure and properties was found in other coelenterates, like actinia and metridium, and in the mammalian brain, localised mainly in the hypothalamus (Schaller, 1975a). In the following experiments and arguments are presented which demonstrate how the head activator acts in hydra.

## II. EFFECTS OF THE HEAD ACTIVATOR AS A MORPHOGEN CONTROLLING HEAD FORMATION IN HYDRA

### *A. Specificity*

The head activator is active at an extremely low concentration, less than $10^{-10}$ molar (Schaller, 1973). This is important, because in embryology a number of substances are known, e.g. the classical "inducer" LiCl, which cause unspecific inductions or inhibitions. All these unspecific substances act at concentrations at least 5 orders of magnitude higher than the head activator. The unspecific effects are easily explainable as interfering with the release of specific morphogenetic substances present in the embryo or animal in an inactive, stored form. In hydra this can best be demonstrated by incubating pieces of tissue in distilled water which leads to a release of e.g. head activator and consequently to an "unspecific" induction.

### *B. Graded Distribution*

The capacity to form a head is graded disto-proximally in hydra. A piece of tissue from the upper gastric region will regenerate a head faster and with more tentacles than an equal-sized piece from the lower gastric region (Schaller, 1973). Similarly in reaggregation experiments a head will form preferentially where cells from head-near regions are located (Gierer *et al.*, 1972). If the head-activator is the morphogen, or one of the morphogens responsible for the gradient in head-forming potential, it should be graded disto-proximally in hydra. That this is so is shown in Fig. 1. In the head (hypostomal) region the concentration of the head activator is highest. There is a continuous decrease towards the foot. This shows that the head activator is present in hydra in its structure-bound form as a gradient from head to foot.

### *C. Specific Release During Head Regeneration*

The head activator is released extensively during head, but not during foot regeneration (Schaller, 1976a). This release which occurs in the first 4–8 hr after

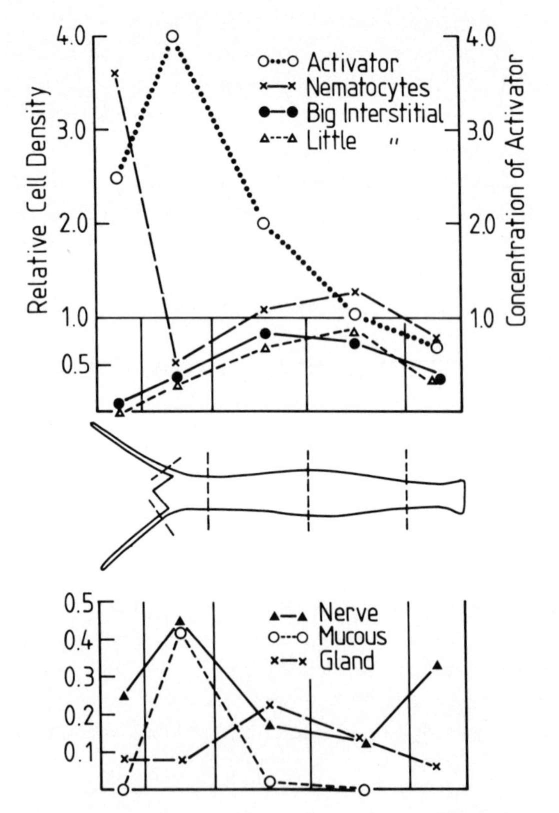


Fig. 1. Distribution of the head activator and of cell types in different body regions of hydra. The concentration of the head activator was determined for each region and is expressed as biological units (BU) per $A_{280}$ nm (Schaller; Schaller and Gierer, 1973). The density of a given cell type is expressed as the ratio of the number of cells of that cell type to the number of epithelial cells (Bode *et al.*, 1973).

cutting (Fig. 2) precedes changes in morphogenetic properties as measured in transplantation experiments e.g. by Webster & Wolpert (1966). If the release of the head activator is prevented (Fig. 2) by incubation of the regenerates in an inhibitor purified by Berking (1974, 1977), head formation is blocked or retarded (Berking, 1974 and unpublished results). This shows that the release of head activator is necessary to initiate head regeneration.

### D. *Action of the Purified Head Activator on Morphogenesis*

Incubation of regenerates in purified head activator accelerates head and bud, but not foot formation (Schaller, 1973). Also the head forming potential of cells in reaggregates is increased, if either the animals before dissociation into cells or the cells are incubated in head activator (Schaller, 1976b). As shown in Fig. 3, reaggregates containing cells treated for 4 hr with head activator form tentacles earlier than untreated cells and the head forms preferentially in that part of the reaggregate where the cells treated with head activator are located. All these arguments favor the notion that the head activator is the morphogen, or one of the morphogens controlling head formation in hydra.

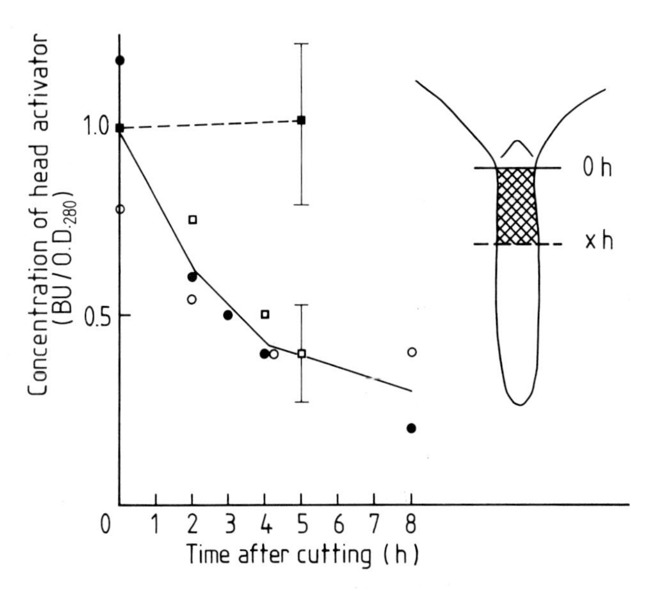


Fig. 2. Decrease in the concentration of head activator in the head regenerating tip. The upper third of a regenerate from which the head (hypostome and tentacles) was removed at time 0, was assayed for its content of head activator at different periods (h) after cutting. Solid line: normal head regeneration; dashed line: regeneration in the presence of inhibitor.

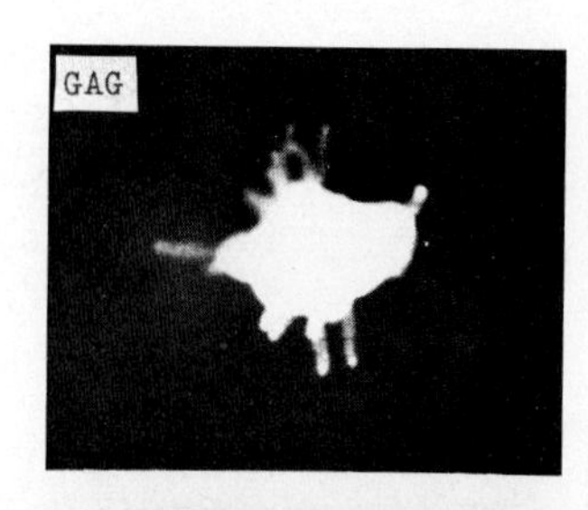

Fig. 3. Linear combinations of aggregates of cells to form triplet reaggregates. 3½ day old triplet reaggregate consisting of cells from the gastric region incubated during disaggregation for 4–5 hr in the presence (A) and absence (G) of head activator (Schaller, 1975b): a) combination GAG, b) AGA, c) GGG.

## III. ACTION OF THE HEAD ACTIVATOR AT THE CELLULAR LEVEL

### A. *Effects as a Growth Hormone*

Steadily growing hydra consist of 5 major cell types: epithelial cells, gland cells and interstitial cells as proliferating cell types, and nerve cells and nematocytes as differentiation products of the interstitial stem cell (Bode *et al.*, 1973; David and Gierer, 1974). The cell cycle of proliferating cells in hydra consists of the following periods (David and Campbell, 1972; Campbell and David, 1974): Mitosis last 1–2 hr, the $G_1$ period is very short or non-existing (0–1 hr), the S phase lasts 10–12 hr for interstitial cells, 12–15 hr for epithelial cells, the $G_2$ period is extremely variable, 4–22 hr for interstitial cells, 24–72 hr for epithelial cells. The $G_2$ period represents the resting phase in the cell cycle of proliferating cells in hydra. It is, therefore, expected that growth control is exerted in $G_2$.

Under starvation conditions cellular proliferation gradually stops (Bode *et al.*, 1973). From this may be concluded that under normal conditions feeding is the trigger for cell division giving rise to the circadian rhythm in cell cycle parameters observed e.g. for epithelial cells (David and Campbell, 1972). In fact, it was found (Schaller, 1976b) that under starvation conditions more cells become arrested in the $G_2$ period. If such starved hydra are incubated in head activator, cells are stimulated to divide. This can be measured either as a more or less immediate increase in the mitotic index occurring 1–2 hr after addition of head activator or, since proliferating cells enter the S phase directly after mitosis, as a more gradual increase in the labelling index of cells in autoradiographs. The effect of the head activator on epithelial cells in 24 hr starved hydra is shown in Fig. 4, on interstitial cells (big interstitial cells in nests of 1 and 2) in Fig. 5. In spite of the missing feeding stimulus epithelial cells of untreated control animals

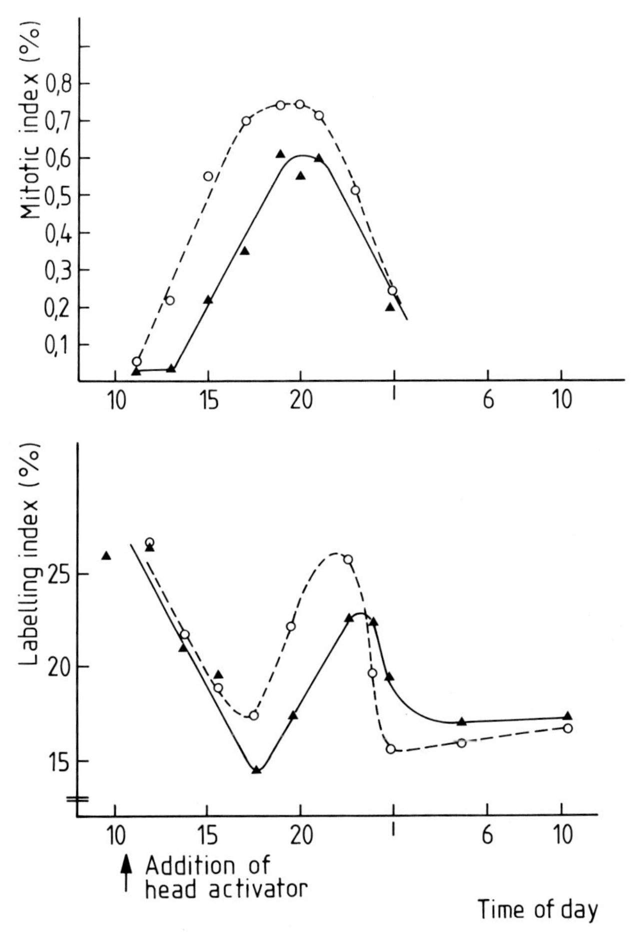


Fig. 4. Effect of the head activator on the mitotic and labelling index of epithelial cells. The animals were starved for 24 hr at the start of the experiment (10 a.m.). Half of the animals were incubated in 20 BU of purified (> $10^6$ fold) head activator (dashed line), half were used as untreated controls (solid line). At the times shown 10 hydra each were macerated to determine the mitotic index and 10 each were pulse-labelled with [$^3$H] thymidine and macerated 30 min later to determine the labelling index.

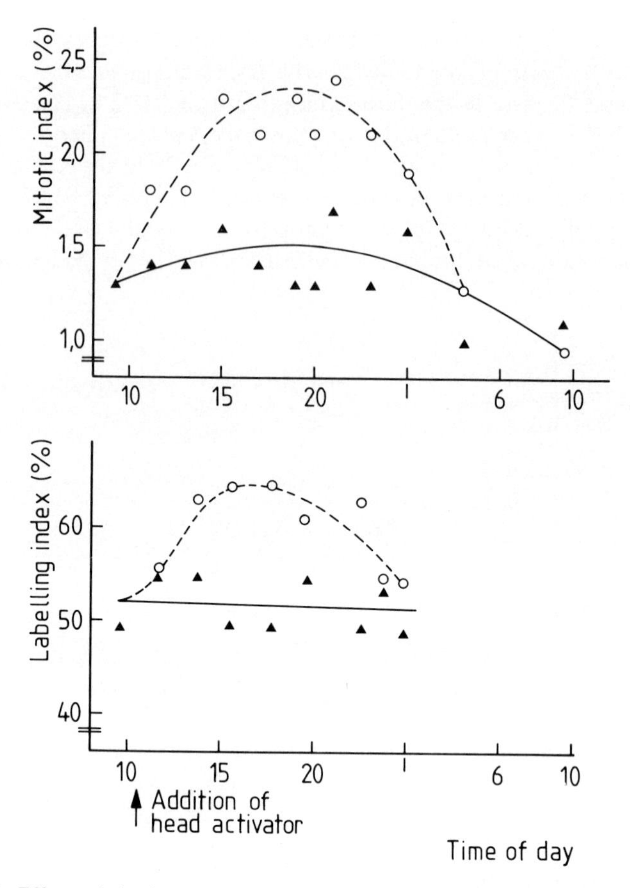


Fig. 5. Effect of the head activator on the mitotic and labelling index of interstitial cells (big interstitial cells in nests of 1 and 2). Solid line: untreated controls, dashed line: head-activator treated animals.

still show somewhat less pronounced fluctuations in cell cycle parameters. No daily rhythm was observed in interstitial cell cycles. With head activator cells re-enter the cell cycle earlier and more cells do so. The head activator is, therefore, able to mimick the effects normally produced by feeding by anticipating and enlarging the fluctuations in case of the epithelial cells and by re-introducing them in case of the interstitial cells. Likewise the labelling index of gland cells and of interstitial cells in nests of 4 and 8 is increased after a 12 hr incubation in head activator (Table I). From this is concluded that the head activator triggers all proliferating cell types in hydra to go through mitosis and start DNA synthesis.

As shown by David and Gierer (1974) in the differentiation pathway of interstitial cells to nerves one stem cell very probably differentiates in a single division step to two nerves, in the pathway to nematocytes in several division

**TABLE 1**

*Effects of the Head Activator on the Labelling Index of Gland Cells and of Interstitial Cells in Nests of 4 and 8*

| Animals | Labelling index (%) | | |
|---|---|---|---|
| | Gland cells | Nest of 4 i-cells | Nest of 8 i-cells |
| Control | 15.2 ± 2.9 | 58.0 ± 6.8 | 72.9 ± 9.2 |
| Animals treated 12 hr with head activator | 22.8 ± 2.6 | 73.4 ± 3.7 | 89.8 ± 6.3 |

steps to 8 or 16 and less frequently into 4 or 32 nematocytes (Fig. 6). In Table I is shown that with head activator more interstitial cells in nests of 4 and 8 become labelled, i.e. that they enter the S phase earlier than untreated controls. This indicates that the intermediate steps in the proliferation of interstitial cells to nematocytes are shortened in the presence of head activator.

In addition to this stimulating effect on cells which continue to proliferate, the head activator also stimulates those cells to divide which differentiate after their last mitosis to nerve cells and to nematocytes (Schaller, 1976b). From the last S phase to recognisable differentiation products at least 18 hr are needed in case of nerve cells, 12 hr in case of nematocytes (David and Gierer, 1974). If head activator is present during this period, differentiation products appear earlier. The outline of such an experiment is given in Fig. 7 and the outcome in Table II.

In all these cell types, in dividing cells and in differentiating cells, the head activator seems to follow the same pattern of action. By triggering cells to go through mitosis the $G_2$ period is shortened. The head activator is thus one of the substances which control cell cycle times and thus growth in hydra.

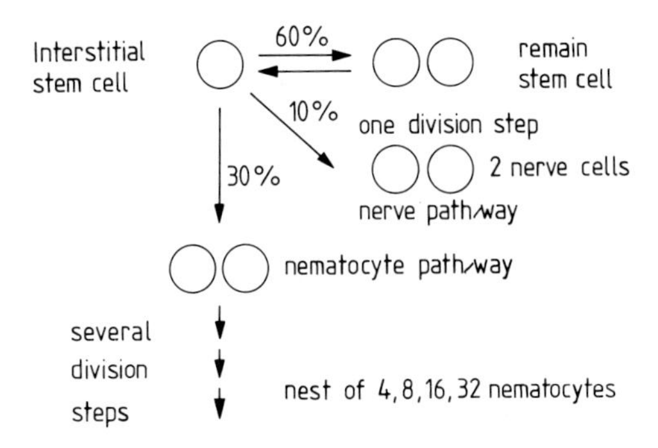


Fig. 6. Cell flow model of the proliferation and differentiation of interstitial stem cells (according to David and Gierer, 1974).

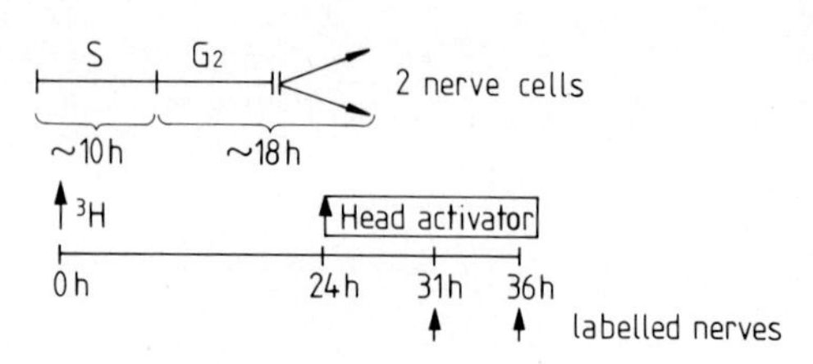


Fig. 7. Differentiation of interstitial cells to nerves. A [$^3$H] thymidine pulse was given at time 0, head activator was added 24 hr later. After a 7 and 12 hr period of incubation in head activator controls and treated animals were dissociated into cells by maceration, and the labelling index of nerve cells was determined.

### *B. Effects on the Determination of Interstitial Stem Cells*

David and Gierer (1974) presented evidence that close to the head the interstitial stem cells preferentially differentiate to nerves, in the gastric region to nematocytes. During head regeneration a reprograming of the interstitial cells of the gastric region occurs with the effect that approximately 28 hr after cutting the number of nerve cells doubles in the head regenerating tip, whereas that of nematoblasts decreases (Bode *et al.*, 1973). Since from the last S phase to recognisable nerve cells 16–18 hr are needed and the S phase lasts 10 hr (David and Gierer, 1974), it can be calculated that the rise in nerve cell density after 28 hr is probably the result of a new determination of pre-S-phase stem cells. That this is so was shown by the experiments outlined in Fig. 8. [$^3$H] thymidine pulses were given either before or after initiation of head regeneration by cutting, nerve cell labelling was measured at least 28 hr later. It was found (Schaller, 1976c) that nerve cells were labelled, if the [$^3$H] thymidine pulse was given later than 4 hr after cutting, i.e. the interstitial precursor cells to the new nerves of the head regenerating tip start to be in the S period 4–6 hr after cutting.

To determine when in the cell cycle the determination of interstitial cells to nerves occurs, the regenerates were incubated for varying periods of time after cutting in head activator. As shown in Table III a further increase in new nerve cells is obtained if animals are incubated from 0–5 hr in head activator. No increase is found, if head activator is present from 0–2 hr only. From this is

**TABLE 2**

*Acceleration of Nerve Cell Differentiation in the Presence of Head Activator*

| | Labelled Nerve Cells (%) | |
|---|---|---|
| Incubation period | 7 hr | 12 hr |
| Control | 2.3 ± 1.4 | 5.5 ± 1.3 |
| Treated with head activator | 5.6 ± 1.6 | 11.0 ± 1.1 |

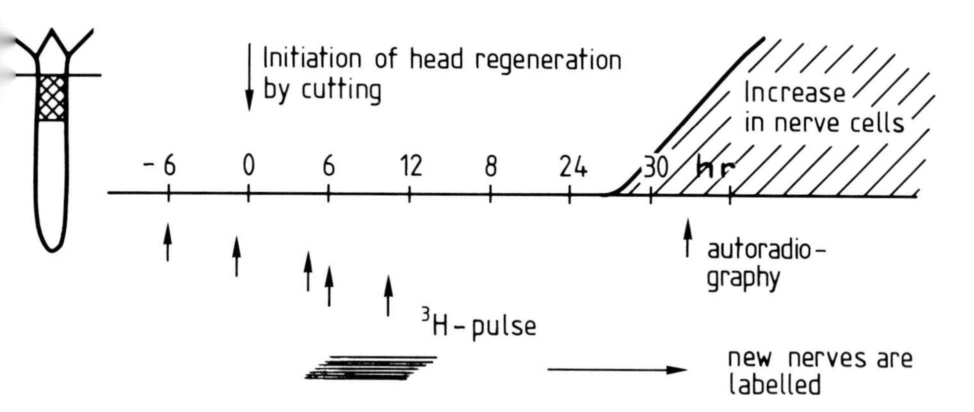


Fig. 8. Determination of interstitial cells to nerves during head regeneration. [$^3$H] thymidine pulses were given as indicated. The head regenerating tip (upper third of regenerate) was dissociated into cells 28–36 hr after initiation of head regeneration by cutting and the labelling index of nerve cells was determined (Schaller, 1976c).

**TABLE 3**

*Effect of the Head Activator on the Determination of Nerve Cells during Head Regeneration. [$^3$H] thymidine was injected into the regenerates between 8 and 10.5 hr after cutting. At 36 hr the head regenerating tip (14,000 ± 2000 cells) was excised and 20 pieces each prepared for autoradiography. (4–5 × 100–150 cells were counted each; the standard deviation was calculated from these individual counts)*

| Time in 20 BU of Head Activator | Nerve Cells Labelling Index % ± S.D. |
|---|---|
| Control | 42.0 ± 1.9 |
| 0–2 hr | 35.7 ± 4.4 |
| 0–5 hr | 51.6 ± 2.3 |
| 0–10.5 hr | 48.4 ± 2.7 |

concluded that the determination of interstitial cells to nerves – or processes leading to that determination – occur before or during the very early S period of the interstitial precursor cells and that presence of head activator is necessary for this determination.

The ratio of interstitial cells which remain stem cells to those which differentiate to nerves or to nematocytes is relatively constant under normal conditions (David and Gierer, 1974). Since with head activator more stem cells are determined to enter the nerve cell pathway a decrease in one of the other pathways is expected. It was found (Schaller, 1976c) that in hydra treated with head activator the number of interstitial cells in nests of 4 and 8 was reduced, whereas in nests of 1 and 2 no change occurred. This suggests that the head activator has a negative effect on the determination of interstitial cells to nematocytes, the proliferation of stem cells is not affected, the determination of interstitial cells to nerves is stimulated.

## CONCLUSION

The head activator has two types of effects at the cellular level: 1) It acts as a growth hormone or mitogen by stimulating cells arrested in the $G_2$ period to go through mitosis. 2) Its presence is necessary for the determination of interstitial cells to nerves. This determination occurs before or in the very early S period of the interstitial precursor cells. Since in hydra no $G_1$ period exists (David and Campbell, 1972; Campbell and David, 1974), this means that the effect on the determination may be exerted in the $G_2$ period as well. If the assumption is made that cells possess some internal mechanism to measure the time spent in the $G_2$ period, the following model is proposed to link the two processes: If an interstitial cell is close to the head, the high head activator concentration leads to a short $G_2$ period and therefore to determination to nerves. Further down the body column the somewhat lower head activator concentration results in longer $G_2$ periods and may allow interstitial cells to remain stem cells. At still lower levels and longer $G_2$ periods determination of interstitial cells to nematocytes may be possible. Whether such a model is applicable for hydra and how many other substances are involved in these control processes needs further investigation.

The effects of the head activator at the cellular level, namely the rapid influence on cell cycles and the more permanent influence on cellular determination may explain its effect or its function as a morphogen. In regions with a high head activator content more nerve cells will form. These nerve cells produce more head activator and thus insure that the head activator potential stays high. The high head activator potential leads to the maintenance of head structures, to the induction of new buds, tentacles or heads. Its action on stimulating growth, on the other hand, is reflected at the morphogenetic level as acceleration of processes like bud outgrowth, tentacle outgrowth or head formation during head regeneration.

## REFERENCES


Berking, S. (1974). Ph.D. Thesis, Tübingen.

Berking, S. (1977). Wilhelm Roux's Archives **181**, 215–225.

Bode, H., Berking, S., David, C. N., Gierer, A., Schaller, H. C. and Trenkner, E. (1973). *Wilhelm Roux's Archives.* **171**, 269–285.

Campbell, R. D. and David, C. N. (1974). *J. Cell Sci.* **16**, 349–358.

David, C. N. and Campbell, R. D. (1972). *J. Cell Sci.* **11**, 557–568.

David, C. N. and Gierer, A. (1974). *J. Cell Sci.* **16**, 359–376.

Gierer, A., Berking, S., Bode, H., David, C. N., Flick, K., Hausmann, G., Schaller, H. C. and Trenkner, E. (1972). *Nature New Biol.* **239**, 98–101.

Schaller, H. C. (1973). *J. Embryol. Exp. Morph.* **29**, 27–38.

Schaller, H. C. (1975a). *J. Neurochem.* **25**, 187–188.

Schaller, H. C. (1975b). *Cell Differentiation* **4**, 265–272.

Schaller, H. C. (1976a). *Wilhelm Roux's Archives* **180**, 287–295.

Schaller, H. C. (1976b). *Cell Differentiation* **5**, 1–11.

Schaller, H. C. (1976c). *Cell Differentiation* **5**, 13–20.

Schaller, H. C. and Gierer, A. (1973). *J. Embryol. Exp. Morph.* **29**, 39–52.

Webster, G. and Wolpert, L. (1966). *J. Embryol. Exp. Morph.* **16**, 91–142.

# Mating-Type Specific Pheromones as Mediators of Sexual Conjugation in Yeast

Vivian L. MacKay

*Waksman Institute of Microbiology*
*Rutgers University*
*New Brunswick, New Jersey 08903*

## I. INTRODUCTION

Simple eukaryotic organisms, such as fungi or ciliates, are often excellent subjects for studies of cellular regulation at the molecular level, since they lack the complexity of higher organisms, yet possess many of the same characteristics. Furthermore, in some species such as the yeast *Saccharomyces cerevisiae*, their genetic systems have been well defined, thereby allowing detailed analysis comparable to bacterial genetics in an organism that behaves remarkably like higher eukaryotes. One cellular regulatory process that has been amenable to both genetic and biochemical investigations is sexual conjugation in *Sacch. cerevisiae*, a function which is under the control of the mating-type locus (*mat*).

Conjugation results from the fusion of two complementary cell types, a phenomenon which is dependent upon the action of several diffusible hormone-like substances (pheromones) and the interaction of surface-bound agglutination factors. Since mutants are available that are altered in one or more aspects of conjugation, the mating process provides an opportunity to investigate questions of hormonal control, cell recognition, and cell fusion at the molecular and organismic level. This paper will summarize the results from a number of laboratories, with particular emphasis on the roles of the mating pheromones as mediators of conjugation. More detail about the genetics, physiology, and biochemistry of the mating-type regulatory system is available in a recent review (Crandall *et al.*, 1977).

### A. *The Life Cycle of Saccharomyces cerevisiae*

Since *Sacch. cerevisiae* has both a stable haploid phase and a diploid phase (Fig. 1), mutations can be induced, isolated, and characterized in haploids and

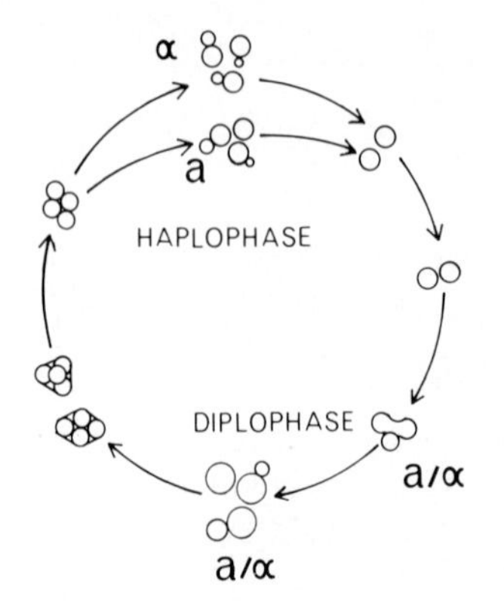


Fig. 1. The life cycle of bakers' yeast, *Saccharomyces cerevisiae.*

then easily crossed into diploids for complementation and recombination analyses. Haploid cells are either mating type *a* or mating type α, a phenotype that is determined by the mating-type locus on chromosome III. (From genetic analysis, *Sacch. cerevisiae* appears to have 17 chromosomes in the haploid state (Mortimer and Hawthorne, 1973).) As long as *a* and α haploids are cultured separately, they will reproduce by a mitotic budding cycle. (See Hartwell, 1974 for a review of the mitotic cycle.) However, if *a* and α cells are mixed, within two hours, the culture contains dumbbell-shaped zygotes that result from the direct fusion of *a* and α cells. Since cell fusion is followed immediately by nuclear fusion, zygotes are diploid and heterozygous (*a*/α) at *mat*. This process of sexual conjugation results in part through the action of the mating-type specific pheromones. Zygotes then bud off diploid cells that also reproduce mitotically by budding. However, if nutritional conditions are altered, diploids can be induced to undergo meiosis and sporulation, resulting in the development of four haploid ascospores enclosed in the wall of the old diploid cell. By enzymatic digestion, the outer wall material can be removed and the four ascospores isolated as a tetrad, with the use of a microneedle. These spores can then be separated with the microneedle and cultured separately, leading to the recovery of all four haploid products of a single meiosis, two of which are mating type *a* and two are α. Thus, the mating type segregates in a Mendelian fashion, as expected for a phenotype specified by a single nuclear gene.

## *B. Regulation by Mating Type*

Before a discussion of the intercellular interactions that lead to conjugation, several other aspects of the *mat* regulatory system should be mentioned. In addition to the difference in mating behavior between *a* and α haploid cells, *mat* also influences processes in the diploid phase. This role of *mat* is disclosed by a comparison of the properties of diploid cells heterozygous at *mat* (*a*/*q*) with those homozygous at *mat* (*a*/*a* or α/α). (Homozygous *mat* diploids do not occur normally in the life cycle as shown in Fig. 1, but can be obtained in the laboratory as mitotic recombinants of *a*/α diploids or as meiotic segregants from tetraploids.) Although the homozygous *mat* diploids mate with the same high frequency seen with haploid cells and also produce their respective sex pheromone (Duntze *et al.*, 1970), *a*/α cells do not mate at all (Gunge and Nakatomi, 1972) nor do they produce or respond to sex pheromones (Duntze *et al.*, 1970; MacKay and Manney, 1974a; Betz *et al.*, 1977). Other properties of *a*/α diploids that distinguish them from homozygous *mat* diploids include the ability to undergo meiosis and sporulation (Lindegren and Lindegren, 1943; Roman and Sands, 1953), increased survival after X-irradiation (Mortimer, 1958; Laskowski, 1960), higher frequencies of mitotic recombination after induction with UV light (Friis and Roman, 1968), and different budding patterns (R. K. Mortimer, personal communication; B. Weinstein and V. L. MacKay, unpublished). Thus, heterozygosity at *mat* in diploids results in the apparent inhibition of certain cellular functions and the implementation or enhancement of others. Because of the multiplicity of effects of *mat* on both haploids and diploids, our working hypothesis is that *mat* itself is a regulatory gene that turns on or off various structural genes located elsewhere in the genome. At present, we have no reason to believe that any of the pheromones, mating factors, or any other identifiable gene products are specified by *mat*.

## II. PHYSIOLOGICAL LANDMARKS LEADING TO CONJUGATION

At the macroscopic level, several physiological events leading ultimately to conjugation have been identified, including synchronization of cells in the G1 stage of the life cycle, initial pair formation, heterosexual agglutination, and formation of the conjugation bridge. As will be discussed in Sections III and IV below, each of these steps in the preconjugational sequence appears to be mediated in part by the mating pheromones and the surface-bound factors.

### *A. G1 Arrest*

Bücking-Throm *et al.* (1973) and Hartwell (1973) have demonstrated that, in mixed *a* and α cultures, cells of both mating types accumulate as unbudded cells in the G1 stage of the cell cycle, prior to the initiation of the next round of DNA synthesis. The observed G1 arrest appears to be the result of the action of the diffusible pheromones. Furthermore, in studies using different temperature-sensitive *c*ell *d*ivision *c*ycle (*cdc*) mutants (B. Reid, cited in Hartwell, 1974), the only mutant exhibiting continued mating ability at the restrictive temperature was *cdc-28*, whose lesion causes the cell to arrest in G1 at the same point as the pheromone-induced block. Finally, pair formation, which is the next step of the preconjugational pathway, occurs only between single, unbudded cells (Campbell, 1973; Sena *et al.*, 1973). All of these results indicate that mating is restricted to the G1 interval when haploid cells contain a single complement of the genome. This mechanism would seem to be optimal for subsequent nuclear fusion, replication and mitosis or meiosis of the resultant diploid.

### *B. Initial Pair Formation*

The second interaction between cells of opposite mating type is cellular recognition by formation of weak bonds between cell pairs (Campbell, 1973; Sena *et al.*, 1973). This weak bonding may occur between boiled cells or cell wall fragments and live cells of opposite mating type (Sakai and Yanagishima, 1972), as well as in cell mixtures incubated in saline instead of growth medium (Campbell, 1973). Therefore, at least one kind of recognition factor is present at all times on the cell surface.

### *C. Heterosexual Agglutination*

Following initial pair formation, the development of larger agglutinated clumps occurs in mating mixtures in growth medium and requires protein synthesis (Sakai and Yanagishima, 1971, 1972; Sena *et al.*, 1973; Yanagishima, 1973). These clumps exhibit tighter intercellular binding and cannot be disrupted by rapid pipetting or mixing (Radin *et al.*, 1973). Thus, heterosexual agglutination may result from the synthesis or activation of additional or more reactive surface-bound agglutination factors.

### *D. Formation of the Conjugation Bridge*

Formation of the conjugation bridge and cell fusion require breakdown of the thick wall of yeast cells. In electron micrographs of conjugational pairs (Osumi *et al.*, 1974), the parental cell walls were observed to become progressively thinner as the area of contact between the cells enlarged. This cell wall dissolution was most pronounced in the electron transparent region thought to be glucan. Changes in cytoplasmic membrane structure occurred simultaneously until finally both the wall material and the membrane became sufficiently fragmented to allow cytoplasmic fusion. Although pronounced alterations of cell walls may require extended physical interaction between cells, the process can be initiated in either mating type by treatment with the pheromone of the opposite mating type (Sections III, A and IV, A).

## III. PROPERTIES OF MATING FACTORS

The existence of diffusible substances associated with mating in *Sacch. cerevisiae* was first suggested by Winge (cited in Levi, 1956). When *a* and α haploid cells are paired close together but not touching, they elongate and eventually fuse at the narrow end of the "copulatory processes" (Ahmad, 1953; Levi, 1956). Levi further supported this hypothesis by demonstrating that *a* cells undergo a distinct morphological change when they are placed on agar on which a mating mixture had previously been incubated. The asymmetric, pear-shaped cells that develop have been called shmoos (H. Roman, cited in MacKay and Manney, 1974a), such as those shown in Fig. 2. These early observations have stimulated a great deal of recent research into the nature and function of the diffusible substances and of other factors correlated with the mating reaction. Before discussing their proposed roles in the preconjugational pathway, I will summarize the available data on their chemical, physical, and biological properties (Table I).

*A. α-Factor*

In Levi's experiments (1956), *a* cells developed into shmoos when incubated on agar medium that had been preconditioned by a mating mixture. The pheromone (α-factor) responsible for the reaction (Fig. 2) is not unique to mating mixtures, however, but is secreted constitutively by α cells into the growth medium (Duntze *et al.*, 1970). Duntze and co-workers (Duntze *et al.*,

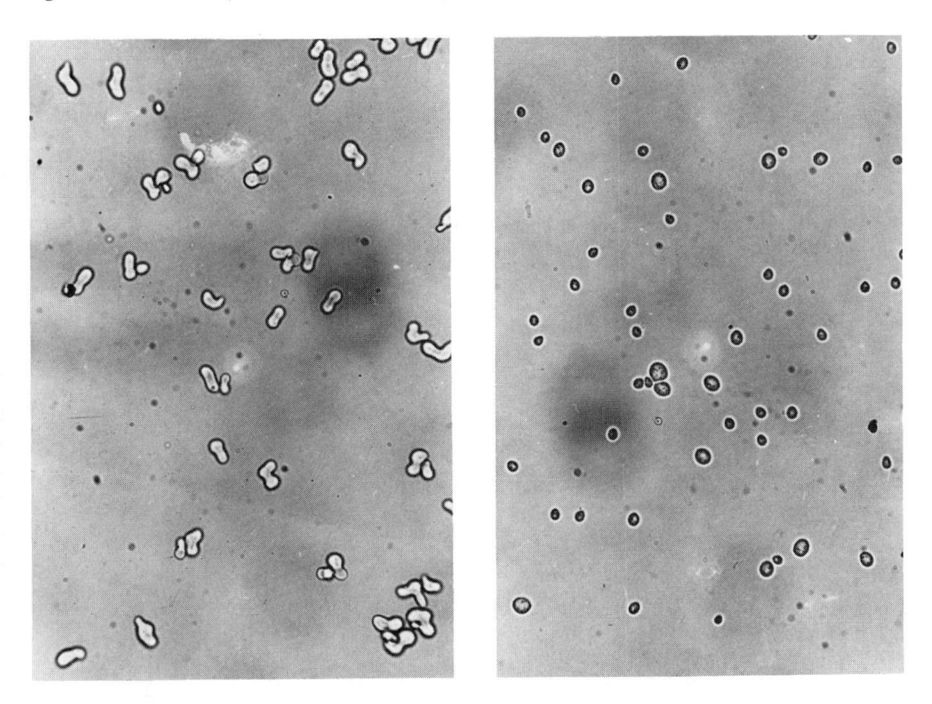

Fig. 2. Formation of shmoos in *a* cells treated with α-factor. Left: Stationary phase cells of mating type *a* exposed for 4 hours to 8 units/ml of α-factor prepared from a culture filtrate of mating type α cells by the method of Duntze *et al.* (1970). Right: Untreated stationary phase cells of mating type *a*. (From MacKay and Manney, 1974a)

**TABLE 1**

*Diffusible and Surface-bound Mating Factors in Sacch. cerevisiae*

| Factor | Made by | Acts on | Chemical nature | Effects |
|---|---|---|---|---|
| α-factor | α cells | *a* cells | Oligo-peptide | G1 arrest<br>Cell wall changes<br>Shmoos<br>Agglutination factors<br>Induction of *a*-factor II |
| *a*-factor I | *a* cells | α cells | Protein | Transient G1 arrest |
| *a*-factor II | *a* cells | α cells | Protein | Cell wall changes<br>Shmoos |
| Barrier factor | *a* cells | *a* cells or α-factor | ? | Inhibition of response to α-factor |
| *a* agglutination factor | *a* cells | α cells | Glyco-protein | Intercellular binding |
| α agglutination factor | α cells | *a* cells | Glyco protein | Intercellular binding |

*See text for details and references.

1973; Duntze, 1974; Stötzler *et al.,* 1976) have purified, characterized and sequenced α-factor, determining that it is a small peptide (*ca.* 1700 daltons) with the following primary amino acid sequence:

$H_2N$-(Trp)-His-Trp-Leu-Gln-Leu-Lys-Pro-Gly-Gln-Pro-Met-Tyr-COOH. (The amino terminal tryptophan is sometimes lacking and the methionine residue can be present as a methionine sulfoxide.) Cupric ion co-purifies with α-factor, but it is not known whether copper is required for activity. Synthesis of α-factor is inhibited immediately by the addition of cycloheximide or by different conditional mutations that prevent protein or RNA synthesis (Scherer *et al.,* 1974), implying that α-factor synthesis requires ribosomal translation of a specific mRNA. (The small size of the molecule suggests that it might result from translation of a larger mRNA or from cleavage of a larger polypeptide.)

In the presence of α-factor, the formation of *a* shmoos is only one of several responses to the pheromone. Within one generation, essentially all of the treated *a* cells arrest in late G1 of the cell cycle (Bücking-Throm *et al.,* 1973). Thus, subsequent nuclear DNA replication and cell division are blocked, but the synthesis of mitochondrial DNA, cellular protein, and bulk RNA are unaffected (Throm and Duntze, 1970; Petes and Fangman, 1973). Cell wall synthesis also continues, but the new wall material differs structurally by containing more glucan and less mannan than the wall of untreated cells (Lipke *et al.,* 1976). Furthermore, the mannan of *a* shmoos contains an increased proportion of shorter side chains and unsubstituted backbone mannose moieties. Shmoos are more susceptible to lysis by glucanases, a result which suggests that, in mating mixtures, breakdown or restructuring of the cell wall may begin as a result of pheromonal action before cells have established contact. Lipke *et al.* (1976) have also noted in electron micrographs that the cell wall at the tip of the shmoo is thinner, as was seen for the walls of mating pairs in the early stages of interaction (Osumi *et al.,* 1974). In addition, small vesicles are visible in the region of the tip (Fig. 3; W. Duntze, personal communication) and might contain β-glucanases as proposed for the vesicles present during budding in vegetatively growing cells (Wiemken, 1975). Thus, it seems possible that the formation of the shmoo morphology is the result of the change in cell wall structure and prolonged inhibition of the cell division cycle in G1, while RNA and protein synthesis continue unimpaired.

The mechanism of action of α-factor has not been elucidated but it could be either transported into the cell where it might trigger the initiation of mating or it may interact with a surface receptor to stimulate *a* cells. The possibility that α-factor acts by altering the intracellular concentration of cyclic 3′,5′-adenosine monophosphate (cAMP) was examined but no differences were found either shortly after treatment of *a* cells with α-factor or after prolonged incubation (V. L. MacKay, unpublished). Furthermore, addition of cAMP, its dibutyryl derivative, or dibutytyl cGMP did not induce the shmoo response (W. Duntze, personal communication). Thus, the mechanism of G1 arrest does not appear to be analogous to that reported for many mammalian cell culture systems in which cAMP levels are elevated in resting G1 cells (Abell and Monahan, 1973).

There are at least two additional aspects of the *a* cell response to α-factor that will be discussed in detail below: 1) the increased agglutinability of *a* cells

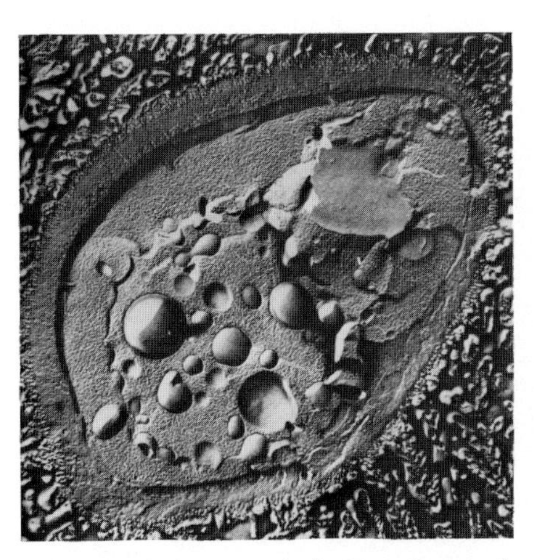

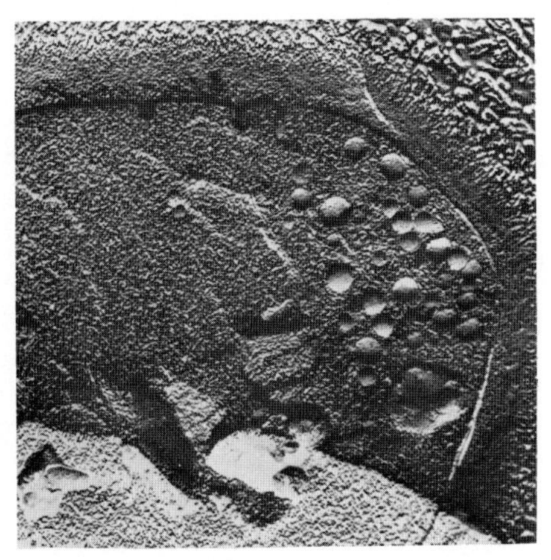

Fig. 3. Freeze-etch electron micrographs of *a* shmoos, formed in response to treatment with α-factor. Left: Whole shmoo, showing thinning of the cell wall at the tip and the presence of small vesicles under the thin area. Right: Higher magnification of the tip area showing small vesicles. (Courtesy of W. Duntze, unpublished.)

treated with α-factor and 2) the appearance of an *a*-factor activity (that acts on α cells) in the medium of *a* cultures incubated with α-factor.

The question of the significance of α-factor to the mating process can be answered indirectly by the properties of nonmating (sterile or *ste*) mutants isolated from both *a* and α parent strains. In a collection of approximately 400 α *ste* strains, 51% had simultaneously become defective for α-factor production, whereas all 66 *a ste* mutants were incapable of responding to α-factor, as assayed by both G1 arrest and shmoo formation (MacKay and Manney, 1974a). Subsequent genetic analysis of many of the mutants established that in each mutant the dual phenotypes resulted from single mutational events, that could occur in one of several loci, either linked to *mat* or unlinked (MacKay and Manney, 1974b).* Recently, Manney and Woods (1976) isolated *a* strains that are resistant to α-factor (i.e., that do not exhibit G1 arrest and shmooing) and found that all of these are also sterile. Therefore, the correlation between mating ability and α-factor properties of the mutants demonstrates that the response to α-factor is essential for conjugation, whereas the synthesis of the pheromone is probably necessary but not sufficient.

### *B. a-Factor(s)*

When α cells are incubated in culture medium from *a* strains, they do not form shmoos, suggesting that *a* cells do not synthesize constitutively an activity analogous to α-factor. However, within two hours after addition of concentrated *a* medium, α cells do arrest in G1 as unbudded cells (Wilkinson and Pringle, 1974). The G1 block seems to be transient, since after four hours the α cells begin to develop and enter into DNA synthesis. The constitutively synthesized diffusible substance that is responsible for G1 arrest has been designated *a*-factor I.

However, a shmoo-causing activity (*a*-factor II) could not be detected in liquid cultures of *a* cells, although the existence of *a*-factor II was suggested by the presence of both *a* and α shmoos in mating mixtures (Ahmad, 1953; Levi, 1956; Bücking-Throm *et al.*, 1973). α Shmoos are also formed in a "confrontation" assay on rich complex agar medium by placing fresh α cells close to a heavy overnight growth of *a* cells, although most of the challenged α cells do not shmoo and the response in those that do is less pronounced than seen with *a* shmoos and α-factor (MacKay and Manney, 1974a). Thus, the minimal response of α cells to *a* cells on agar may reflect a low level of constitutive *a*-factor II synthesis, poor diffusion of the substance, instability of the *a*-factor II, or a

*Because these nonmating mutants had been obtained using a strongly selective method that could select against certain mutant phenotypes, Hicks (1975) isolated more nonmating mutants by a nonselective screening of clones derived from mutagenized cells. The phenotypic classes of mutants that he obtained, however, were essentially identical to the original ones.

reduced or transient response by α cells. Taken together, the above results could indicate that significant *a*-factor II activity is not constitutive, but induced in mating mixtures. As discussed below, this interpretation may be in part correct.

When supernatant fluids from *a* cultures are concentrated by rotary evaporation, *a*-factor II can be detected only when the *a* cells have been grown in the presence of α cells or α-factor, suggesting an induction mechanism (Fig. 4, V. L. MacKay, unpublished). Yet chromatographic fractionation of uninduced culture supernates from at least two *a* strains indicates that some activity is present constitutively, as well (Betz *et al.*, 1977). Thus, α-factor may "induce" by increasing *de novo* synthesis or activity of *a*-factor II, by removing an inhibitor, or by a combination of the two.

Recently, *a*-factor II has been partially purified and characterized chemically to some extent (Betz *et al.*, 1977). The pheromone is a protein that may contain carbohydrate moieties and appears to be either a large molecule (greater than 600,000 daltons by gel filtration) or a complex of firmly aggregated subunits, yet the activity is resistant to boiling. Although α cells shmoo in response to *a*-factor II only between pH 4 and 6, the pheromone is stable from pH 2.5 to 7.5, at least. Since the activity has been purified only approximately 200-fold as yet, further purification may resolve the questions of molecular weight, subunit composition, etc. If *a*-factor II is actually a large glycoprotein (*ca.* $10^6$ daltons), it could be a component of the *a* cell wall that is assembled and/or released in response to α-factor.

Throughout the purification steps developed to date, the shmooing activity of *a*-factor II and G1 arrest activity of *a*-factor I have co-purified (Betz *et al.*, 1977), suggesting that both functions may reside on the same molecule, as found

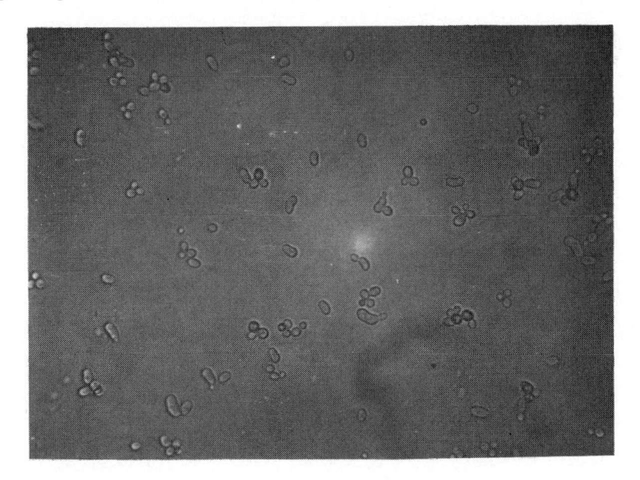

Fig. 4. Cells of mating type α exposed for 5 hours to a crude preparation of *a*-factor. (V. L. MacKay, unpublished.)

with α-factor. Alternatively, α cells might shmoo only when arrested in G1, preventing the detection of *a*-factor II in the absence of *a*-factor I. Analogous to G1 arrest by α-factor, *a*-factor I blocks α cells in G1, preventing DNA replication, but does not impair the synthesis of RNA and protein. Whether the similarity between α-factor and *a*-factor(s) extends to changes in the cell wall structure has not been determined, although studies on the binding of fluorescent concanavalin A to both *a* and α shmoos indicate that the same type of wall alteration occurs in both cases (J. Tkacz and V. L. MacKay, submitted; see Section IV, A). Thus, at least at the qualitative level, *a* and α cells appear to be brought to the same stage in the preconjugational pathway *via* α-factor and *a*-factor(s), respectively.

In order to determine if *a*-factor has any function in conjugation, the nonmating *a* and α mutants described in the previous section (Section III, A) were assayed for the production of and response to the pheromone. In the confrontation assay system using a rich growth medium, only five of the 66 *a ste* strains still produced any detectable activity that could cause α cells to shmoo (MacKay and Manney, 1974a). Using the *in vitro* procedures recently developed for *a*-factor II studies, none of the six *a ste* mutants that were tested could be induced by α cells or α-factor to produce detectable *a*-factor II (V. L. MacKay, unpublished). (This result is not surprising since none of the mutants responds to α-factor by shmooing or arresting in G1.) However, at least one of the *a ste* strains appears to make a constitutive level of *a*-factor II that can be demonstrated if the culture medium is fractionated by column chromatography (V. L. MacKay, unpublished); this mutant is one of the five originally classified as having retained the ability to produce *a*-factor (MacKay and Manney, 1974a). Of 107 α *ste* strains studied, only one showed even a slight, probably insignificant, response to *a*-factor, whether by confrontation assay or with partially purified *a*-factor II (MacKay and Manney, 1974a; V. L. MacKay, unpublished). Thus, as with α-factor, the ability of an α cell to respond to *a*-factor is required for mating ability, whereas secretion of the pheromone is probably, but not necessarily, involved.

### *C. Barrier factor*

When *a* cells are placed on agar near a source of α-factor (i.e., α cells), they form shmoos unless "barrier" cells are interposed between the α cells and the *a* assay cells (Hicks and Herskowitz, 1976). Only *a* cells and some *a ste* mutants exhibit this barrier effect on the further diffusion of α-factor through the agar to the assay *a* cells. Furthermore, at least part of the barrier phenotype results from the constitutive production of a diffusible factor that inhibits the response of *a* cells to α-factor, either by interacting with and protecting *a* cells or by inactivating α-factor. Hicks and Herskowitz concluded that the barrier factor is distinct from *a*-factor II, since some mutant strains that produce the inhibitor are *a*-factor deficient and *vice versa*. However, with better assay and purification techniques now available for *a*-factor(s), this question should be re-examined.

The recovery of *a* cells from α-factor induced G1 arrest was also investigated by Chan (1977) who likewise found that recovery was associated with a loss of α-factor from the medium. However, in this case, the removal of α-factor appeared to be completely cell-mediated, perhaps by uptake, rather than by the action of a diffusible uactor. Furthermore, *a* cells that had recovered retained some immunity up to 4.5 hours after transfer to fresh medium. Removal of α-factor from the medium was correlated with mating ability, since neither of two nonmating *a* mutants was capable of this depletion. (One of these mutants was determined by Hicks and Herskowitz (1976) to be barrier-positive.) Chan has suggested that differences in growth media and experimental techniques might be sufficient to account for the lack of agreement with the results of Hicks and Herskowitz (1976).

What function these resistance and recovery mechanisms might serve in mating or the normal life cycle is not immediately apparent. In mating mixtures, unmated *a* cells that can recover quickly after exposure to α-factor would have an obvious selective advantage over other unmated *a* cells. Hicks and Herskowitz (1976) have also pointed out that a high concentration of the "anti-α-factor" activity would develop in mating mixtures with a high ratio of *a* to α cells. If mating were delayed by the action of the inhibitor until further growth of the α cells, such an imbalance could be corrected, resulting in a higher mating frequency in the population as a whole.

### *D. Agglutination Factors*

In *a* cells, two patterns of heterosexual agglutinability have been described: a constitutive mode and an inducible one (Sakai and Yanagishima, 1972). In contrast, all of the α strains studied were determined to be constitutively agglutinative. In mixtures of constitutive *a* strains with α cells, rapid agglutination takes place, even in the presence of cycloheximide or after heat killing. When inducible *a* cells are used, however, agglutination of *a* and α cells occurs only after a lag period and can be blocked by cycloheximide or heat killing. Sakai and Yanagishima (1972) have proposed that the responsible agglutination factors are present on the surface of α cells and some *a* strains, but their synthesis, activation or accessibility must be induced in other *a* strains (Yanagishima, 1973). By treating cells with Glusulase (a crude glucanase fraction from snail digestive juices), Yanagishima *et al.* (1975) have succeeded in releasing the sex-specific agglutination factors. When cells of each mating type are pretreated with the crude agglutination factor from the opposite mating type, subsequent mixed cell agglutination is inhibited. The factors are thought to be univalent, because they do not cause agglutination of the opposite mating type, but do inhibit the reaction (Shimoda *et al.*, 1975). Preparations of both factors contain sugar and both biological activities are susceptible to certain proteases, both *in vivo* and *in vitro*. These results suggest that the surface agglutination factors in *Sacch. cerevisiae* may be glycoproteins like those in *Hansenula wingei* (reviewed in Crandall *et al.*, 1977). Furthermore, it is possible that *a*-factor II and the *a* agglutination factor are similar molecules and that a certain portion of the surface bound factor is released into the medium.

Inducible *a* strains develop agglutinability after a lag period when mixed with α cells, after incubation of the *a* cells with filtrates of α cultures or after exposure to a partially purified fraction from α culture filtrates (Sakai and Yanagishima, 1972; Sakurai *et al.*, 1973; Yanagishima, 1973). The activity (α substance-I) present in these filtrates that induces the agglutinability may be a peptide (Sakurai *et al.*, 1975). However, D. Radin (personal communication) has determined that the inducer of agglutinability produced by α cells (presumably α substance-I) shows the same response as α-factor to heat, pH changes and proteases. Since highly purified α-factor is also a strong inducer of agglutinability in *a* cultures and since nonmating α mutants which fail to produce detectable α-factor also do not make the agglutinability inducer (D. Radin, personal communication), α substance-I and α-factor appear to be identical peptides.

## IV. INVOLVEMENT OF THE MATING FACTORS IN CONJUGATION

### A. *Fluorescence Microscopy*

The chemical and biological properties of the various mating-related substances strongly indicate that they have essential functions in the preconjugational pathway. With the use of fluorescence microscopy, we have recently obtained additional evidence for their physiological activities that supports the scheme for conjugation proposed in Section IV, B. *a* and α Cells were incubated for various times with the appropriate pheromone to yield shmoos and then treated with FITC-concanavalin A (concanavalin A conjugated with fluorescein isothiocyanate). As shown at the left in Fig. 5, both *a* and α cells exhibit a round morphology with uniform fluorescence over the surface at the beginning of incubation. After two hours, cells of both mating types have formed shmoos, but, following treatment with FITC-concanavalin A, the distribution of fluorescence is no longer uniform (Fig. 5, middle). The portion of the shmoo synthesized in response to pheromones has a greater affinity for the lectin, demonstrating that the alteration in the cell surface occurs at a defined region and not randomly over the whole cell.* By four hours in this experiment, many of the shmoos had recovered from the pheromone effects and were beginning to

*It can be seen in Fig. 5 that the change in the *a* cell surface in response to α-factor is more pronounced than the *a*-factor mediated alteration of the α cell surface, at least as assayed by the binding of FITC-concanavalin A. It seems that all aspects of the response of α cells to *a*-factor are qualitatively similar, but less dramatic, than the effects of α-factor on *a* cells.

bud, with the progeny buds having the intensity of fluorescence of untreated cells (Fig. 5, right).†

The roles of the pheromones, agglutination factors, and shmoos become apparent when zygotes are stained (Fig. 6). These were taken from a mating mixture of *a* and α cells without the addition of any exogenous pheromones. The appearance of a more intense fluorescent band at the conjugation bridge demonstrates that zygotes are formed by the fusion of two cells at the region of altered wall structure, i.e., at the tips of the developing shmoos. Again, it should be noted that the developing diploid bud at the conjugation bridge has the intensity of fluorescence of untreated cells.

### *B. Scheme for Interactions Leading to Conjugation*

Bücking-Throm *et al.* (1973) first proposed a model for conjugation in which the first step in mating was reciprocal arrest of cells of the opposite mating type in G1, caused by α-factor and *a*-factor. With recent additional observations on the biological activities of the mating factors, this viewpoint of the role of the pheromones can be expanded. The proposed scheme for the interactions leading to conjugation as illustrated in Fig. 7 represents an attempt to integrate the properties of the factors with the cellular events of the pathway.

Mating type α cells constitutively secrete the oligopeptide α-factor into the culture medium. In mating mixtures, α-factors triggers the multiple *a* cell response. (1) *a* Cells are arrested in the G1 phase of the cell cycle, preventing the initiation of any new rounds of DNA synthesis. (2) The structure of the newly synthesized cell wall is altered; I would propose that this alteration is responsible for the appearance of the inducible agglutination factors and ultimately for the development of the shmoo morphology. (3) *a* Cells are "induced" to synthesize, activate, or unmask the inducible *a*-factor II activity.

While *a* cells are responding to α-factor, the α cells in the mating mixture are being affected initially by *a*-factor I, which is secreted constitutively by *a* cells, and, shortly thereafter, by the *a*-factor II activity. The combined effects of the pheromone(s) are analogous to those induced by α-factor. (1) α Cells are arrested in G1, although perhaps only transiently. (2) The nature of the newly synthesized cell surface is altered, as evidenced by the binding of fluorescent concanavalin A and by the development of shmoos, although the chemical

†The budding pattern of *a* shmoos was apparently random, with approximately half of the buds developing at the old cell end and the others at the shmoo tip (Fig. 5, top). In some *a* shmoos, limited synthesis of "cellular" wall material appeared to have occurred at the shmoo end before budding. With α shmoos, however, the budding pattern was extremely restricted, with buds developing only at the shmoo end (Fig. 5, bottom). As yet, these studies have been conducted with only one *a* and one α strain, so any differences in budding pattern may be due to strain differences, rather than mating type regulation.

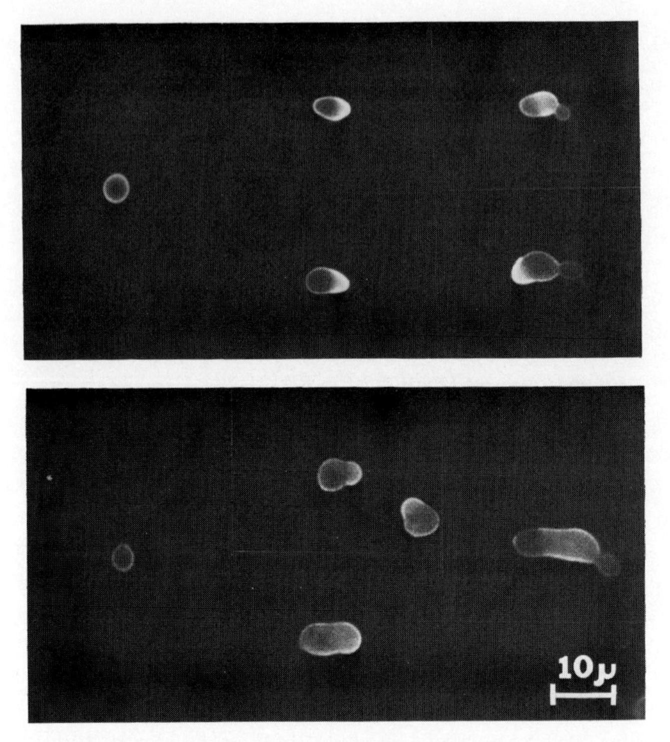


Fig. 5. Cells incubated with the appropriate pheromone for 0 hours (left), 2 hours (middle), and 4 hours (right), then stained with FITC-concanavalin A by the method of Tkacz *et al.* (1971). Top panel: *a* cells incubated with α-factor. Bottom panel: α cells incubated with *a*-factor. (J. Tkacz and V. L. MacKay, submitted.)

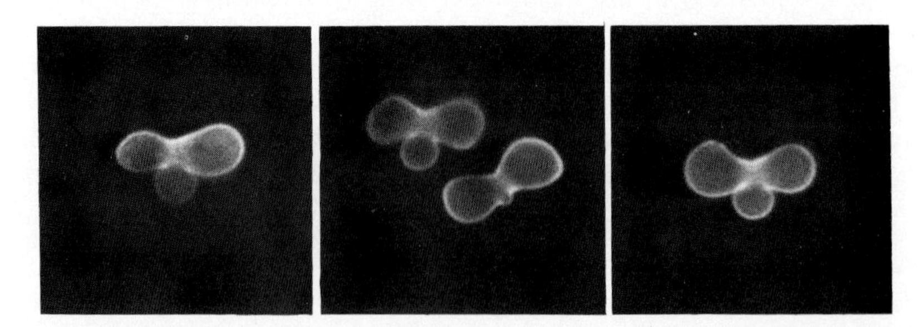

Fig. 6. Zygotes stained with FITC-concanavalin A. A mating mixture of *a* and α cells was incubated without the addition of exogenous mating pheromones for approximately 4 hours and then stained by the method of Tkacz *et al.* (1971). (J. Tkacz and V. L. MacKay, submitted.)

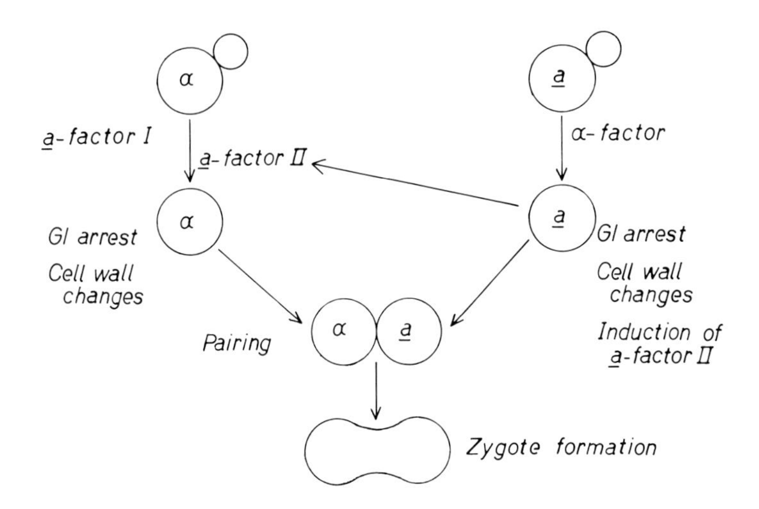


Fig. 7. Proposed scheme for cell and pheromonal interactions leading to conjugation in *Sacch. cerevisiae*. See text for details.

changes are not yet known. The surface alteration may also enhance the agglutinability of the α cells with *a* cells (D. Radin, personal communication).

At this point then, both *a* and α cells are poised in the G1 stage with complementary surface factors that recognize and bind the cells together at the area of new wall material. As determined by fluorescence and electron microscopy, this fusion occurs at the tips of the shmoos. The small vesicles present in the region may then release cell wall degrading enzymes (e.g., β-glucanases) that would complete dissolution of the wall, forming a conjugation bridge between the two haploid cells.

## V. SUMMARY

The sequence of events leading to sexual conjugation in *Sacch. cerevisiae* is mediated by the action and interaction of several diffusible pheromones and surface-bound factors for recognition and agglutination. Although we have learned a great deal about the process in recent years, many questions remain to be answered. Do the mating pheromones bind to surface receptors or are they taken up into the cell? What receptors, regulatory proteins, DNA binding proteins, etc. are involved? What genes or gene products are activated during the preconjugational period? What is the nature of the interactions between pheromones and between the surface-bound agglutination factors? These questions are not unique to the problem of sexual conjugation in *Sacch. cerevisiae*, but represent areas of interest in many systems of eukaryotic regulation, such as the immune response, hormonal control, and malignant transformation. With the availability of defective mutants, purified pheromones, and assayable responses, future studies of conjugation in yeast may provide some specific answers for these general problems common to all eukaryotic organisms.

## ACKNOWLEDGMENTS

I wish to thank my colleagues who provided unpublished data and manuscripts prior to publication. Our work was supported in part by a grant from the United States Public Health Service (GM22149) and by the Biomedical Sciences Support Grant awarded to Rutgers University, The State University of New Jersey.

## REFERENCES

Abell, C. W. and Monahan, T. M. (1973). *J. Cell Biol.* **59,** 549–558.
Ahmad, M. (1953). *Ann. Bot.* **17,** 329–342.
Betz, R., MacKay, V. L. and Duntze, W. (1977). *J. Bacteriol.,* in press.
Bücking-Throm, E., Duntze, W., Hartwell, L. H. and Manney, T. R. (1973). *Exp. Cell Res.* **76,** 99–110.
Campbell, D. A. (1973). *J. Bacteriol.* **116,** 323–330.
Chan, R. K. (1977). *J. Bacteriol.* **130,** 766–774.
Crandall, M., Egel, R. and MacKay, V. L. (1977). *Adv. Microbial Physiol.* **15,** 307–398.
Duntze, W. (1974). *Postepy Mikrobiologii* **13,** 41–51.
Duntze, W., MacKay, V. L. and Manney, T. R. (1970). *Science* **168,** 1472–1473.
Duntze, W., Stötzler, D., Bücking-Throm, E. and Kalbitzer, S. (1973). *Eur. J. Biochem.* **35,** 357–365.
Friis, J. and Roman, H. (1968). *Genetics* **59,** 33–36.
Gunge, N. and Nakatomi, Y. (1972). *Genetics* **70,** 41–58.
Hartwell, L. H. (1973). *Exp. Cell Res.* **76,** 111–117.
Hartwell, L. H. (1974). *Bacteriol. Rev.* **38,** 164–198.
Hicks, J. B. (1975). Ph.D. Thesis, University of Oregon, Eugene.
Hicks, J. B. and Herskowitz, I. (1976). *Nature* **260,** 246–248.
Laskowski, W. (1960). *Z. Naturf.* **15b,** 495–506.
Levi, J. D. (1956). *Nature* **177,** 753–754.
Lindegren, C. C. and Lindegren, G. (1943). *Proc. Nat. Acad. Sci. U.S.A.* **29,** 306–308.
Lipke, P. N., Taylor, A. and Ballou, C. E. (1976). *J. Bacteriol.* **127,** 610–618.
MacKay, V. L. and Manney, T. R. (1974a). *Genetics* **76,** 255–271.
MacKay, V. L. and Manney, T. R. (1974b). *Genetics* **76,** 273–288.
Manney, T. R. and Woods, V. (1976). *Genetics* **82,** 639–644.
Mortimer, R. K. (1958). *Rad. Res.* **9,** 312–326.
Mortimer, R. K. and Hawthorne, D. C. (1973). *Genetics* **74,** 33–54.
Osumi, M., Shimoda, D. and Yanagishima, N. (1974). *Arch. Mikrobiol.* **97,** 27–38.
Petes, T. D. and Fangman, W. L. (1973). *Biochem. Biophys. Res. Comm.* **55,** 603–609.
Radin, D., Sena, E. and Fogel, S. (1973). *Genetics* **74,** Supplement, No. 2/Part 2, p. s222 (abstract).
Roman, H. and Sands, S. M. (1953). *Genetics* **39,** 171–179.
Sakai, K. and Yanagishima, N. (1971). *Arch. Mikrobiol.* **75,** 260–265.
Sakai, K. and Yanagishima, N. (1972). *Arch. Microbiol.* **84,** 191–198.

Sakurai, A., Tamura, S., Yanagishima, N., Shimoda, C., Hagiya, M. and Takao, N. (1973). *Proc. 8th Intern. Congr. Plant Growth Sub.*, 185–192.

Sakurai, A., Tamura, S., Yanagishima, N. and Shimoda, C. (1975). *Proc. Jap. Acad.* **51,** 291–294.

Scherer, G., Haag, G. and Duntze, W. (1974). *J. Bacteriol.* **199,** 386–393.

Sena, E. P., Radin, D. N. and Fogel, S. (1973). *Proc. Nat. Acad. Sci. U.S.A.* **70,** 1373–1377.

Shimoda, C., Kitano, S. and Yanagishima, N. (1975). *Antonie van Leeuwenhoek J. Microbiol. Serol.* **41,** 513–519.

Stötzler, D., Kiltz, H-H. and Duntze, W. (1976). *Eur. J. Biochem.* **69,** 397–400.

Throm, E. and Duntze, W. (1970). *J. Bacteriol.* **104,** 1388–1390.

Tkacz, J., Cybulska, B. and Lampen, J. O. (1971). *J. Bacteriol.* **105,** 1–5.

Wiemken, A. (1975). *Meth. Cell Biol.* **12,** 99–109.

Wilkinson, L. E. and Pringle, J. R. (1974). *Exp. Cell Res.* **89,** 175–187.

Yanagishima, N. (1973). *Curr. Adv. in Plant Sci.: Commentaries in Plant Sci.* #7, 55–66.

Yanagishima, N., Shimoda, C. and Kitano, S. (1975). *Proc. 1st Intersect. Congr.* Int. Assoc. Microbiol. Soc., vol. 1, pp. 265–268. Hasegawa, T., ed. Science Council of Japan.

# Subject Index

**D**



**E**



**F**



**G**



**H**

R



S



T



U



Y



A B C 8 D 9 E 0 F 1 G 2 H 3 I 4 J 5